Structure and Health Effects of Natural Products on Diabetes Mellitus

Haixia Chen • Min Zhang
Editors

Structure and Health Effects of Natural Products on Diabetes Mellitus

Springer

Editors
Haixia Chen
School of Pharmaceutical Science and Technology, Tianjin University
Tianjin Key Laboratory for Modern Drug Delivery and High-Efficiency
Tianjin, China

Min Zhang
State Key Laboratory of Food Nutrition and Safety
Tianjin Agricultural University
Tianjin, China

ISBN 978-981-15-8790-0   ISBN 978-981-15-8791-7 (eBook)
https://doi.org/10.1007/978-981-15-8791-7

This Springer imprint is published by the registered company Springer Nature Singapore Pte Ltd.
The registered company address is: 152 Beach Road, #21-01/04 Gateway East, Singapore 189721, Singapore

Preface

We are pleased to present a book on *Structure and Health Effects of Natural Products on Diabetes Mellitus*. Diabetes is an important public health problem, one of the four priority noncommunicable diseases (NCDs) targeted for action by world leaders. Both the number of cases and the prevalence of diabetes have been steadily increasing over the past few decades. It has been estimated that about 463 million adults between the ages of 20 and 79 suffered from diabetes globally in 2019 (that is 1 in 11 persons), according to the International Diabetes Federation. Diabetes mellitus is a common endocrine and metabolic disease, which not only causes physiological damage to patients' kidneys, cardiovascular and cerebrovascular vessels, peripheral blood vessels, nerves, and eyes but also causes mental and psychological pressure to patients. Currently, most treatments for diabetes and its complications are chemical and biological drugs, which have the obvious short-term hypoglycemic effects, but many adverse reactions and high price. Due to the evidence that traditional medicine and natural herbal formula have advantages in treating diabetes, natural products with hypoglycemic activity have been studied extensively in recent years and have been accepted by many scholars all over the world.

The purpose of this book is to introduce the classified chemical components of hypoglycemic components in natural products, summarize the recent research progress of natural products with hypoglycemic activity in the past 20 years, and provide the original analysis and development opinions of relevant scholars.

This book focuses on the progress on the study of the structure, hypoglycemic activities, structure–activity relationships, and mechanism of a wide range of polysaccharides, flavonoids, saponins, alkaloids, terpenoids, polyphenols, and other constituents. It will help graduate students and researchers to understand current approaches and progress in the treatment of diabetes with natural products, which may also be beneficial to develop new hypoglycemic drugs. This book will be useful to all those working in the field of botany, phytochemistry, pharmacy, drug delivery, molecular biology, forestry, biotechnology, industrial food, and medical products. This work is arranged in 17 well-illustrated chapters.

This project was spread over almost 2 years, from concept to print. We would like to acknowledge the cooperation, patience, and support of our contributors who have put their serious efforts to ensure the high scientific quality of this book with up-to-date information. We are thankful to the staff at Springer, for their professional support in this project.

Tianjin, China Haixia Chen
Tianjin, China Min Zhang

Contents

Chapter 1
Introduction of Diabetes Mellitus and Future Prospects of Natural Products on Diabetes Mellitus

Haixia Chen and Ruilin Li

Abstract Diabetes mellitus is a chronic disease caused by inherited and/or acquired deficiency or due to ineffective insulin production by the pancreas. There is a great need for its medication, and natural products play an important role in the treatments of diabetes. This chapter mainly summarizes the introduction, the progress in the medication of diabetes mellitus, and use of natural products in the treatment of diabetes mellitus.

Keywords Diabetes mellitus · Medication · Natural products · Properties · Bioactivities

1 Diabetes Mellitus

With the improvement of human living standards and the accelerated development of population aging, the incidence of diabetes has continued to increase rapidly. It is the third major chronic disease endangering human health after tumors, cardiovascular, and cerebrovascular diseases, and its seriousness lies in its high incidence and mortality. Diabetes mellitus is a considerable chronic metabolic disorder characterized by chronic hyperglycemia due to insufficient insulin secretion and/or deficiency of insulin action [1]. The estimate for 2040 is that the numbers concerning the prevalence of diabetes mellitus will rise to 642 million worldwide [2]. In addition to having major effects, diabetes also raises risks such as high blood sugar, high blood pressure, dyslipidemia, atherosclerosis, and platelet aggregation. Without enhanced control, diabetic patients will be accompanied by diabetic complications such as kidney disease, encephalopathy, and retinopathy, which may trigger cardiovascular and cerebrovascular events [3, 4]. Diabetes has become a serious public health problem that greatly affects patients' quality of life and longevity. The complexity

H. Chen (✉) · R. Li
Tianjin Key Laboratory for Modern Drug Delivery and High-Efficiency, School of Pharmaceutical Science and Technology, Tianjin University, Tianjin, China
e-mail: chenhx@tju.edu.cn

H. Chen, M. Zhang (eds.), *Structure and Health Effects of Natural Products on Diabetes Mellitus*, https://doi.org/10.1007/978-981-15-8791-7_1

of treatment places a huge burden on health, economy, and society. Diabetes mellitus has become a stubborn disease that physicians and pharmacists need to overcome.

Diabetes mellitus is a chronic disease caused by inherited and/or acquired deficiency or ineffective insulin production by the pancreas. This deficiency leads to increased glucose concentrations in the blood, which further harm our physical health, especially the blood vessels and nervous system. Diabetes mellitus is a metabolic disorder with abnormally high blood glucose levels, which is called as hyperglycemia [5]. It is worth noting that diabetes is a broad term for a class of diseases that cause long-term hyperglycemia. There are many types of diabetes, and the different mechanisms by which diabetes mellitus is formed will be the basis for classifying it. Diabetes can be divided into two categories: insulin-dependent diabetes mellitus (IDDM) or type 1 (insulin-deficient type, hyperglycemia, and ketosis due to autoimmune diseases) and noninsulin-dependent diabetes mellitus (NIDDM) or type 2 (metabolic disorders and elevated blood sugar due to inadequate production of insulin or effective use of insulin, but resistance to ketosis) [6]. According to the statistics, over 90% of diabetic patients are having type 2 diabetes [7].

Diabetes mellitus is a metabolic disorder that is mainly caused by insufficient insulin secretion or resistance to its effects. In the case of diabetes mellitus, side effects caused by the accumulation of glucose are one of the most serious problems [8]. Studies have shown that excess glucose can be converted into various compounds and free radicals, which can damage vital organs and organelles. In human body, plenty of pathways and systems work synergistically to achieve and keep our physiological state in a healthy environment [9]. The ability of organisms to maintain a steady state or have homeostasis in the body is the core of these processes. Disturbance of homeostasis will cause damage to various organs and accelerate the development of pathological conditions. The ability of the body to regulate glucose levels in the blood is reduced in persons with diabetes mellitus and causes numerous complications. Clinical antidiabetic medications show that there are many types of antidiabetic medications . However, there are currently no specific drugs in the market to treat these complications. Prolonged use of antidiabetic drugs can cause a variety of both mild and severe adverse reactions, including hypoglycemia, weight gain, gastrointestinal discomfort, headache, nausea, and abdominal pain [10]. Therefore, there is an urgent need to find safer and more effective antidiabetic drugs to counteract the aforesaid problems related to the current treatment options.

2 Status of Medication

Many researchers have worked on the improvement of insulin on its target tissues and found compounds that are capable of improving insulin secretion from β-cells. Over the past decade, several new oral medicines have also been discovered to control blood sugar in patients with type 2 diabetes [11]. These agents work through different mechanisms. Varieties of natural products have been acquired, and formulations that were especially used for the treatment of diabetes and its related complications have been prepared. As natural products are usually a mixture of multiple ingredients, the active ingredient among them is unclear and is not defined well. This has become one of the major drawbacks with natural products. Therefore, more effort is needed to seek out the active ingredients and their molecular interactions, which will not only contribute to analyze the mechanism of product action but also can normalize formulated products. Natural products play an indelible role in the treatment of diabetes in people around the world, which has driven us to deepen our knowledge of natural products [12]. For searching of the feasible treatment plan for diabetes and its complications, complex pathogenesis requires extensive research. These drugs have their own adverse effects, and the complexity of treatment places a huge burden on health, economy, and society. The need for effective treatment will continue to increase with the increase in the population of diabetic patients.

The increasing prevalence of diabetes mellitus has stimulated the development of many new treatments for hyperglycemia in order to keep blood glucose concentrations as close to normal as possible and prevent complications. To date, many different classes of antidiabetics and mechanisms of action are available as oral drugs, such as biguanides, meglitinides, sulfonylureas, PPAR-γ agonists (glitazones), thiazolidinediones, α-glucosidase inhibitors, dopamine D2-receptor agonists, DPP-4 inhibitors [13], and glucagon-like peptide-1 receptor agonists, and injectable therapies include GLP-1 agonists, amylin analogs, insulin, insulin analogues, etc. [14]. However, many of these drugs have undesirable adverse effects and insufficient efficacy that greatly affect the rehabilitation of diabetic patients. Metformin may cause some mild adverse effects, like mild abdominal discomfort, diarrhea, anorexia, and nausea, most of which are gastrointestinal diseases. Dipeptidyl peptidase-IV inhibitors (DPP-4 inhibitors) can also cause other side effects like nausea, nasopharyngitis, headache, hypersensitivity, or skin reactions [15]. For incretin-based drugs, in spite of many kinds of benefits have been certificated, they are still accompanied by severe gastrointestinal troubles, for instance, nausea, vomiting, sour stomach, indigestion, and diarrhea [16]. Diabetes mellitus is a chronic disease, which demands treatment throughout life. As a result, the cost of antidiabetic drugs has become a major concern for patient compliance. Based on the patient's drug list, researchers calculated and analyzed the cost of drug therapy, and the results revealed that the price of the drugs per prescription was expensive, which was considered to be one of the main reasons of patient's nonadherence to treatment [17]. Therefore, safer and more effective therapeutic drugs still need to be urgently developed.

3 Natural Products

During diabetes, chronic hyperglycemia causes glycosylation of human proteins, which in turn leads to secondary complications that affect the eyes, kidneys, nerves, and arteries. At present, the development of chemically synthesized diabetes drugs is increasingly difficult, and the problem of adverse drug reactions is prominent. In addition to the fact that synthetic antidiabetic drugs have serious side effects, currently they are also expensive for patients [18]. Even with many treatment options, many natural products have been recommended to treat diabetes mellitus.

Based on the high abundance and diversity of marine organisms and terrestrial organisms, natural products have become sources of drugs for treating many disorders and diseases successfully. Nowadays, they remain to be a repository of potential drugs. The discovery of natural products is re-emerging as a reputable source of current drugs in the market. At the same time, the World Health Organization is also aware of the importance of biodiversity, which would be able to offer affordable, therapeutic solutions to the majority of the world population. Today, almost half of the existing medicines are inspired by natural products. We believe that along with this trend, a large number of natural products derived (lead drugs or extracts) will continue to be produced and successfully enter the market in the future. The chemical structure of many pharmaceuticals commonly used today is inspired by natural compounds found in traditional medicinal plants. Natural products have been used in the treatment of diabetes mellitus worldwide. Natural products for the treatment of diabetes have accumulated rich experience in thousands of years of theoretical and clinical practice and have obvious advantages in the treatment of diabetes and chronic complications. Natural products with hypoglycemic properties and other beneficial therapeutic properties are the main reasons mentioned above, and modern pharmacological studies have also proven that natural products can lower blood sugar through multiple pathways and multiple targets. Fortunately, natural antidiabetic drugs have never been outdated in the management of diabetes and still play a major role, many of which are known to be effective against diabetes [19]. The active ingredients of hypoglycemic activities found in natural products are alkaloids, sugars, polyphenols and flavonoids, terpenes, saponins, unsaturated fatty acids, and so on. The first medicinal plant with obvious antidiabetic effect was *Galega officinalis* L. (Fabaceae), which has been used as a prescription medicine to treat diabetes mellitus since the Middle Ages. From this plant the guanidine derivative galactose was extracted. The chemical structure of this compound is very similar to that of the metformin, an antidiabetic drug, which the plant extracts has a blood sugar-lowering effect and is one of the most prescribed drugs in the USA for diabetes mellitus treatment [20]. Various phenolic compounds, for example, flavonoids and anthocyanins, have a mitigation effect on the development of diabetes mellitus [21]. For instance, different anthocyanins from *Ipomoea batatas* effectively inhibit α-glucosidase/maltase activity in the intestine, which can reduce postprandial blood glucose levels. Inhibitory effects of anthocyanins were decided by their structure, as their potency as α-glucosidase inhibitors is higher in acylated anthocyanins than in deacylated derivatives [22, 23]. Some of these drugs have been derived from plants or microorganisms, such as galegine isolated from officinalis, which is very similar to the antidiabetic drug metformin. Picnogenol, acarbose, miglitol, and voglibose are other antidiabetic products of natural origin [24]. In the past few years, natural products including fruits, vegetables, spices, mushrooms, or natural beverages have shown good therapeutic effects on diabetes mellitus by inhibiting α-amylase and α-glucosidase, sodium-dependent glucose transporter, glycogenase, aldose reductase, advanced glycation end products (AGEs), and other mechanisms [25]. Besides, natural products can increase insulin secretion and enhance its activity, promote glucose uptake, protect pancreatic β-cellion; regulate glucose transporter-4 (GLUT-4), reduce oxidative stress, and mimic the effects of insulin [26]. In addition to being used as drugs, natural products are frequently identified disease-relevant targets in the form of molecular probes, which also greatly promote the research and application of natural products at the industrial level [11].

Because of the obvious outcome, compared with synthetic drugs, natural drugs have lower side effects and less cost effectiveness, attracting the attention of the scientific community. However, these natural products still have their own disadvantages and limitations. Complex physical and chemical properties, low yields, unclear adverse events, and uncertain mechanisms are still obstacles in the clinical research process [27]. As we all know, biodiversity is diverse. It is known to date that there are an estimated two million species of plants, animals, fungi, and microorganisms [28]. How to effectively acquire and utilize this natural chemical diversity is one of the major challenges. The main reasons for these challenges include: extremely small quantities of lead compounds obtained, difficult to source/harvest samples, extensive synthetic routes, and development times resulting in poor yields, unfeasible scale-up, complicated isolation and/or purification procedures, high toxicity of the active compound, in addition to ecological and legal aspects, government policy considerations, and its lack of infrastructure and insufficient capital investment. Therefore, it is considered that this area of natural product research should be vigorously expanded.

3.1 *Toxicity*

Natural products in a broad sense include any substance produced by life, usually containing toxic and active ingredients. Before using natural products as a potential therapeutic drug, it is important to research on safety and minimal toxicity to human cells to identify possible bioactive phytochemical and selective ingredients. Previously, coumarin was a constituent of a wide range of plants, and various researches implemented in rodents have manifested that it had hepatotoxicity [29]. Application of whole grass extracts or fractions consisting of numerous compounds can cause different biological effects in the human body. These toxic effects are caused by some special parts of the plant [30].

Natural products and related various strategies that will be used to control diabetes mellitus in the future. Natural products are a good source of medicinal compounds, and there are still many active substances extracted and isolated from natural products that need to be fully characterized and studied. More active ingredient studies on antidiabetic and mimicking insulin action mechanisms are expected in the future. Choosing an appropriate therapeutic dose is extremely necessary for proper treatment of diabetes mellitus. Whether natural products and medicines have a therapeutic effect or cause adverse reactions depends largely on the dosage taken. It must be believed in the future that plants are safe, but not all ingredients are safe, which is why it is important to pay close attention to the toxicology of these plant compounds before giving treatment. Therefore, prior to the use of natural products in diabetic patients, a comprehensive understanding of all aspects of the drug can effectively cut down the possible risks of drug interactions and maximize its therapeutic efficacy. We must objectively understand the causes of adverse events of natural products and draw on modern toxicology research ideas, methods, and means to construct natural product toxicology evaluation models and methods that meet the clinical characteristics of natural products and clarify the material basis for the occurrence of adverse reactions of natural products [31]. Establish a natural product safety evaluation and risk management system that complies with the characteristics of natural products. What is more, we need standardize the toxicology research standards of natural products to ensure the safety of patients' medication and promote the sustainable development of natural products.

3.2 *Solubility and Stability*

Solubility is one of the factors that must be considered in determining whether the drug can become a candidate, and it is also one of the most important physicochemical properties. Ultimately, the vast majority of new chemical entities (NCEs) will eventually be administered to patients via oral or intravenous injection, and for any of these pathways, compound solubility is a key parameter. However, low water solubility is a significant limitation in some natural products that develop into drugs [32, 33]. Poor physicochemical properties, especially poor solubility, are part of the reasons for these failures. In terms of medicinal chemistry, one of the most effective strategies to enhance the water solubility of natural products is chemical modification. A review of the literature [32] over the past 15 years showed that strategies to increase aqueous solubility could be summarized as follows: (1) adding solubilizing/polar groups; (2) increasing sp^3 fraction; (3) deteriorating the flatness, and hence destroying crystal packing forces; and (4) using the prodrug approach.

Natural products often face difficult challenges in terms of in vivo stability, which imposes significant restrictions on the structure of natural products. Certain natural products with high in vitro activity may be limited due to their sensitivity to metabolism. Reducing the binding or reactive structural modification of a compound at an unstable site will enhance metabolic stability. Solutions that are commonly employed include adding other blocking groups, removing unstable functional groups, and replacing unstable groups, thereby blocking the metabolic site [34]. For instance, studies illustrated that [35] since the group was easily hydrolyzed, the lactone oxygen was replaced with nitrogen to obtain a lactam compound. This single atom modification enhanced metabolic stability, which is not susceptible to esterase-mediated hydrolysis. In addition, this single atom change increased its water solubility. Although the solubilizer is allowed to be added, it is as small as possible. The improved formulation reduced the amount of solubilizing agent. The lipophilicity of the compound is adjusted by adding a hydrophilic group to the methylazetidine, which has good activity and metabolic stability [36].

Because drug discovery, research, and registration are expensive and time-consuming processes with low success rates, the pharmaceutical industry is focusing its efforts on repositioning withdrawn or approved drugs. The costs for bringing such a drug to market are about 60% lower [37]. Exploring established drugs for new bioactivity offers opportunities to advance therapeutic strategies quickly into clinical trials. In order to develop more indications for existing drugs, drug repositioning is performed, also known as drug profiling [38]. In addition to reducing drug development costs and shortening approval time, new drug applications can help extend patent life and increase return on drug development investment.

The specific method is as follows: first, exploiting binding similar site for drug repositioning. There have been a plenty of approaches to computational drug repositioning, including similarity of side effects mined from the literature [39, 40], similarity of gene expression profiles of different diseases [41], and structural similarity of binding sites [42]. And clarifying protein–drug interactions helps to better understand how drugs work and helps cut down drug doses. Second, identification of new drug targets. In order to systematically find new drug targets so that repositioning drugs can be found, a choice can be made between various techniques. For example, non-in silico approaches, which involve noninvasive imaging, disease animal models in vivo, improved bioanalytical procedures, and in vitro screening processes. Not only can in silico approaches avoid animal testing, it can also help with screening and significantly reduce drug development costs. The large amount of data is accessible and processable, reducing the large-scale experimental research.

3.3 Combinations of Drugs

Although the most important thing is to achieve blood glucose control as soon as possible to minimize the toxic effects of glucose on the body, it is also necessary to provide treatments to control other related risk factors, including dyslipidemia, oxidative stress, mitochondrial dysfunction, and vascular complications [43]. As the current antidiabetic drugs have more adverse effects and higher costs, the acceptance rate and use rate of natural products as alternative therapies are increasing. In addition, it is reported that natural products may interact synergistically with antidiabetic drugs, which may be the target of combination therapy [44]. Therefore, a combination therapy must be adopted to combat the multiple risk factors in diabetics.

Historically, drug combinations have been used for treating diseases and reducing suffering [45]. Reasonable drug combinations include sulfonylurea plus metformin, sulfonylurea plus glucosidase inhibitor, sulfonylurea plus thiazolidinedione, metformin plus repaglinide, metformin plus α-glucosidase inhibitor, and metformin plus thiazolidinedione [5]. This suggests that care should be taken in selecting suitable agents for combination therapy, which can provide most metabolic benefits to the patients with type 2 diabetes. Thus, it is necessary that there must be a synergistic combination therapy to treat diabetes mellitus conditions. When the pharmacokinetics and pharmacodynamics of these two compounds match, the maximum benefit can be obtained by combining the drugs. Synergy may also be helpful in reducing the dose of antidiabetic drugs in the treatment of diabetes mellitus while minimizing adverse reactions related to the drugs. Hence, combination therapy may be a new and highly effective treatment strategy for hyperglycemia management. While in order to select phytochemicals that exert drug synergy, the molecular mechanism must be fully understood whether or not a synergistic drug is added, laying a solid foundation for combined therapy.

3.4 Lifestyle

Lifestyle changes or treatment medications can prevent or at least delay the progression of diabetes mellitus in patients with impaired glucose tolerance. Initially, some patients with mild conditions do not need survival insulin therapy, and in some cases weight loss, exercise, and/or oral hypoglycemic drugs can achieve adequate glycemic control [46].

For centuries, it has been thought that people who are more sedentary are more likely to develop diabetes. And overweight people experience more than tenfold increase in the risks of diabetes mellitus compared with thin individuals [47]. Many prospective studies [48] have demonstrated that obesity and lack of exercise are risk factors for diabetes. Weight loss and exercise can improve insulin sensitivity and insulin secretion in the short term. Studies have shown that changes in diet and exercise habits may delay the onset of diabetes [49]. Whether in different places, in the work environment or at home, modern life in developed countries has significantly reduced the opportunity for people to consume energy. As a result of urbanization, availability has increased, such as convenient transportation and shopping kept indoors, people's physical activity has also decreased substantially, and motorized transport has replaced walking and cycling and has strengthened the mechanization of the labor force. However, regular physical exercise is a key factor in controlling weight and preventing obesity. In addition to the key role in retaining a healthy weight, regular physical activity reduces the risk of stroke, coronary heart disease, osteoporotic fractures, osteoarthritis, and depression. In our daily life, walking and cycling as a healthy form of transportation are worth promoting.

Walking or cycling can be an effective and practical means of taking both transport and sports into account and remains the most common way of transport.

With the development of society, people increasingly realize that bioactive food ingredients are important to human health, thus diet may play a very important role in the pathogenesis and prevention of chronic diseases. Functional foods have more powerful functions than basic diets, which can promote health while preventing diseases. Their food ingredients usually contain biologically active, and these ingredients work in conjunction with conventional nutrients to prevent or delay the onset of chronic diseases by imparting beneficial physiological effects. However, the inclusion of vegetables, fruits, legumes, nuts, and whole grains have the effect in reducing the incidence of diabetes mellitus. Because a diet rich in nutrients and phytochemicals could modulate genetic functions, silence the phenotypic expression, affect the heritability of variant phenotypes, and reduce the incidence of disease [50].

For our daily lives, we can make positive adjustments from our eating habits: (1) Make sure you eat plenty of fruits and vegetables. Strong evidence suggests that high intake of fruits and vegetables will reduce diabetes risk. (2) Consume cereal products in their whole-grain, high-fiber form. Consuming grains in a whole-grain, high-fiber form has double benefits. First, fiber in edible cereal products has consistently been linked to reducing the risk of diabetes, probably because of both the fiber itself and the vitamins and minerals naturally present in whole grains. High intake of refined starches exacerbates the metabolic syndrome and increases the risk of diabetes. Second, high dietary fiber consumption also seems to facilitate weight control and helps prevent constipation [51]. (3) Sugar has a negative impact on the health of people at risk of being overweight. Consumption of sugar and sugar-based beverages should be limited. In addition, sugar increases the blood sugar load in the diet, which exacerbates the metabolic syndrome and is closely related to the risk of diabetes. (4) Limit excessive calorie intake from any source and decrease consumption of foods and beverages high in calories. Given the importance of obesity and overweight in causing diabetes, it is important to avoid excessive energy consumption from any source. Because the calories consumed as beverages are poorly regulated compared to calories from solid foods, it is especially important to limit the consumption of sugary beverages. In this sense, lifestyle changes have proven to be the most convenient and cost-effective method of prevention with almost no side effects.

4 Outlook of Natural Products on Diabetes Mellitus

Natural products have been used in traditional medicine to treat various diseases affecting humans and animals since ancient times, and many natural products have been used worldwide to treat diabetes. Natural products possess a long history, especially with regard to direct use or development into antidiabetic drugs. As an endocrine disorder, in hyperglycemic conditions, the entire metabolic homeostasis of our body is affected by elevated blood glucose levels caused by pancreatic defects or insufficient insulin secreted by insulin-resistant cells. Diabetes remains one of the leading fatal diseases with many concomitant complications. Prediabetes, diabetes, and their complications easily lead to various organ failures [52, 53], which affect people worldwide.

We have been relying on plants or other natural resources to maintain or restore health since ancient times. According to the World Health Organization (WHO), there are more than 100 million users of traditional medicine and natural products, including people in Asia, Africa, Europe, North America, etc. Natural products are usually the main source of health care, especially in developing countries [54]. Along with herbal extracts and natural products, many other medicinal plants are used around the world to relieve cardiovascular and metabolic disorders. There is no doubt that the use of medicinal plants and natural products for the prevention or treatment of cardiovascular and metabolic diseases has become increasingly popular in Western society.

It is worth noting that natural products provide many potential mechanisms of actions to improve glucose homeostasis, which could reduce and/or eliminate diabetic complications. Aloe, banaba, bitter melon, cinnamon, and other medicinal plants or natural product-derived antidiabetic compounds have entered the stage of clinical research [55]. Among them, natural products treat diabetes through various routes and mechanisms, including inhibition of α-glucosidase and α-amylase, regulation of PPAR receptors, inhibition of protein tyrosine phosphatase 1B activity, regulation of hormones involved in glucose homeostasis such as adiponectin pharmacokinetics, amelioration of intestinal insulinotropic activity, suppression of oxidative stress, etc. [55].

Natural products have provided a rich source of compounds that have a wide range of applications in the fields of pharmacy, medicine, biology, and health sciences. Natural products have a reliable track record in drug discovery and uncontested unique structural diversity [56]. Although natural active substances are good lead compounds for discovering new drugs, there are still several problems associated with natural products and most of them are plagued by various deficiencies or shortcomings, such as poor solubility, weak stability, complex structures, and so on. In the process of developing novel drugs from natural products, typical restrictions such as low solubility, unstable metabolism, and unclear mechanism of action limit the development of natural products, which especially hinder the development of similar drugs [57, 58]. Therefore, it is necessary to modify the structure of natural products in order to achieve the purpose of developing new compounds with specific properties. Modifying the structure of natural products can not only improve their dynamics in vivo by changing the physical and chemical properties of the compounds such as solubility and stability but also increase their activity and selectivity as drugs and reduce human toxicity. Natural products as a semisynthetic precursor of drugs and templates for chemical synthesis of drugs provide a broad idea for drug design.

Natural products have been widely used for treating complex chronic diseases favorable by a huge number of patients owing to abundant experience and expectative efficacy. In order to provide a better experimental basis for the clinic, plentiful experimental researches were carried out to study the mechanisms of natural products. The results revealed that natural products exert a beneficial action on diabetes including reducing oxidative stress, regulating metabolic disorders, and inhibiting AGEs activity [59]. In the future, we still need to gather more studies correlated with the regulatory mechanisms of natural products on diabetes. In addition, the absorption, distribution, metabolism, elimination, and toxicity of natural products should be clarified along with their pharmacological activity. The pathophysiology of diabetes mechanisms is diverse and still is not clearly understood. More studies are warranted to identify the active compounds from natural products as well as to evaluate their activity and efficacy in vitro and in vivo. There is a need for well-designed, large sample, long-term, randomized, controlled, clinical trials to verify the efficacy and safety of natural products in patients with diabetes. In addition, because the development of medicines from natural compounds or plants is very time consuming, a faster and more advanced approach for natural product collection is needed, and it is necessary to screen through bioassays to isolate and develop lead compound to shorten the time [60]. For those targets of enzyme characteristics of diabetes, hopefully, increasing applications of updated technologies, such as omics detection, tandem analyzer, X-ray, and cryo-electron microscopy, can greatly promote the analysis of natural products and bring a better understanding of structural and functional proteins in the pathogenesis of diabetes. Some emerging technologies have also attracted great attention, such as metabolomic, genomic, or in silico studies [16]. Therefore, natural products with antidiabetic effects are very promising as diabetes treatment agents pave the way for promising therapeutic and promote the progress of diabetes treatment. In conclusion, natural products, especially natural products of plant origin, are an important source of compounds with different chemical structures. They act through a variety of mechanisms to become an alternative therapy for the treatment of diabetes. Since the research on these compounds is not sufficient, it is necessary to continue working on their research and development before using them as new antidiabetic drugs.

There is a limited potential with the use of any of the available drugs in patients with diabetes because it is difficult to balance the glucose-lowering efficacy, the side effect profiles, the anticipation of additional benefits, cost, and other practical aspects of care. Moreover, data on the side effects of most of the possible drug combinations are lacking. Hence, the demand for new compounds that can supplement current therapies for diabetes and their comorbidities is expanding [61]. Compared with chemical drugs specific to a single molecular target, natural products containing different ingredients exhibit distinct advantages with synergistic effects for treatment of diabetes. Hence, screening and discovery of active ingredients for lowering blood glucose from natural products have become an important direction for the development of new diabetes drugs. The market for natural products is booming, and the evidence for their effectiveness is growing, but insufficient regulations and lack of appropriate standards have limited their use. Therefore, product standardization needs to be properly evaluated to fully understand the efficacy, safety, and treatment risks associated with the use of natural products. Natural products standardization is another point. In order to ensure quality, safety, and reproducibility, in-depth analysis of each step during the preparation process of samples is necessary. With the development of new and complementary strategies to improve traditional natural products, they have the potential to continue making significant contributions to the discovery and development of natural products.

Natural products come from a wide range of sources, which possess invaluable biodiversity. Countless secondary metabolites, animals, plants, fungi, microorganisms, and other natural resources have become a rich source for finding new medicines [28, 62]. Natural products have a complex structure, which contains a wealth of activity and a variety of medicinal possibilities. As a natural gift to humans, even if natural products cannot be directly used as medicines, they can often inspire researchers' thinking and guide them. Perspectives provide important help for scientific research. Natural product-derived compounds obtained through structural modification, chemical synthesis, and other methods are also commonly used to obtain ideal drugs. The research process from goat beans to metformin is a good example. Diabetes is a highly metabolic disease caused by a variety of factors. There are many types of hypoglycemic drugs and different mechanisms of action, which determine its different clinical characteristics and adverse reactions. As more and more glucose-lowering drugs are available to clinicians, they need to invest more energy in researching the progress of diabetes treatment, grasp the characteristics of glucose-lowering drugs of different glucose-lowering drugs, and consider both adverse reactions and drug costs, and patient's compliance to provide patients with scientific, personalized hypoglycemic solutions. In the subsequent research, we also need to find natural antidiabetic drugs for hot research, develop targeted drugs, and comprehensively research them to achieve the ultimate goal of natural products for the treatment of diabetes.

References

1. Gordin D, Shah H, Shinjo T, St-Louis R, Qi W, Park K, Paniagua SM, Pober DM, Wu IH, Bahnam V, Brissett MJ, Tinsley LJ, Dreyfuss JM, Pan H, Dong Y, Niewczas MA, Amenta P, Sadowski T, Kannt A, Keenan HA, King GL (2019) Characterization of glycolytic enzymes and pyruvate kinase M2 in type 1 and 2 diabetic nephropathy. Diabetes Care 42(7):1263–1273. https://doi.org/10.2337/dc18-2585
2. Munhoz A, Frode TS (2018) Isolated compounds from natural products with potential antidiabetic activity-A systematic review. Curr Diabetes Rev 14(1):36–106
3. Kahn SE, Cooper ME, Del Prato S (2014) Pathophysiology and treatment of type 2 diabetes: perspectives on the past, present, and future. Lancet 383(9922):1068–1083
4. Miranda PJ, DeFronzo RA, Califf RM, Guyton JR (2005) Metabolic syndrome: evaluation of pathological and therapeutic outcomes. Am Heart J 149(1):20–32
5. Lin Y, Sun Z (2010) Current views on type 2 diabetes. J Endocrinol 204:1):1
6. Zhou Y, Huang Y, Ji X, Wang X, Shen L, Wang Y (2020) Pioglitazone for the primary and secondary prevention of cardiovascular and renal outcomes in patients with or at high risk of type 2 diabetes mellitus: a meta-analysis. J Clin Endocrinol Metab 105(5). https://doi.org/10.1210/clinem/dgz252

7. Prabhakar PK, Doble M (2011) Mechanism of action of natural products used in the treatment of diabetes mellitus. Chin J Integr Med 17(8):563
8. Li S, Yang H, Chen X (2019) Protective effects of sulforaphane on diabetic retinopathy: activation of the Nrf2 pathway and inhibition of NLRP3 inflammasome formation. Exp Anim 68(2):221–231. https://doi.org/10.1538/expanim.18-0146
9. Liu J, Nile SH, Xu G, Wang Y, Kai G (2019) Systematic exploration of Astragalus membranaceus and Panax ginseng as immune regulators: insights from the comparative biological and computational analysis. Phytomedicine:153077–153077. https://doi.org/10.1016/j.phymed.2019.153077
10. Chen X-W, He Z-X, Zhou Z-W, Yang T, Zhang X, Yang Y-X, Duan W, Zhou S-F (2015) An update on the clinical pharmacology of the dipeptidyl peptidase 4 inhibitor alogliptin used for the treatment of type 2 diabetes mellitus. Clin Exp Pharmacol Physiol 42(12):1225–1238. https://doi.org/10.1111/1440-1681.12469
11. Schmitt EK, Moore CM, Krastel P, Petersen F (2011) Natural products as catalysts for innovation: a pharmaceutical industry perspective. Curr Opin Chem Biol 15(4):497–504
12. Moradi-Marjaneh R, Paseban M, Sahebkar A (2019) Natural products with SGLT2 inhibitory activity: possibilities of application for the treatment of diabetes. Phytother Res 33 (10):2518–2530. https://doi.org/10.1002/ptr.6421
13. Tong L, Adler S (2018) Glycemic control of type 2 diabetes mellitus across stages of renal impairment: information for primary care providers. Postgrad Med 130(4):381–393. https://doi.org/10.1080/00325481.2018.1457397
14. Braga T, Kraemer-Aguiar LG, Docherty NG, Le Roux CW (2019) Treating prediabetes: why and how should we do it? Minerva Med 110(1):52–61. https://doi.org/10.23736/s0026-4806.18.05897-4
15. Gooßen K, Gräber S (2012) Longer term safety of dipeptidyl peptidase-4 inhibitors in patients with type 2 diabetes mellitus: systematic review and meta-analysis. Diabetes Obes Metab 14 (12):1061–1072
16. He J-H, Chen L-X, Li H (2019) Progress in the discovery of naturally occurring anti-diabetic drugs and in the identification of their molecular targets. Fitoterapia 134:270–289
17. Acharya KG, Shah KN, Solanki ND, Rana DA (2013) Evaluation of antidiabetic prescriptions, cost and adherence to treatment guidelines: a prospective, cross-sectional study at a tertiary care teaching hospital. J Basic Clin Pharm 4(4):82–87. https://doi.org/10.4103/0976-0105.121653
18. Donga J, Surani V, Sailor G, Chauhan S, Seth A (2011) A systematic review on natural medicine used for therapy of diabetes mellitus of some Indian medicinal plants. Pharm Sci Monit 2(1):36–72
19. Xu L, Li Y, Dai Y, Peng J (2018) Natural products for the treatment of type 2 diabetes mellitus: pharmacology and mechanisms. Pharmacol Res 130:451–465
20. Perla V, Jayanty SS (2013) Biguanide related compounds in traditional antidiabetic functional foods. Food Chem 138(2–3):1574–1580
21. Dzydzan O, Bila I, Kucharska AZ, Brodyak I, Sybirna N (2019) Antidiabetic effects of extracts of red and yellow fruits of cornelian cherries (Cornus mas L.) on rats with streptozotocin-induced diabetes mellitus. Food Funct 10(10):6459–6472. https://doi.org/10.1039/c9fo00515c
22. Matsui T, Ueda T, Oki T, Sugita K, Terahara N, Matsumoto K (2001) α-Glucosidase inhibitory action of natural acylated anthocyanins. 1. Survey of natural pigments with potent inhibitory activity. J Agric Food Chem 49(4):1948–1951
23. Bedekar A, Shah K, Koffas M (2010) Natural products for type II diabetes treatment. In: Advances in applied microbiology, vol 71. Elsevier, Amsterdam, pp 21–73
24. Luis Ros J, Francini F, Schinella GR (2015) Natural products for the treatment of type 2 diabetes mellitus. Planta Med 81(12–13):975–994. https://doi.org/10.1055/s-0035-1546131
25. Carter P, Gray LJ, Troughton J, Khunti K, Davies MJ (2010) Fruit and vegetable intake and incidence of type 2 diabetes mellitus: systematic review and meta-analysis. BMJ 341:c4229
26. Alam F, Islam A, Khalil I, Hua Gan S (2016) Metabolic control of type 2 diabetes by targeting the GLUT4 glucose transporter: intervention approaches. Curr Pharm Des 22(20):3034–3049

27. Bunel V, Ouedraogo M, Nguyen AT, Stévigny C, Duez P (2014) Methods applied to the in vitro primary toxicology testing of natural products: state of the art, strengths, and limits. Planta Med 80(14):1210–1226
28. David B, Wolfender J-L, Dias DA (2015) The pharmaceutical industry and natural products: historical status and new trends. Phytochem Rev 14(2):299–315
29. Ezuruike UF, Prieto JM (2014) The use of plants in the traditional management of diabetes in Nigeria: pharmacological and toxicological considerations. J Ethnopharmacol 155(2):857–924
30. Sun X, Yu W, Hu C (2014) Genetics of type 2 diabetes: insights into the pathogenesis and its clinical application. Biomed Res Int 2014:926713
31. Moloney MG (2016) Natural products as a source for novel antibiotics. Trends Pharmacol Sci 37(8):689–701
32. Ahmad NM (2016) Solubility-driven lead optimisation: recent examples and personal perspectives. Bioorg Med Chem Lett 26(13):2975–2979
33. Lee W-H, Loo C-Y, Young PM, Traini D, Mason RS, Rohanizadeh R (2014) Recent advances in curcumin nanoformulation for cancer therapy. Expert Opin Drug Deliv 11(8):1183–1201
34. Lepsik M, Sommer R, Kuhaudomlarp S, Lelimousin M, Paci E, Varrot A, Titz A, Imberty A (2019) Induction of rare conformation of oligosaccharide by binding to calcium-dependent bacterial lectin: X-ray crystallography and modelling study. Eur J Med Chem 177:212–220. https://doi.org/10.1016/j.ejmech.2019.05.049
35. Hunt JT (2009) Discovery of ixabepilone. Mol Cancer Ther 8(2):275–281
36. Hubbs JL, Fuller NO, Austin WF, Shen R, Creaser SP, McKee TD, Loureiro RM, Tate B, Xia W, Ives J (2012) Optimization of a natural product-based class of γ-secretase modulators. J Med Chem 55(21):9270–9282
37. Haupt VJ, Schroeder M (2011) Old friends in new guise: repositioning of known drugs with structural bioinformatics. Brief Bioinform 12(4):312–326
38. Novac N (2013) Challenges and opportunities of drug repositioning. Trends Pharmacol Sci 34 (5):267–272
39. Kuhn M, Campillos M, Letunic I, Jensen LJ, Bork P (2010) A side effect resource to capture phenotypic effects of drugs. Mol Syst Biol 6(1):343
40. Plake C, Schroeder M (2011) Computational polypharmacology with text mining and ontologies. Curr Pharm Biotechnol 12(3):449–457
41. Vilar S, Hripcsak G (2017) The role of drug profiles as similarity metrics: applications to repurposing, adverse effects detection and drug-drug interactions. Brief Bioinform 18 (4):670–681. https://doi.org/10.1093/bib/bbw048
42. Ab Ghani NS, Ramlan EI, Firdaus-Raih M (2019) Drug ReposER: a web server for predicting similar amino acid arrangements to known drug binding interfaces for potential drug repositioning. Nucleic Acids Res 47(W1):W350–W356. https://doi.org/10.1093/nar/gkz391
43. Spallone V (2019) Update on the impact, diagnosis and management of cardiovascular autonomic neuropathy in diabetes: what is defined, what is new, and what is unmet. Diabetes Metab J 43(1):3–30. https://doi.org/10.4093/dmj.2018.0259
44. Prabhakar P, Kumar A, Doble M (2014) Combination therapy: a new strategy to manage diabetes and its complications. Phytomedicine 21(2):123–130
45. Bijnsdorp IV, Giovannetti E, Peters GJ (2011) Analysis of drug interactions. In: Cancer cell culture. Springer, Berlin, pp 421–434
46. Association AD (2014) Diagnosis and classification of diabetes mellitus. Diabetes Care 37 (Suppl 1):S81–S90
47. Willett WC, Koplan JP, Nugent R, Dusenbury C, Puska P, Gaziano TA (2006) Prevention of chronic disease by means of diet and lifestyle changes. In: Disease control priorities in developing countries, 2nd edn. The International Bank for Reconstruction and Development/ The World Bank
48. van den Berk-Clark C, Secrest S, Walls J, Hallberg E, Lustman PJ, Schneider FD, Scherrer JF (2018) Association between posttraumatic stress disorder and lack of exercise, poor diet,

obesity, and co-occuring smoking: a systematic review and meta-analysis. Health Psychol 37 (5):407–416. https://doi.org/10.1037/hea0000593
49. Allende-Vigo MZ (2015) Diabetes mellitus prevention. Am J Ther 22(1):68–72
50. Anuradha CV (2013) Phytochemicals targeting genes relevant for type 2 diabetes. Can J Physiol Pharmacol 91(6):397–411
51. Choi C-Y, Cho S-S, Yoon I-S (2018) Hot-water extract of the branches of Hovenia dulcis Thunb. (Rhamnaceae) ameliorates low-fiber diet-induced constipation in rats. Drug Des Dev Ther 12:695–703. https://doi.org/10.2147/dddt.s150284
52. Wise J (2014) A third of adults in England have "prediabetes," study says. BMJ 348:g3791
53. Mukai N, Doi Y, Ninomiya T, Hirakawa Y, Nagata M, Yoshida D, Hata J, Fukuhara M, Nakamura U, Kitazono T (2014) Trends in the prevalence of type 2 diabetes and prediabetes in community-dwelling Japanese subjects: the Hisayama study. J Diabet Invest 5(2):162–169
54. WHO (2013) WHO traditional medicine strategy: 2014–2023. World Health Organization, Geneva
55. Ríos JL, Francini F, Schinella GR (2015) Natural products for the treatment of type 2 diabetes mellitus. Planta Med 81(12/13):975–994
56. Yao H, Liu J, Xu S, Zhu Z, Xu J (2017) The structural modification of natural products for novel drug discovery. Expert Opin Drug Discovery 12(2):121–140
57. Bauer A, Brönstrup M (2014) Industrial natural product chemistry for drug discovery and development. Nat Prod Rep 31(1):35–60
58. Chen J, Li W, Yao H, Xu J (2015) Insights into drug discovery from natural products through structural modification. Fitoterapia 103:231–241
59. Di S, Han L, Wang Q, Liu X, Yang Y, Li F, Zhao L, Tong X (2018) A network pharmacology approach to uncover the mechanisms of Shen-qi-di-huang decoction against diabetic nephropathy. Evid Based Complement Alternat Med 2018:7043402
60. Alam F, Islam MA, Kamal MA, Gan SH (2018) Updates on managing type 2 diabetes mellitus with natural products: towards antidiabetic drug development. Curr Med Chem 25 (39):5395–5431
61. Clarence O, Donatus B, Samuel E (2014) Prophylaxis and treatment of types 1 and 2 diabetes mellitus. Int J Dis Disord 2:65–73
62. Newman DJ, Cragg GM (2016) Natural products as sources of new drugs from 1981 to 2014. J Nat Prod 79(3):629–661

Chapter 2
An Overview of Hypoglycemic Modern Drugs

Haixia Chen and Yangpeng Lu

Abstract Modern hypoglycemic drugs mainly include insulin sensitizers, secretagogues (increase insulin output from the pancreas), which take effect by reducing blood glucose sources and increasing blood glucose pathways, including insulin and its analogs, sulfonylurea secretions, metformin, α-glucosidase enzyme inhibitors, thiazolidinediones derivative sensitizers, phenanthrene derivatives secretion enhancers, GLP-1 receptor agonists, and DPP-4 enzyme inhibitors. This article mainly describes the classification of hypoglycemic drugs, mechanism of action, characteristics, representative drugs, and their limitations.

Keywords Hypoglycemic · Blood glucose · Sulfonylurea secretions · Metformin · α-Glucosidase enzyme inhibitors · Thiazolidinediones derivative sensitizers · Phenanthrene derivatives secretion enhancers · GLP-1 receptor agonists · DPP-4 enzyme inhibitors · Mechanism · Characteristics

1 Introduction

Diabetes is a group of metabolic diseases characterized by high blood sugar. Hyperglycemia is caused by defective insulin secretion, impaired biological effects, or both. Chronic hyperglycemia in diabetes leads to chronic damage and dysfunction of various tissues, especially eyes, kidneys, heart, blood vessels, and nerves. In the past, drug treatment, diet control, and exercise were recognized by the medical community as the "troika" for the treatment of diabetes. If a diabetic patient's blood sugar is still not under control after diet and exercise treatment, he or she may need to use drug treatment [1–3].

Drugs used for diabetes can treat diabetes by reducing the level of glucose in the blood. In addition to insulin, exenatide, liraglutide, and pramlintide, all drugs are

H. Chen (✉) · Y. Lu
Tianjin Key Laboratory for Modern Drug Delivery and High-Efficiency, School of Pharmaceutical Science and Technology, Tianjin University, Tianjin, China
e-mail: chenhx@tju.edu.cn

H. Chen, M. Zhang (eds.), *Structure and Health Effects of Natural Products on Diabetes Mellitus*, https://doi.org/10.1007/978-981-15-8791-7_2

oral, so they are also called oral hypoglycemic drugs or oral hypoglycemic drugs. There are different types of antidiabetic drugs, and their choice depends on the nature of diabetes, people's age and condition, and other factors [4–7].

Type 1 diabetes, formerly known as insulin-dependent diabetes, occurs mostly in children and adolescents and can also occur at various ages. They must be given insulin treatment due to the absolutely insufficient insulin in the body; otherwise, it will be life-threatening [8, 9]. While type 2 diabetes is a disease in which cells are resistant to insulin, it is the most common type of diabetes whose treatment includes: (1) drugs to increase the amount of insulin secreted by the pancreas, (2) drugs to increase the sensitivity of target organs to insulin, and (3) drugs to reduce the absorption rate of glucose in the gastrointestinal tract [10].

It is mainly divided into six types according to the classification of pathophysiological mechanism, including insulin sensitizers, secretagogues, GLP-1 receptor agonist, DPP-4, α-glucosidase inhibitor, and SGLT-2 inhibitor. This article mainly describes the classification of hypoglycemic modern drugs, mechanism of action, characteristics, representative drugs, and their limitations [11–14].

2 Insulin Sensitizers

Insulin sensitizers solve the core problem of insulin resistance in type 2 diabetes including the following.

2.1 Biguanides

Biguanides work by reducing glucose output from the liver and increasing the absorption of glucose by surrounding tissues, including skeletal muscles [15].

Take metformin, the first-line drug for the treatment of type 2 diabetes for example. Its A1C value is typically reduced to 1.5–2.0%. It has quick-release and slow-release preparations, which are usually reserved for patients with gastrointestinal side effects. It can also be used in combination with other oral diabetes drugs [16].

Metformin hydrochloride is the first choice of hypoglycemic drugs. This kind of drugs does not stimulate β cells of islets of Langerhans and has little effect on normal people but has obvious hypoglycemic effect on type 2 diabetic patients. It does not affect insulin secretion, mainly by promoting the uptake of glucose by peripheral tissues, inhibiting gluconeogenesis, reducing the output of liver glycogen, and delaying the absorption of glucose in the intestinal tract, so as to achieve the effect of lowering blood glucose. Commonly used drug is metformin [17]. Metformin (Fortamet, Glucophage, Glumetza): its hypoglycemic effect is weaker than that of phenformin, but its toxicity is smaller, and it has no hypoglycemic effect on normal people. When compared with sulfonylurea, this product does not stimulate insulin secretion, so it rarely causes hypoglycemia. In addition, this product has the effect of increasing insulin receptor and reducing insulin resistance, as well as improving fat metabolism and fibrinolysis, and reducing insulin resistance and platelet aggregation. Metformin, which is beneficial to alleviate the occurrence and development of cardiovascular complications, is the first choice drug for children and overweight and obese patients with type 2 or type 1 diabetes. It can reduce the amount of insulin and also be used in the treatment of insulin resistance syndrome. Because of its large response to the gastrointestinal tract, it should be taken during or after meals. It is forbidden for patients with renal function impairment [18, 19].

Metformin is the most common type 2 diabetes drug in children and adolescents, although metformin must be used with caution in patients with impaired liver or kidney function. Among the common diabetes drugs, metformin is the only widely used oral drug that does not cause weight gain.

However, the use of metformin is now limited due to the increased risk of involving lactic acidosis [20].

2.2 *Thiazolidinediones*

Thiazolidinediones (TZDs), also known as "glitazones," bind to PPARγ, a type of nuclear regulatory protein involved in transcription of genes regulating glucose and fat metabolism. These PPARs act on peroxisome proliferator responsive elements (PPREs). The PPREs influence insulin-sensitive genes, which enhance the production of mRNAs of insulin-dependent enzymes. The final result is better use of glucose by the cells [21].

Typically, its reductions in glycated hemoglobin (A1C) values are 1.5–2.0%, including:

Rosiglitazone (Avandia): The European Medicines Agency recommended in September 2010 that it will be suspended from the EU market due to elevated cardiovascular risks.

Pioglitazone (Actos): It remains on the market but has also been associated with increased cardiovascular risks.

Troglitazone (Rezulin): It was used in 1990s and later withdrawn due to hepatitis and liver damage risks.

However, pioglitazone may decrease the overall incidence of cardiac events in people with type 2 diabetes who have already had a heart attack [22–25].

2.3 *Lyn Kinase Activators*

The LYN kinase activator tolimidone has been reported to potentiate insulin signaling in a manner that is distinct from the glitazones. The compound has demonstrated positive results in a Phase 2a clinical study involving 130 diabetic subjects [26–30].

3 Secretagogues

Insulin secretagogues are the first-line antidiabetic agents, which are sulfonylureas and non-sulfonylureas. It mainly plays a role by promoting insulin secretion, inhibiting ATP-dependent potassium channel, making K^+ outflow, β cell depolarization, Ca^{2+} inflow, and inducing insulin secretion. In addition, it can strengthen the binding between insulin and its receptor, relieve the effect of insulin resistance after receptor, and enhance the effect of insulin, including sulfonylureas and non-sulfonylurea secretagogues [31–35].

3.1 Sulfonylureas Secretagogues

Glipizide (Glucotrol, Glucotrol XL) As the second generation of sulfonylureas, it has a fast onset, and its efficacy can last for 6–8 h in human body, especially effective in reducing postprandial hyperglycemia; because of its inactive metabolites and rapid excretion, it causes less hypoglycemia than glibenclamide, and it is suitable for the elderly.

Gliclazide (Diamicron, Diaprel, Azukon) [14] As the second generation of sulfonylurea, its efficacy is more than 10 times stronger than that of the first generation of tolbutamide; in addition, it has the effect of inhibiting platelet adhesion and aggregation, which can effectively prevent the formation of microthrombosis, thus preventing the microvascular disease of type 2 diabetes. It is suitable for adult type 2 diabetes, with obesity or vascular disease. It should be used with caution in the elderly and those with renal dysfunction.

Glibenclamide (Diabeta, Flycron, Glyburide) [36, 37] It is the second generation of sulfonylureas. It has the strongest hypoglycemic effect among all sulfonylureas, 200–500 times of tolbutamide, and its effect can last for 24 h. It can be used in mild and moderate noninsulin-dependent type 2 diabetes, but it is prone to hypoglycemia. The elderly and those with renal insufficiency should use it with caution.

Glimepiride (Amaryl) [36] It is the third generation of oral sulfonylureas. Its mechanism of action is the same as other sulfonylureas, but it can increase the intake of cardiac glucose through a way unrelated to insulin, less affecting the cardiovascular system than other oral hypoglycemic drugs. Its half-life can be up to 9 h, only once a day. It is suitable for noninsulin-dependent type 2 diabetes.

Gliquidone (Glurenorm) It is the second generation of oral sulfonylurea hypoglycemic agents, which is highly active islet beta cell friendly agent, can be combined with the specific receptor on the islet beta cell membrane to induce the production of appropriate insulin to reduce blood glucose concentration. After 2–2.5 h of oral administration, the drug reached the highest blood concentration and was absorbed completely. The half-life of plasma is 1.5 h, the metabolism is complete, the metabolites do not have hypoglycemic effect, and most of the metabolites are excreted through the digestive system of bile duct. It is suitable for mild and moderate noninsulin-dependent type 2 diabetes mellitus with unsatisfactory effect of diet control alone, and the β cells of the islets of Langerhans have certain insulin secretion function without serious complications.

All these drugs are commonly used sulfonylurea antidiabetic drugs with antidiabetic intensity glibenclamide > glipizide > gliquidone > gliclazide [38–40].

3.2 *Non-sulfonylurea Secretagogues*

This kind of drug binds to ATP-dependent K^+ (K_{ATP}) channel on cell membrane of pancreatic beta cells with similar manner to sulfonylurea secretagogues, but they tend to have weaker binding affinity and faster dissociation from the SUR1-binding site. They increase the concentration of intracellular potassium and cause the electric potential over the membrane to become more positive. This depolarization opens voltage-gated Ca^{2+} channels. The rise in intracellular calcium leads to increased fusion of insulin granula in the cell membrane and therefore increased secretion of (pro)insulin.

The representative drugs include repaglinide (Prandin), which gained FDA approval in 1997, nateglinide (Starlix), and mitiglinide (Glufast).

However, their side effects include both weight gain and hypoglycemia. While the potential for hypoglycemia is less than for those on sulfonylureas, it is still a serious potential side effect that can be life threatening. Patients on this medication should know the signs and symptoms of hypoglycemia and appropriate management.

Repaglinide (Prandin) can cause an increased incidence in male rats of benign adenomas (tumors) of the thyroid and liver. While no such effect is seen with another drug of this class, nateglinide (Starlix) [41–43].

4 α-Glucosidase Inhibitor

α-Glucosidase inhibitors are the first-line antidiabetic drugs. Although they have no direct effect on insulin secretion or sensitivity, they are not antidiabetic drugs in essence [44]. But these drugs slow down the digestion of starch in the small intestine, making glucose in meal powder enter the blood stream more slowly, and can be matched more effectively by impaired insulin response or sensitivity. These drugs are only effective in the early stages of impaired glucose tolerance but may be helpful in combination with other type 2 diabetes drugs [45–47].

The typical reduction in A1C was 0.5–1.0%. These drugs competitively inhibit maltase, glucoamylase, and sucrase, block the hydrolysis of 1,4-glycoside bond, delay the decomposition of starch, sucrose, and maltose into glucose in the small intestine, and reduce postprandial blood glucose. Its commonly used drugs are: sugar-100, acarbose, and voglibose.

Sugar-100 The main ingredient btd-1 is a soybean fermentation extract, which regulates the rapid increase of blood sugar after meals. It is produced by *Bacillus subtilis* Mori using defatted soybean meal. 1-Deoxynojirimycin (DNJ) produced by *Bacillus subtilis* Mori has a very good inhibitory activity on α-glucosidase in small intestinal villi.

Acarbose (Betacarbose) It does not cause hypoglycemia or affect body weight when used alone; it can be combined with other oral hypoglycemic drugs and insulin. It can be used in various types of diabetes mellitus to improve postprandial blood sugar of type 2 diabetes patients, and it can also be used in patients on other oral hypoglycemic drugs with no obvious effect.

Voglibose It is a new generation of α-glucosidase inhibitor. The inhibitory effect of the drug on α-glucosidase (maltase, isomaltase, glucosidase) in small intestinal mucosa was stronger than that of acarbose, and the inhibitory effect on α-amylase from pancreas was weaker. It can be used as the first choice medicine for type 2 diabetes and can also be combined with other oral hypoglycemic drugs and insulin.

These drugs are rarely used in the USA because they have serious side effects (e.g., flatulence and bloating). They are more common in Europe. They do have the potential to lose weight by reducing glucose metabolism [48–52].

5 SGLT-2 Inhibitors

Sodium-dependent glucose transporters 2, SGLT-2 inhibitors, can inhibit the renal reabsorption of glucose, so that excess glucose is excreted through the urine, thus reducing blood sugar. This is a new type of antidiabetic drug.

SGLT-2 was mainly expressed in the kidney, while SGLT-1 was partially expressed in the kidney and mainly expressed in the intestine. About 90% of glucose was reabsorbed by SGLT-2 in S1 segment and about 10% by SGLT-1 in S3 segment. That is to say, SGLT-2 plays a major role in glucose reabsorption. SGLT-2 transports 90% of glucose to the kidney, while SGLT-1 only accounts for 10% of the rest. Therefore, SGLT-2 inhibitor can block the reabsorption of glucose by proximal convoluted tubules and excrete excess glucose through urine, so as to reduce blood glucose.

SGLT-2 inhibitor has high selectivity and specificity. It can reduce the reabsorption of glucose by proximal convoluted tubules, increase the excretion of urine glucose, and thus reduce the blood glucose level. Several studies have shown that SGLT-2 inhibitors have different degrees of decrease in HbA1c in patients with type 2 diabetes. Fasting plasma glucose (FPG) also decreased significantly.

Kidney plays a very important role in glucose metabolism. Glucose is filtered in glomerulus and reabsorbed in renal proximal convoluted tubules. Glucose cannot freely pass through lipid bilayer of cell membrane in organism, and it must rely on glucose transporter on cell membrane. Sodium-dependent glucose transporters (SGLTs) are a family of transporters found in small intestinal mucosa and renal proximal convoluted tubules. The process of renal reabsorption of glucose is mainly mediated by SGLTs. Among them, SGLT-1 and SGLT-2 are the most important, and SGLT-2 plays a leading role. SGLT-1 is mainly distributed in the brush border of the small intestine and the S3 segment of the renal proximal convoluted tubule, which is expressed in the heart and trachea, and it is a kind of transporter with high affinity and low transport capacity. SGLT-2 is mainly distributed in S1 segment of renal proximal convoluted tubules, which is a kind of transporter with low affinity and high transport capacity. Its main physiological function is to complete the reabsorption of 90% glucose in glomerular filtration fluid in renal proximal convoluted tubules, and the rest 10% is completed by SGLT-1. It was found that SGLT-1 gene mutation can lead to severe diarrhea, even life threatening, while SGLT-2 gene mutation can lead to 140 g·D-1 renal sugar excretion and no obvious adverse reactions.

SGLT-2 selective inhibitors, as a new target of hypoglycemic drugs, have no significant effect on other tissues and organs due to their specific distribution in the kidney; diabetic patients with insulin resistance can still benefit. They are not prone to hypoglycemia and do not increase the weight of diabetic patients. SGLT-2 inhibitor has become a new research hotspot at home and abroad. The first SGLT-2 inhibitor was natural product phlorizin, but it was easy to be hydrolyzed into glucosides and phlorizin by glycosidase in vivo. Moreover, it had poor selectivity to SGLT-1 and SGLT-2 and had large adverse reactions, so it did not become a treatment drug for diabetes. At present, there are six kinds of SGLT-2 inhibitors listed in the world, namely canagliflozin, dapagliflozin, empagliflozin, ipragliflozin, luseogliflozin, and tofogliflozin. Among them, dapagliflozin has been approved by the State Food and Drug Administration (CFDA) to be listed in China.

It was found that SGLT-2 inhibitor can protect the function of islet β cells. By comparing with control group, there were small decrease of MAGE in the dapagliflozin group (17.1% [95% CI – 0.7, 30.8], $p = 0.042$) and a small, non-significant, reduction in the exercise group (15.3% [95% CI – 1.2, 29.1], $p = 0.067$), while MAGE kept unchanged in the metformin group (0.1% [95% CI – 16.1, 19.4], $p = 0.991$) [53].

Therefore, SGLT-2 inhibitor is a new type of hypoglycemic drug at present, which has a unique hypoglycemic pathway independent of insulin secretion. The existing clinical research shows that SGLT-2 inhibitor has a very definite and effective hypoglycemic effect no matter it is used alone or in combination. In 2015, the American Diabetes Association (ADA) and the European Association for the Study of Diabetes (EASD) both recommended SGLT-2 inhibitors as second-line and third-line drugs for type 2 diabetes, which can be used in combination with dimethyl biguanide or other hypoglycemic drugs [54–58]. To conclude, we list them in Table 2.1.

6 Conclusions and Future Prospects

Different modern drugs in the treatment of diabetes can have differnt action mechanism and they may have differnt side effects. GLP-1 agonists, such as exenatide, can cause acute pancreatitis, pancreatic cancer risk, and renal function changes. SGLT2 inhibitors, such as canglitazine, can cause severe reproductive tract infection, ketoacidosis risk, leg and foot amputation risk, pancreatitis, and so on. With further research, these problems may be solved in the future, and more new drugs will be found.

At the same time, it is very important for diabetes prevention and treatment to change diet, exercise, and weight loss. According to relevant statistics, type 2 diabetes mellitus (85–90% of all cases in the world) can usually be prevented or delayed by maintaining normal weight, exercising, and eating healthy. A higher level of physical activity (more than 90 min per day) reduces the risk of diabetes by 28%. Dietary changes known to be effective in preventing diabetes include maintaining a diet rich in whole grains and fiber, selecting high-quality fats, such as nuts, vegetable oil, and polyunsaturated fats found in fish, limiting sugary drinks, and eating less red meat and other saturated fats, which can also help prevent diabetes. Smoking is also associated with an increased risk of diabetes and its complications, so quitting smoking is also an important preventive measure [60]. Weight loss can prevent the development from prediabetes to type 2 diabetes, reduce the risk of cardiovascular disease, or lead to partial remission of diabetic patients [61, 62]. There is no single diet for all diabetics. Although dietary evidence does not support other diets, it is generally recommended to adopt healthy eating habits, such as Mediterranean diet, low carbohydrate diet, or DASH diet [61, 62]. According to ADA, "reducing the total carbohydrate intake of diabetic patients has shown the best evidence for improving blood glucose." It is a feasible method for type 2 diabetic patients who cannot reach the blood glucose target or give priority to reducing antiblood glucose drugs by low carbohydrate diet [62]. For overweight type 2 diabetic patients, any diet that can reduce weight is effective [63, 64]. Beside the healthy management, the new drugs can be discovered based on the new action mechanism with high efficacy and low side effects.

Table 2.1 Summary of antidiabetic drugs

Type [59]	Representative drugs	Mechanism [10]	Advantages [59]	Disadvantages [59]
Sulfonylurea secretagogues	Glyburide, glimepiride, glipizide	Stimulating insulin release by pancreatic beta cells by inhibiting the K_{ATP} channel	Low cost, fast in effect, no effect on blood pressure or low-density lipoprotein Lower risk of gastrointestinal problems than with metformin and dosing is more convenient	Causes an average of 5–10 pounds weight gain, increases risk of hypoglycemia, glyburide, increases risk of hypoglycemia slightly more as compared with glimepiride and glipizide
Non-sulfonylurea secretagogues	Meglitinides, Repaglinide, Nateglinide	Help pancreas produce insulin, act on the same potassium channels as sulfonylureas, but at a different binding site [33]. By closing the potassium channels of the pancreatic beta cells, they open the calcium channels, thereby enhancing insulin secretion [34]	Absorbed rapidly by oral administration, with the onset time of 0–30 min. The plasma drug concentration reaches the peak value within 1 h after taking the medicine, then drops rapidly, and is cleared within 4–6 h. The half-life of plasma is about 1 h, which is suitable for reducing postprandial blood sugar	Weight gain and hypoglycemia
Sensitizers: Thiazolidinediones	Pioglitazone, Rosiglitazone	Reduce insulin resistance by activating PPAR-γ in fat and muscle	Lower risk of hypoglycemia Slight increase in high-density lipoprotein Actos linked to decreased triglycerides Convenient dosing	Increased risk of heart failure, triglycerides, and risk of heart attack, bladder cancer Causes weight gain, higher risk of edema, anemia, limb fractures Increases low-density lipoprotein Slower onset of action Expensive

(continued)

Table 2.1 (continued)

Type [59]	Representative drugs	Mechanism [10]	Advantages [59]	Disadvantages [59]
Sensitizers: Biguanides	Metformin, Phenformin, Buformin	Acts on the liver to reduce gluconeogenesis and causes a decrease in insulin resistance via increasing AMPK signaling	Not associated with weight gain Low risk of hypoglycemia as compared to alternatives Good effect on LDL cholesterol Decreases triglycerides No effect on blood pressure Low cost	Increased risk of gastrointestinal problems contraindicated for people in shock; with kidney disease, or at risk for impaired kidney function from intravenous dye, and with acute or chronic metabolic acidosis Risk of lactic acidosis also is increased for people with unstable or acute heart failure, liver disease, or alcoholism, or who are recovering from major surgery Increased risk of vitamin B12 deficiency [10] Less convenient dosing Metallic taste [10]
Alpha-glucosidase inhibitor	Acarbose, Miglitol, Voglibose	Reduces glucose absorbance by acting on small intestine to cause decrease in production of enzymes needed to digest carbohydrates	Slightly decreased risk of hypoglycemia as compared to sulfonylurea Not associated with weight gain Decreases triglycerides No effect on cholesterol	Less effective than most other diabetes pills in decreasing glycated hemoglobin Increased risk of GI problems than other diabetes pills except metformin Inconvenient dosing Expensive

(continued)

Table 2.1 (continued)

Type [59]	Representative drugs	Mechanism [10]	Advantages [59]	Disadvantages [59]
Vanadium	Sodium Orthovanadate	Protein tyrosine phosphatase (PTP1B) inhibitors plays an important role in the pathogenesis of diabetes	Both regulating glucose metabolism and also altering lipid metabolism	Safety yet unknown
Peptide analogs: SGLT-2 inhibitors	Dapagliflozin, Canagliflozin, Empagliflozin	Amylin agonist analogues slow down gastric emptying and suppress glucagon. They have all the incretin actions except stimulation of insulin secretion	SGLT-2 inhibitors block the reuptake of glucose in the renal tubules, promoting loss of glucose in the urine, causes mild weight loss and reduces blood sugar levels with little risk of hypoglycemia [55]. Oral preparations may be available alone or in combination with other agents [56]. Along with GLP-1 agonists, they are considered preferred second or third agents for type 2 diabetics suboptimally controlled with metformin alone, according to the most recent clinical practice guidelines	The side effects of SGLT-2 inhibitors are derived directly from their mechanism of action; these include an increased risk of: ketoacidosis, urinary tract infections, candidal vulvovaginitis, and hypoglycemia [58]

References

1. Nield L, Summerbell CD, Hooper L, Whittaker V, Moore H (2008) Dietary advice for the prevention of type 2 diabetes mellitus in adults. Cochrane Database Syst Rev (3):CD005102. https://doi.org/10.1002/14651858.CD005102.pub2.hdl:10149/92337. Nield L (ed.). PMID 18646120. (Retracted)

2. Schwingshackl L, Hoffmann G, Lampousi A-M, Knüppel S, Iqbal K, Schwedhelm C, Bechthold A, Schlesinger S, Boeing H (2017) Food groups and risk of type 2 diabetes mellitus: a systematic review and meta-analysis of prospective studies. Eur J Epidemiol 32(5):363–375. https://doi.org/10.1007/s10654-017-0246-y. ISSN 0393-2990. PMC 5506108. PMID 28397016
3. Xi B, Li S, Liu Z, Tian H, Yin X, Huai P, Tang W, Zhou D, Steffen LM (2014) Intake of fruit juice and incidence of type 2 diabetes: a systematic review and meta-analysis. PLoS One 9(3): e93471. Bibcode:2014PLoSO...993471X. doi:10.1371/journal.pone.0093471. PMC 3969361. PMID 24682091
4. Reynolds A, Mann J, Cummings J, Winter N, Mete E, Te Morenga L (2019) Carbohydrate quality and human health: a series of systematic reviews and meta-analyses. Lancet 393 (10170):434–445. https://doi.org/10.1016/S0140-6736(18)31809-9. PMID 30638909
5. Haw JS, Galaviz KI, Straus AN, Kowalski AJ, Magee MJ, Weber MB, Wei J, Narayan KM, Ali MK (2017) Long-term sustainability of diabetes prevention approaches: a systematic review and meta-analysis of randomized clinical trials. JAMA Intern Med 177(12):1808–1817. https://doi.org/10.1001/jamainternmed.2017.6040. PMC 5820728. PMID 29114778
6. Seida JC, Mitri J, Colmers IN, Majumdar SR, Davidson MB, Edwards AL, Hanley DA, Pittas AG, Tjosvold L, Johnson JA (2014) Clinical review: effect of vitamin D3 supplementation on improving glucose homeostasis and preventing diabetes: a systematic review and meta-analysis. J Clin Endocrinol Metab 99(10):3551–3560. https://doi.org/10.1210/jc.2014-2136. PMC 4483466. PMID 25062463
7. McBrien K, Rabi DM, Campbell N, Barnieh L, Clement F, Hemmelgarn BR, Tonelli M, Leiter LA, Klarenbach SW, Manns BJ (2012) Intensive and standard blood pressure targets in patients with type 2 diabetes mellitus: systematic review and meta-analysis. Arch Intern Med 172 (17):1296–1303. https://doi.org/10.1001/archinternmed.2012.3147. PMID 22868819
8. Atkinson MA, Eisenbarth GS, Michels AW (2014) Type 1 diabetes. Lancet 383(9911):69–82
9. Daneman D (2006) Type 1 diabetes. Lancet 367(9513):847–858
10. DeFronzo RA (1999) Pharmacologic therapy for type 2 diabetes mellitus. Ann Intern Med 131 (4):281–303
11. Boussageon R, Bejan-Angoulvant T, Saadatian-Elahi M, Lafont S, Bergeonneau C, Kassaï B, Erpeldinger S, Wright JM, Gueyffier F, Cornu C (2011) Effect of intensive glucose lowering treatment on all cause mortality, cardiovascular death, and microvascular events in type 2 diabetes: meta-analysis of randomised controlled trials. BMJ 343:d4169. https://doi.org/10.1136/bmj.d4169. PMC 3144314. PMID 21791495
12. Webster MW (2011) Clinical practice and implications of recent diabetes trials. Curr Opin Cardiol 26(4):288–293. https://doi.org/10.1097/HCO.0b013e328347b139. PMID 21577100
13. Inzucchi SE, Bergenstal RM, Buse JB, Diamant M, Ferrannini E, Nauck M, Peters AL, Tsapas A, Wender R, Matthews DR (2015) Management of hyperglycaemia in type 2 diabetes, 2015: a patient-centred approach. Update to a position statement of the American Diabetes Association and the European Association for the Study of diabetes. Diabetologia 58(3):429–442. https://doi.org/10.1007/s00125-014-3460-0. PMID 25583541
14. Joint Formulary Committee (2015) British national formulary (bnf) 69
15. Yaribeygi H, Ashrafizadeh M, Henney NC, Sathyapalan T, Jamialahmadi T, Sahebkar A (2020) Neuromodulatory effects of anti-diabetes medications: a mechanistic review. Pharmacol Res 152:104611. https://doi.org/10.1016/j.phrs.2019.104611
16. Warncke K, Kummer S, Kann PH, Bergis D, Bollow E, Hummel M, Zoicas F, Wernert S, Holl RW (2020) Cardivascular risk profile in patients with diabetes and acromegaly or Cushing's disease—analysis from the DPV database. Exp Clin Endocrinol Diabetes 128(2):104–110. https://doi.org/10.1055/a-0600-9649
17. Maruthur NM, Tseng E, Hutfless S, Wilson LM, Suarez-Cuervo C, Berger Z et al (2016) Diabetes medications as monotherapy or metformin-based combination therapy for type 2 diabetes: a systematic review and meta-analysis. Ann Intern Med 164(11):740–751

18. Kameda T, Kumamaru H, Nishimura S, Kohsaka S, Miyata H (2020) Use of oral antidiabetic drugs in Japanese working-age patients with type 2 diabetes mellitus: dosing pattern for metformin initiators. Curr Med Res Opin. https://doi.org/10.1080/03007995.2020.1729710
19. Ihana-Sugiyama N, Sugiyama T, Tanaka H, Ueki K, Kobayashi Y, Ohsugi M (2020) Comparison of effectiveness and drug cost between dipeptidyl peptidase-4 inhibitor and biguanide as the first-line anti-hyperglycaemic medication among Japanese working generation with type 2 diabetes. J Eval Clin Pract 26(1):299–307. https://doi.org/10.1111/jep.13171
20. Ena J, Carretero-Gomez J, Zapatero-Gaviria A, Carrasco Sanchez FJ, Del Romero-Sanchez M, Gonzalez-Becerra C, Blazquez-Encinar JC, Iguzquiza-Pellejero MJ, de Escalante Yanguela B, Gomez-Huelgas R, en representacion del Grupo de Trabajo de Diabetes OyN, de la Sociedad Espanola de Medicina I, Investigadores del Grupo de Trabajo de Diabetes OyNd, la Sociedad Espanola de Medicina I (2020) Use of antihyperglycaemic therapy with cardiovascular benefit in patients with type 2 diabetes who require hospitalisation: a cross-sectional study (Uso de terapia antihiperglucemiante con beneficio cardiovascular en pacientes con diabetes tipo 2 que requieren hospitalizacion: un estudio transversal.). Revista clinica espanola. https://doi.org/10.1016/j.rce.2019.12.009
21. Yuan Z, DeFalco F, Wang L, Hester L, Weaver J, Swerdel JN, Freedman A, Ryan P, Schuemie M, Qiu R, Yee J, Meininger G, Berlin JA, Rosenthal N (2020) Acute pancreatitis risk in type 2 diabetes patients treated with canagliflozin versus other antihyperglycemic agents: an observational claims database study. Curr Med Res Opin:1–8. https://doi.org/10.1080/03007995.2020.1761312
22. Rigato M, Avogaro A, Vigili de Kreutzenberg S, Fadini GP (2020) Effects of basal insulin on lipid profile compared to other classes of Antihyperglycemic agents in type 2 diabetic patients. J Clin Endocrinol Metab 105(7):dgaa178. https://doi.org/10.1210/clinem/dgaa178
23. Kim SG, Kim KJ, Yoon KH, Chun SW, Park KS, Choi KM, Lim S, Mok J-O, Lee HW, Seo JA, Cha B-S, Kim MK, Shon HS, Choi DS, Kim DM (2020) Efficacy and safety of lobeglitazone versus sitagliptin as an add-on to metformin in type 2 diabetes with metabolic syndrome over 24 weeks. Diabetes Obes Metab. https://doi.org/10.1111/dom.14085
24. Khunti K, Davies MJ, Seidu S (2020) Cardiovascular outcome trials of glucose-lowering therapies. Expert Rev Pharmacoecon Outcomes Res:1–13. https://doi.org/10.1080/14737167.2020.1763796
25. Alzhrani ZMM, Alam MM, Neamatallah T, Nazreen S (2020) Design, synthesis and invitro antiproliferative activity of new thiazolidinedione-1,3,4-oxadiazole hybrids as thymidylate synthase inhibitors. J Enzyme Inhib Med Chem 35(1):1116–1123. https://doi.org/10.1080/14756366.2020.1759581
26. Song W, Li D, Tao L, Luo Q, Chen L (2020) Solute carrier transporters: the metabolic gatekeepers of immune cells. Acta Pharm Sin B 10(1):61–78. https://doi.org/10.1016/j.apsb.2019.12.006
27. Nair A, Chakraborty S, Banerji LA, Srivastava A, Navare C, Saha B (2020) Ras isoforms: signaling specificities in CD40 pathway. Cell Commun Signal 18(1). https://doi.org/10.1186/s12964-019-0497-1
28. Lipinski CA, Reaume AG (2020) High throughput in vivo phenotypic screening for drug repurposing: discovery of MLR-1023 a novel insulin sensitizer and novel Lyn kinase activator with clinical proof of concept. Bioorg Med Chem 28(9). https://doi.org/10.1016/j.bmc.2020.115425
29. Lee M-K, Kim SG, Watkins E, Moon MK, Rhee SY, Frias JP, Chung CH, Lee S-H, Block B, Cha BS, Park HK, Kim BJ, Greenway F (2020) A novel non-PPARgamma insulin sensitizer: MLR-1023 clinical proof-of-concept in type 2 diabetes mellitus. J Diabetes Complicat 34(5). https://doi.org/10.1016/j.jdiacomp.2020.107555
30. Severin F, Frezzato F, Visentin A, Martini V, Trimarco V, Carraro S, Tibaldi E, Brunati AM, Piazza F, Semenzato G, Facco M, Trentin L (2019) In chronic lymphocytic Leukemia the JAK2/STAT3 pathway is constitutively activated and its inhibition leads to CLL cell death

unaffected by the protective bone marrow microenvironment. Cancers 11(12). https://doi.org/10.3390/cancers11121939
31. Sovetkina A, Nadir R, Fung JNM, Nadjarpour A, Beddoe B (2020) The physiological role of ghrelin in the regulation of energy and glucose homeostasis. Cureus 12(5). https://doi.org/10.7759/cureus.7941
32. Peterson QP, Veres A, Chen L, Slama MQ, Kenty JHR, Hassoun S, Brown MR, Dou H, Duffy CD, Zhou Q, Matveyenko AV, Tyrberg B, Sorhede-Winzell M, Rorsman P, Melton DA (2020) A method for the generation of human stem cell-derived alpha cells. Nat Commun 11 (1):2241–2241. https://doi.org/10.1038/s41467-020-16049-3
33. Elabadlah H, Hameed R, D'Souza C, Mohsin S, Adeghate EA (2020) Exogenous ghrelin increases plasma insulin level in diabetic rats. Biomol Ther 10(4). https://doi.org/10.3390/biom10040633
34. Cunha V, Cotrim HP, Rocha R, Carvalho K, Lins-Kusterer L (2020) Metformin in the prevention of hepatocellular carcinoma in diabetic patients: a systematic review. Ann Hepatol 19(3):232–237. https://doi.org/10.1016/j.aohep.2019.10.005
35. Abusamaan MS, Klonoff DC, Mathioudakis N (2020) Predictors of time-to-repeat point-of-care glucose following hypoglycemic events in hospitalized patients. J Diabetes Sci Technol 14 (3):526–534. https://doi.org/10.1177/1932296819883332
36. British national formulary: BNF 76 (76 ed.)
37. Makrilakis K (2007) Diabetes in clinical practice: questions and answers from case studies. Wiley, Hoboken, p 342
38. Wu J-X, Ding D, Wang M, Chen L (2020) Structural insights into the inhibitory mechanism of insulin Secretagogues on the pancreatic ATP-sensitive potassium channel. Biochemistry 59 (1):18–25. https://doi.org/10.1021/acs.biochem.9b00692
39. Wang J-S, Chen H, Tang F, Sheu WH-H (2020) Associations of fear of hypoglycemia with second-line use of insulin secretagogues or insulin and subsequent glycemic control in patients with type 2 diabetes: an analysis using data from the DISCOVER study. Int J Clin Pract. https://doi.org/10.1111/ijcp.13485
40. Mekuria AN, Ayele Y, Tola A, Mishore KM (2019) Monotherapy with metformin versus sulfonylureas and risk of cancer in type 2 diabetic patients: a systematic review and meta-analysis. J Diabetes Res 2019. https://doi.org/10.1155/2019/7676909
41. Rendell M (2004) Advances in diabetes for the millennium: drug therapy of type 2 diabetes. Med Gen Med 6(3 Suppl):9–9
42. Rendel M (2004) Advances in diabetes for the millennium: nutritional therapy of type 2 diabetes. Med Gen Med 6(3 Suppl):10–10
43. Einhorn D (2004) Advances in diabetes for the millennium: insulin treatment and glucose monitoring. Med Gen Med 6(3 Suppl):8–8
44. Benalla W, Bellahcen S, Bnouham M (2010) Antidiabetic medicinal plants as a source of alpha glucosidase inhibitors. Curr Diabetes Rev 6(4):247–254
45. Choi M, Kang Y-R, Zu HD, Lim I-S, Jung SK, Chang YH (2020) Effects of time on Phenolics and in vitro bioactivity in autoclave extraction of Graviola (Annona muricata) leaf. Biotechnol Bioprocess Eng 25(1):9–15. https://doi.org/10.1007/s12257-019-0259-3
46. Primo Fernandes KR, Bittercourt PS, Leao de Souza AD, Lima de Souza AQ, Araujo da Silva FM, Lima ES, Rosales Acho LD, Saraiva Nunomura RC, Teixeira AF, Ferreira Koolen HH (2019) Phenolic compounds from Virola venosa (Myristicaceae) and evaluation of their antioxidant and enzyme inhibition potential. Acta Amazon 49(1):48–53. https://doi.org/10.1590/1809-4392201800832
47. Park KH (2019) α-Glucosidase inhibitory caged xanthones from the resin of Garcinia hanburyi. J Appl Biol Chem 62(1):81–86
48. Oh TJ, Yu JM, Min KW, Son HS, Lee MK, Yoon KH, Song YD, Park JY, Jeong IK, Cha BS, Kim YS, Baik SH, Kim IJ, Kim DM, Kim SR, Lee KW, Park JH, Lee IK, Park TS, Choi SH, Park SW (2019) Efficacy and safety of Voglibose plus metformin in patients with type 2 diabetes

mellitus: a randomized controlled trial. Diabetes Metab J 43(3):276–286. https://doi.org/10.4093/dmj.2018.0051

49. Kim J-I, Baek H-J, Han D-W, Yun J-A (2019) Autumn olive (*Elaeagnus umbellata* Thunb.) berry reduces fasting and postprandial glucose levels in mice. Nutr Res Pract 13(1):11–16. https://doi.org/10.4162/nrp.2019.13.1.11
50. Jeon Y-H, Choi S-W (2019) Isolation, identification, and quantification of Tyrosinase and alpha-Glucosidase inhibitors from UVC-irradiated mulberry (Morus alba L.) leaves. Prev Nutr Food Sci 24(1):84–94. https://doi.org/10.3746/pnf.2019.24.1.84
51. de Almeida TS, Machado Araujo ME, Rodriguez LG, Julio A, Mendes BG, Borges dos Santos RM, Martinho Simoes JA (2019) Influence of preparation procedures on the phenolic content, antioxidant and antidiabetic activities of green and black teas. Braz J Pharm Sci 55. https://doi.org/10.1590/s2175-97902019000117695
52. Abbasi MA, Riaz S, Aziz ur R, Siddiqui SZ, Shah SAA, Ashraf M, Lodhi MA, Khan FA (2019) Synthesis of new 2-{2,3-dihydro-1,4-benzodioxin-6-yl (4-methylphenyl) sulfonyl amino}-N-(un/substituted-phenyl) acetamides as alpha-glucosidase and acetylcholinesterase inhibitors and their in silico study. Braz J Pharm Sci 55. https://doi.org/10.1590/s2175-97902019000117032
53. Færch K, Blond MB, Bruhn L, Amadid H, Vistisen D, Clemmensen KK et al (2020) The effects of dapagliflozin, metformin or exercise on glycaemic variability in overweight or obese individuals with prediabetes (the PRE-D Trial): a multi-arm, randomised, controlled trial. Diabetologia:1–14
54. Vaduganathan M, Fonarow GC, Greene SJ, DeVore AD, Kavati A, Sikirica S, Albert NM, Duffy CI, Hill CL, Patterson JH, Spertus JA, Thomas LE, Williams FB, Hernandez AF, Butler J (2020) Contemporary treatment patterns and clinical outcomes of comorbid DiabetesMellitus and HFrEF: the CHAMP-HF registry. JACC Heart Failure. https://doi.org/10.1016/j.jchf.2019.12.015
55. Ikonomidis I, Pavlidis G, Thymis J, Birba D, Kalogeris A, Kousathana F, Kountouri A, Balampanis K, Parissis J, Andreadou I, Katogiannis K, Dimitriadis G, Bamias A, Iliodromitis E, Lambadiari V (2020) Effects of glucagon-like Peptide-1 receptor agonists, sodium-glucose Cotransporter-2 inhibitors, and their combination on endothelial Glycocalyx, arterial function, and myocardial work index in patients with type 2 diabetes mellitus after 12-month treatment. J Am Heart Assoc 9(9):e015716. https://doi.org/10.1161/jaha.119.015716
56. Hess DA, Terenzi DC, Verma S (2020) SGLT-2 inhibitors and regenerative cell exhaustion. Cell Metab 31(5):884–885. https://doi.org/10.1016/j.cmet.2020.04.003
57. Haraguchi A, Shigeno R, Horie I, Morimoto S, Ito A, Chiba K, Kawazoe Y, Tashiro S, Miyamoto J, Sato S, Yamamoto H, Osaki M, Kawakami A, Abiru N (2020) The effect of luseogliflozin on bone microarchitecture in older patients with type 2 diabetes: study protocol for a randomized controlled pilot trial using second-generation, high-resolution, peripheral quantitative computed tomography (HR-pQCT). Trials 21(1):379–379. https://doi.org/10.1186/s13063-020-04276-4
58. Fadini GP (2020) SGLT-2 inhibitors and circulating progenitor cells in diabetes. Cell Metab 31(5):883–883. https://doi.org/10.1016/j.cmet.2020.04.002
59. Carter P, Gray LJ, Troughton J, Khunti K, Davies MJ (2010) Fruit and vegetable intake and incidence of type 2 diabetes mellitus: systematic review and meta-analysis. BMJ 341:c4229. https://doi.org/10.1136/bmj.c4229. PMC 2924474. PMID 20724400
60. Santaguida PL, Balion C, Hunt D, Morrison K, Gerstein H, Raina P, Booker L, Yazdi H (2005) Diagnosis, prognosis, and treatment of impaired glucose tolerance and impaired fasting glucose. Evid Rep Technol Assess (128):1–11. PMC 4780988. PMID 16194123. Archived from the original (PDF) on 2008-09-10
61. Qaseem A, Wilt TJ, Kansagara D, Horwitch C, Barry MJ, Forciea MA (2018) Hemoglobin A1c targets for Glycemic control with pharmacologic therapy for nonpregnant adults with type 2 diabetes mellitus: a guidance statement update from the American college of physicians. Ann Intern Med 168(8):569–576. https://doi.org/10.7326/M17-0939

62. Seaquist ER, Anderson J, Childs B, Cryer P, Dagogo-Jack S, Fish L, Heller SR, Rodriguez H, Rosenzweig J, Vigersky R (2013) Hypoglycemia and diabetes: a report of a workgroup of the American Diabetes Association and the Endocrine Society. Diabetes Care 36(5):1384–1395. https://doi.org/10.2337/dc12-2480. PMC 3631867. PMID 23589542
63. Makam AN, Nguyen OK (2017) An evidence-based medicine approach to Antihyperglycemic therapy in diabetes mellitus to overcome overtreatment. Circulation 135(2):180–195. https://doi.org/10.1161/CIRCULATIONAHA.116.022622. PMC 5502688. PMID 28069712
64. Simpson TC, Weldon JC, Worthington HV, Needleman I, Wild SH, Moles DR, Stevenson B, Furness S, Iheozor-Ejiofor Z (2015) Treatment of periodontal disease for glycaemic control in people with diabetes mellitus. Cochrane Database Syst Rev 11:CD004714. https://doi.org/10.1002/14651858.CD004714.pub3. PMC 6486035. PMID 26545069

Chapter 3
An Overview of Hypoglycemic Biological Drugs

Haixia Chen and Qirou Wang

Abstract Biological drugs are very important to treat diabetes. The hypoglycemic biological drugs mainly include insulin and its analogues, glucagon-like peptide-1 (GLP-1) receptor agonists, islet amyloid peptide (IAPP) analogues, and some others. This chapter mainly describes the categories of hypoglycemic biological drugs, their characteristics, pharmacological activities, representative drugs, and the advantages and disadvantages.

Keywords Hypoglycemic biological drugs · Insulin · GLP-1 receptor agonists · IAPP analogues

1 Introduction

Diabetes is a kind of chronic and metabolic diseases characterized by hyperglycemia, which is caused by deficiency in insulin secretion or insulin activity [1–4].

Currently, modern chemical drugs and traditional drugs are always used in the treatment of diabetes [5]. But most chemical hypoglycemic drugs have side effects, including severe hypoglycemia, weight gain, lactic acidosis, liver damage, gastrointestinal disturbance, edema, anemia, headache, dizziness, nausea, and even death [6]. Numerous common herbs and some foods have good antidiabetic activities and are used to reduce blood glucose level [6–11]. But due to incorrect prescription, overdosage, improper preparation, and inherent toxicity, the usage of traditional drugs still has risks. And most traditional drugs take effect slowly [12].

Apart from modern chemical drugs and traditional drugs, biological drugs are also a very important kind of drugs to treat diabetes.

Biological drugs are a kind of drugs made from living organisms, biological tissues, cells, and body fluids for prevention, treatment, and diagnosis.

H. Chen (✉) · Q. Wang
Tianjin Key Laboratory for Modern Drug Delivery and High-Efficiency, School of Pharmaceutical Science and Technology, Tianjin University, Tianjin, China
e-mail: chenhx@tju.edu.cn

H. Chen, M. Zhang (eds.), *Structure and Health Effects of Natural Products on Diabetes Mellitus*, https://doi.org/10.1007/978-981-15-8791-7_3

Biopharmaceuticals need to use the research achievements of biology, medical science, and biochemistry, and the principles and methods of physics, chemistry, biochemistry, and pharmaceutics, as well as biotechnologies, such as gene engineering, cell engineering, protein engineering, and fermentation engineering. Normally, our body can stay healthy and resist and overcome the diseases. The reason is that our body can constantly produce many types of regulated substances, which are closely related to metabolism, such as proteins, enzymes, nucleic acids, hormones, antibodies, cytokines, and so on. Our body can maintain normal function by the regulating effects of these substances. Based on this feature, we can extract these substances from living organisms and use them as medicines. The raw materials of biological drugs are mainly derived from natural biological materials, including humans, animals, plants, microorganisms, and marine organisms. And with the development of biotechnology, by using genetic engineering, cell engineering, protein engineering, and fermentation engineering, the artificial raw biological materials have also become the important sources of biological drugs.

Biological drugs have the advantages of strong therapeutic pertinence, high pharmacological activity, low toxicity, slight side effects, and high nutritional value. But the differences between species and the individual differences may cause immune reactions or allergic reactions. And the stability of biological drugs is always poor. Some biological drugs are easy to decompose or go bad because of the room temperature or bacteria. So the requirements of production and storage are higher.

According to the different molecular structures, biological drugs can be divided into amino acids and their derivatives, peptides and proteins, enzymes and coenzymes, nucleic acids and their degradants and derivatives, carbohydrate drugs, lipids drugs, cell growth factors, and biologics.

Biological drugs can always be used to treat many diseases, such as tumors, rheumatoid arthritis, psoriasis, cardiovascular and cerebrovascular diseases, autoimmune diseases, etc. They can also prevent some infectious diseases rapidly and accurately diagnose some diseases. Now biological drugs are also used in the treatment of diabetes. With the deepening understanding of diabetes, new pathogenesis is being discovered constantly. Compared to chemical drugs, biological drugs are more safe and efficient.

Currently, the most common biological drug to lower blood glucose level is insulin and its analogues. Their molecular mechanisms have been studied for decades. And insulin has been used as a classic drug for treating diabetes for decades. In addition, there are also some other kinds of biological drugs for lowering blood glucose now, such as glucagon-like peptide-1 (GLP-1) receptor agonists, islet amyloid peptide (IAPP) analogues, interleukin-1 receptor antagonist (IL-1Ra), IL-1β antibodies, and some others. GLP-1 receptor agonist is the focus of current researchs, and IL-1β signaling pathway is a new target for the treatment of Type 2 Diabetes Mellitus (T2DM) in the recent years [13].

2 Insulin and Its Analogues

Insulin is a protein hormone secreted by pancreatic islet β cells in the pancreas. The secretion of insulin is stimulated by endogenous or exogenous substances such as glucose, lactose, ribose, arginine, glucagon, etc. Insulin is the only hormone that can lower the blood glucose in the body and it can promote the synthesis of glycogen, fat, and protein.

2.1 The History of Insulin

According to the different sources, insulin can be divided into animal insulin, human insulin, and insulin analogues. Before the discovery of insulin, diabetes was a fatal disease due to the inevitable ketoacidosis in the late stages of the disease process. In 1921 Frederick Banting and Charles Best first extracted insulin from the pancreas of cows and pigs and turned diabetes from a lethal condition into a manageable disease. But animal insulin has some adverse effects including insulin allergy, lipodystrophy, hypoglycemia, and insulin resistance. In 1978, Genentech first used recombinant DNA technology to synthesize the A and B chains of insulin in *Escherichia coli*, making insulin the first protein ever synthesized in vitro. The development of recombinant DNA technology allowed the large-scale synthesis of insulin. It makes great advances in insulin production and diabetes treatment, but there are still some problems with glycemic control and hypoglycemia. By the early 1990s, great strides had been made in the development of insulin. Insulin analogues were designed to provide either a basal or bolus option to stimulate normal insulin secretion, reduce hypoglycemia, and allow for improved absorption. The substitution of amino acids at specific locations along either of these chains forms the basis of insulin analogue production. The substitution of specific amino acids and other modifications to the peptide chains makes insulin analogues more suitable for human physiological needs. Insulin treatment vastly increases the life expectancy of patients with diabetes and allows them to meet treatment goals [14, 15].

2.2 The Structure of Insulin

Insulin is a protein composed of two peptide chains, A and B. The function of insulin in different animals (humans, cattle, sheep, pigs, etc.) is basically the same, but the composition is slightly different. The human insulin molecule (in Fig. 3.1), a disulfide-bonded heterodimer of 51 amino acids, is the result of numerous years of evolution [16]. The human insulin is composed of a 21-amino acid A chain and a 30-amino acid B chain [15]. The amino acids in A chain are 11 species and the amino acids in B chain are 15 species. There are 16 kinds of amino acids in human insulin

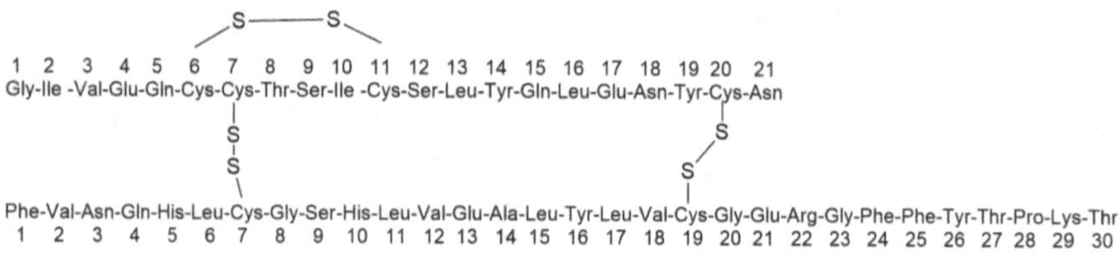

Fig. 3.1 The amino acid sequence of human insulin

in total. The sulfhydryl groups in cysteine (Cys), A7 (Cys) and B7 (Cys), A20 (Cys) and B19 (Cys), form two disulfide bonds that link the A and B chains. There is also a disulfide bond between A6 (Cys) and A11 (Cys) in the A chain. The disulfide-linked monomer is the bioactive form of insulin. When insulin is released into the bloodstream, the hexamer dissociates into monomers that bind to the insulin receptor. But monomers can associate into dimers, and dimers can further associate into hexamers, which are resistant to degradation and fibrillation in the presence of zinc ions [15, 17]. To achieve its many biological activities (blood glucose lowering is only one of these), this protein has not only a defined primary structure but also a well-defined secondary and tertiary structure [16].

2.3 *The Activities of Insulin*

Insulin is the only hormone in the body that can lower blood glucose and promote the synthesis of glycogen, fat, and protein. Endogenous insulin is secreted directly into circulation where it can act on target tissues, primarily muscle and adipose [15]. Insulin rises after the ingestion of food. It can promote the uptake and utilization of glucose in the histocytes of the whole body, promote the synthesis of glycogen, suppress endogenous glucose production, and inhibit gluconeogenesis and the decomposition of glycogen. It has the effect of lowering blood glucose. Insulin can also promote the synthesis and storage of fat, reduce the free fatty acids in the blood, inhibit the decomposition of fat, promote the uptake of amino acids and the synthesis of protein, and inhibit the decomposition of protein. These actions of insulin maintain plasma glucose levels within a fairly narrow range [18].

The insufficient secretion of insulin or a lack of insulin receptors would result in higher blood glucose. If the blood glucose level is more than renal glucose threshold, it will cause diabetes. And due to excessive amounts of glucose in the blood, it may lead to high blood pressure, coronary heart disease, retinal vascular disease, and some other diseases. In addition, the insufficient insulin can also cause the decomposition and the metabolic disorder of fat and higher blood lipids. The body may generate a large number of ketone bodies and lead to ketoacidosis. It may also lead to arteriosclerosis and some more serious diseases in heart and cerebral vessels.

Insulin therapy is the most effective method of lowering blood glucose. The treatment of mimicking normal insulin secretion may be an optimal way to achieve good blood glucose control in patients with diabetes. The key features of the physiologic pattern of insulin secretion by β cells are a meal-stimulated peak in insulin secretion that slowly decays over 2–3 h and a sustained basal level of insulin that remains constant throughout the day [18]. This pattern includes bolus insulin secretion (food-related) in response to food intake and basal insulin secretion (nonfood-related) that maintains a minimal level of insulin throughout the day [18–20]. Insulin therapy is absolutely essential in the treatment of Type 1 Diabetes Mellitus (T1DM) and indispensable in many cases of T2DM as the disease progresses and glycemic control is lost [19–21].

2.4 The Categories of Insulin and Its Analogues

According to the different time of onset, peak, and duration of action, insulin and its analogues can be divided into five different types: fast-acting, intermediate-acting, long-acting, ultra-long-acting, and insulin analogue premixes. Fast-acting insulin and its analogues are mainly used to control bolus insulin (food-related), while long-acting insulin and its analogues provide basal insulin (nonfood-related) [14, 18, 22]. The key pharmacodynamic properties of different insulin and analogues are listed in (Table 3.1) [14, 18, 22–24].

2.4.1 Fast-Acting Insulins and Analogues

Fast-acting insulins and analogues are mainly used to approximate the normal physiologic responses to meal consumption (i.e., the bolus of insulin secretion). Fast-acting insulins and analogues used for bolus therapy, including regular insulin, insulin lispro, insulin aspart, and insulin glulisine. Lispro, aspart, and glulisine, are monomeric insulin analogues [18].

Regular Insulin With the development of recombinant DNA technology, scientists can synthesize the A and B chains of insulin in *Escherichia coli* in a large scale. Regular human insulin is injected subcutaneously before mealtimes, and it has a slower onset and a longer duration action than endogenous insulin and the three fast-acting insulin analogues (lispro, aspart, glulisine). But it may lead to hypoglycemia, weight gain, or allergic reaction [14].

Insulin Lispro It is the first insulin analogue designed to speed up absorption. It is marketed as Humalog developed by Eli Lilly. Insulin lispro exchanges the positions of the proline of B28 and the lysine of B29 (in Fig. 3.2a). The inversion eliminates hydrophobic binding between insulin monomers and weakens the hydrogen bonds of stable hexamers, inhibiting the formation of dimers and hexamers. After subcutaneous injection, the monomers can be rapidly absorbed by vascular epithelial cells [13, 15, 23–25].

Table 3.1 The key pharmacodynamic properties of different insulin and its analogues

Type	Generic name	Brand name	Manufacturer	Approval year	Onset of action	Peak Action	Duration of action
Fast-acting insulin	Regular insulin	Humulin R	Eli Lilly	1982	30 min	2–3 h	8 h
		Novolin R	Novolin Nordisk	1991			
Fast-acting insulin analogue	Insulin lispro	Humalog	Eli Lilly	1996	15 min	1 h	2–4 h
	Insulin aspart	Novolog	Novo Nordisk	2000			
	Insulin glulisine	Apidra	Sanofi	2004			
Intermediate-acting insulin	Neutral protamine Hagedorn (NPH)	Humulin N	Eli Lilly	1982	2–4 h	4–12 h	12–18 h
		Novolin N	Novolin Nordisk	1991			
	Lente insulin	/	/	/	1–3 h	4–8 h	13–20 h
Long-acting insulin analogue	Insulin glargine	Lantus	Sanofi	2000	1–2 h	Flat	16–24 h
	Insulin detemir	Levemir	Novo Nordisk	2005			
Ultra-long-acting insulin analogue	Insulin degludec	Tresiba	Novo Nordisk	2015	1–2 h	Flat	>24 h

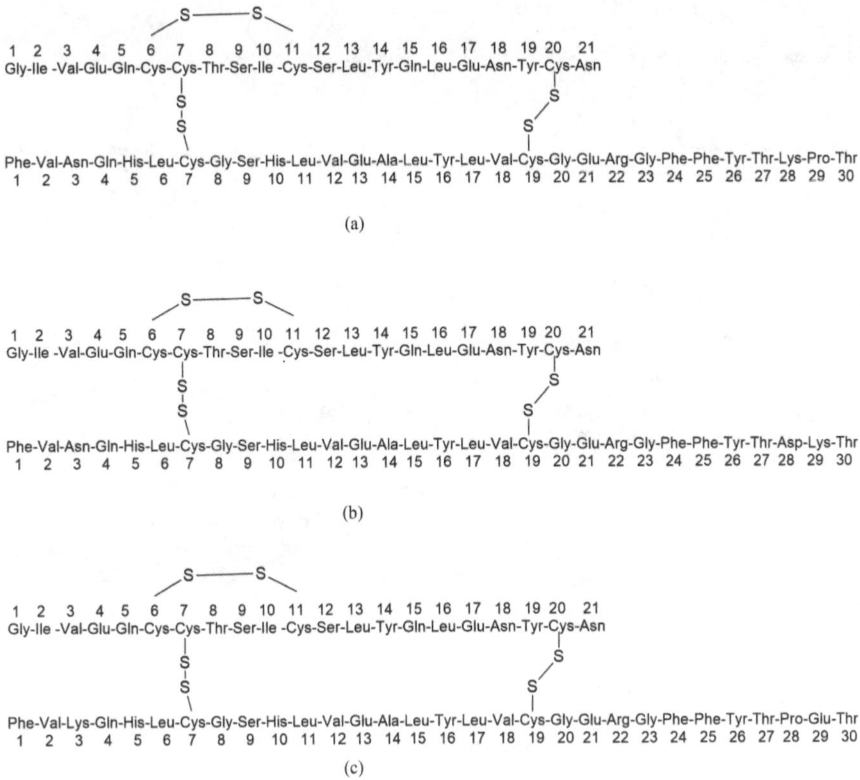

Fig. 3.2 The amino acid sequences of six different insulin analogues. (**a**) Insulin lispro (**b**) Insulin aspart (**c**) Insulin glulisine (**d**) Insulin glargine (**e**) Insulin detemir (**f**) Insulin degludec

Insulin Aspart It is the second rapid-acting insulin analogue branded as NovoLog. The structure of insulin aspart is different from regular human insulin because the proline in position B28 is substituted with aspartate (in Fig. 3.2b). It can also weaken the tendency toward self-association between insulin monomers and inhibit the aggregation into dimmers and hexamers like insulin lispro [15, 26–28].

Insulin Glulisine Insulin glulisine, named as Apidra in the market, is the third fast-acting insulin analogue created via recombinant DNA technology using strains of *Escherichia coli* (K12). The aspargine at position B3 is substituted by lysine, and the lysine at B29 is substituted by glutamic acid (in Fig. 3.2c). The substitutions in glulisine allow it to exist as more stable dimmers and monomers, allowing glulisine to be suspended in a zinc-free buffer and improving the rate of absorption [29–33].

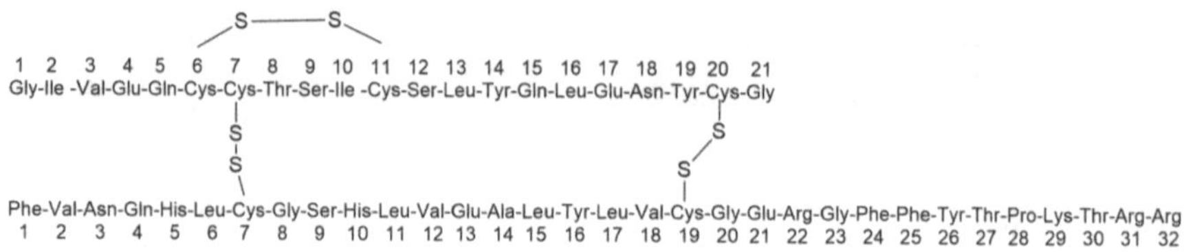

(d)

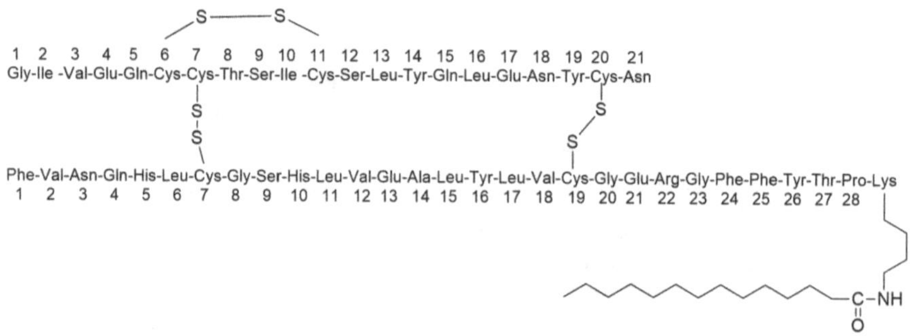

(e)

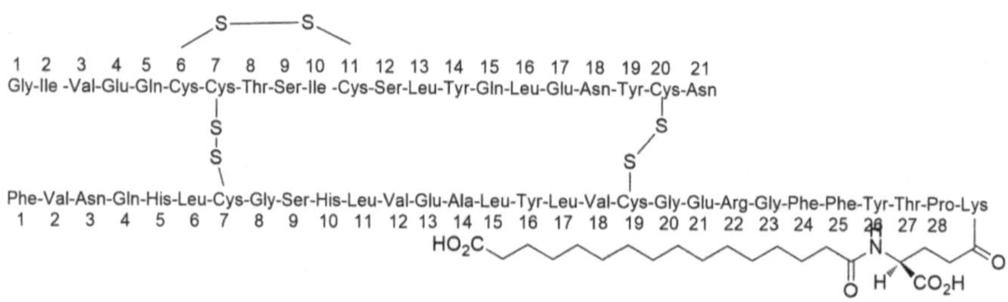

(f)

Fig. 3.2 (continued)

2.4.2 Intermediate-Acting Insulins

Neutral Protamine Hagedorn (NPH) In 1946, Hagedorn and colleagues developed crystalline neutral protamine Hagedorn (NPH) insulin in Denmark. It is a complex of insulin, protamine, and a small amount of zinc. Protamine is a protein from fish sperm, which can reduce the solubility of insulin and zinc. Before the development of long-acting basal insulin analogues, NPH was considered as a long-acting basal agent [14, 34–36].

Lente Insulin In 1952, Hallas-MØller, Petersen, and Schlichtkrull developed three new insulin preparations: semilente, lente, and ultralente. Semilente insulin is fast acting and ultralente insulin is long acting. They are all zinc suspension. Lente insulin is composed of amorphous porcine insulin and bovine crystalline particles in a ratio of 3 to 7. It has a similar intermediate timing of action to NPH insulin [34, 37].

2.4.3 Long-Acting Insulins and Analogues

The long-acting insulins and analogues are used to provide consistent and flat insulin levels mimicking physiologic basal insulin secretion. The long-acting insulins include protamine zinc insulin (PZI) and ultralente insulin, while the common long-acting insulin analogues are insulin glargine and insulin detemir.

Insulin Glargine In 2000, insulin glargine, marketed as Lantus, became the first long-acting insulin analogue introduced into clinical practice. It is made by recombinant DNA technology and has two modifications: the first one is replacement of asparagine with glycine at position A21, and the latter one is the addition of two arginine residues to the B-chain at the C-terminus (position B30) (in Fig. 3.2d). The modifications provide better stability of insulin at acidic pH and make insulin glargine less soluble at neutral pH. Insulin glargine can generate precipitate after injection and thus slow down the absorption [15, 18, 34, 38].

Insulin Detemir It is the second long-acting insulin analogue available for clinical use in 2005 by the use of recombinant DNA technology. In the structure of insulin detemir, the amino acid threonine is removed from position B30 and a 14-carbon myristic fatty acid chain is attached to the amino acid lysine at B29 (in Fig. 3.2e). The myristic fatty acid residue makes insulin detemir have a higher binding affinity to albumin, and insulin detemir can form more stable hexamers, resulting in a prolonged time of onset and duration [15, 19, 23, 34].

2.4.4 Ultra-Long-Acting Insulin Analogues

Insulin Degludec Insulin degludec, branded as "Tresiba," is an ultra-long-acting insulin analogue, which received approval in 2015 by the U.S. Food and Drug Administration (FDA). In the structure of insulin degludec, the amino acid threonine is removed from position B30 and a 16-carbon fatty diacid is attached to lysine at B29 (in Fig. 3.2f). The modifications can make insulin degludec form stable multihexamers after subcutaneous injection at neutral pH, resulting in a duration of action over 24 h [15, 23, 24, 34].

2.4.5 Insulin Analogue Premixes

Insulin analogue premixes are the mixture of intermediate-acting and fast-acting insulins to improve glycemic control. For example, biphasic human insulin 30 (BHI 30) consists of 30% short-acting human insulin and 70% intermediate-acting human insulin. Insulin analogue premixes can meet both basal and bolus insulin needs of diabetic patients. They can also reduce daily injections than classic basal-bolus therapy [14].

2.5 The Adverse Reactions of Insulin Therapy

Hypoglycemia is the main adverse reaction of insulin therapy, which is related to excessive use of insulin. Besides, there may be some other adverse reactions such as allergic reactions, weight gain, insulin resistance, local skin redness and swelling, subcutaneous fat atrophy, etc.

2.6 The Different Dosage Forms of Insulin and Analogues

Subcutaneous Injection of Insulin At present, most insulin therapies are subcutaneous injection. But long-term subcutaneous injection may bring lots of inconvenience and pain to the diabetic patients, even result in local tissue necrosis or infection. Thus, the recent research target of insulin therapy is to develop noninjection delivery of insulin and study noninvasive and more acceptable methods for administering insulin. The newer insulin therapies under research include oral insulin, inhaled insulin, nasal insulin, transbuccal insulin, transdermal insulin, artificial pancreas, etc.

Oral Insulin It is a painless and convenient method to improve glycemic control for diabetic patients. But insulin is a protein with big molecular weight. So there are some barriers for insulin to be absorbed. First barrier is the intestinal epithelium. And different digestive enzymes in the stomach and small intestine can break down insulin into amino acids. In addition, the low pH of the stomach can also break down the structure of insulin. Therefore, liposomes, nanoparticles, micelles, hydrogels, or some other carriers are needed to reduce the destruction and degradation of insulin molecules in gastrointestinal environment. They can promote effective absorption of insulin and improve the bioavailability. The representative oral insulin analogue under investigation is IN-105 [14, 15, 20, 21, 39].

Inhaled Insulin It is a noninvasive alternative for patients with diabetes, especially helpful for patients who are afraid of injections. The lung has large absorption surface area and rich vascularity, which is conducive to drug absorption. Inhaled insulins fall into two different groups: the dry powder preparations and liquid aerosol. Afrezza is a dry power recombinant human insulin approved by the FDA in 2014 [14, 15, 20, 21, 39].

Transmucosal Insulin It can be divided into two groups: nasal insulin and transbuccal insulin. After absorption, insulin can directly enter the systemic circulation, avoiding the liver's first-pass effect. The nasal mucosal absorption surface area is about 150 cm^2. And there are abundant blood vessels and lymphatic vessels under the mucosal cells. However, the drug clearance and the enzyme degradation in the nasal cavity may affect the absorption and utilization of drugs [39, 40]. And transbuccal insulin is absorbed through the inside of the cheeks and in the back of the mouth [15, 20, 21, 40].

Transdermal Insulin It is convenient and can avoid the liver's first-pass effect. But the stratum corneum, the outermost layer of the skin, is the main barrier for the drug to be absorbed and limits the penetration of macromolecular substances. Several chemical and physical enhancement techniques, such as iontophoresis (the use of electric current), sonophoresis (use of ultrasound), microneedles (formation of microchannels in the stratum corneum), laser ablation, and chemical enhancers, have been explored to overcome the stratum corneum barrier to increase skin permeability [20, 21].

Artificial Pancreas It is closed-loop insulin delivery. The goal of artificial pancreas is to achieve good glycemic control with the use of an algorithm that directs insulin delivery without the need for human intervention, and it can also reduce the risk of hypoglycemia [15, 21, 39].

3 Glucagon-Like Peptide-1 (GLP-1) Receptor Agonists

3.1 The Introduction of GLP-1 and GLP-1 Receptor Agonists

Incretins are hormones secreted from the gastrointestinal tract, which are responsible for increasing the insulin secretion and inhibition of glucagon secretion to maintain glucose homeostasis. The incretin hormones include glucagon-like peptide-1 (GLP-1) and glucose-dependent insulinotropic polypeptide (GIP). GLP-1 is a hormone released by endocrine L cells of the ileum in response to a meal stimulus. Normally, GLP-1 can result in a suppression of appetite and food intake, the reduction of blood glucose, increasing insulin secretion, and inhibiting gastric emptying.

GLP-1 is a peptide. In the intestine, GLP-1 is synthesized from inactive 37-amino acid product, GLP-1(1–37) or GLP-1(1–36)-NH2 (in Fig. 3.3a, b), and then further processed to generate biologically active GLP-1, GLP-1(7–37), and GLP-1(7–36)-NH2 (in Fig. 3.3c, d). The latter one is more abundant in humans. But the half-life of bioactive GLP-1 is less than 2 min, due to inactivation by the enzyme dipeptidyl peptidase-4 (DPP-4) and rapid renal clearance. DPP-4 cleaves the peptide bond between Ala8 and Glu9, and the resulting metabolite GLP-1(9–36)-NH2 (in Fig. 3.3e) has a very low affinity for the GLP-1 receptors.

GLP-1 receptors can bind to GLP-1, and the receptors are present not only in the islet β-cells of pancreas but also in the GI tract, brain, heart, and kidney.

His-Asp-Glu-Phe-Glu-Arg-His-Ala-Glu-Gly-Thr-Phe-Thr-Ser-Asp-Val-Ser-Ser-Tyr-Leu-Glu-Gly-Gln-Ala-Ala-Lys-Glu-Phe-Ile-Ala-Trp-Leu-Val-Lys-Gly-Arg-Gly
1 2 3 4 5 6 7 8 9 10 11 12 13 14 15 16 17 18 19 20 21 22 23 24 25 26 27 28 29 30 31 32 33 34 35 36 37

(a)

His-Asp-Glu-Phe-Glu-Arg-His-Ala-Glu-Gly-Thr-Phe-Thr-Ser-Asp-Val-Ser-Ser-Tyr-Leu-Glu-Gly-Gln-Ala-Ala-Lys-Glu-Phe-Ile-Ala-Trp-Leu-Val-Lys-Gly-Arg—NH_2
1 2 3 4 5 6 7 8 9 10 11 12 13 14 15 16 17 18 19 20 21 22 23 24 25 26 27 28 29 30 31 32 33 34 35 36

(b)

DPP4

His-Ala-Glu-Gly-Thr-Phe-Thr-Ser-Asp-Val-Ser-Ser-Tyr-Leu-Glu-Gly-Gln-Ala-Ala-Lys-Glu-Phe-Ile-Ala-Trp-Leu-Val-Lys-Gly-Arg-Gly
7 8 9 10 11 12 13 14 15 16 17 18 19 20 21 22 23 24 25 26 27 28 29 30 31 32 33 34 35 36 37

(c)

DPP4

His-Ala-Glu-Gly-Thr-Phe-Thr-Ser-Asp-Val-Ser-Ser-Tyr-Leu-Glu-Gly-Gln-Ala-Ala-Lys-Glu-Phe-Ile-Ala-Trp-Leu-Val-Lys-Gly-Arg—NH_2
7 8 9 10 11 12 13 14 15 16 17 18 19 20 21 22 23 24 25 26 27 28 29 30 31 32 33 34 35 36

(d)

Glu-Gly-Thr-Phe-Thr-Ser-Asp-Val-Ser-Ser-Tyr-Leu-Glu-Gly-Gln-Ala-Ala-Lys-Glu-Phe-Ile-Ala-Trp-Leu-Val-Lys-Gly-Arg—NH_2
9 10 11 12 13 14 15 16 17 18 19 20 21 22 23 24 25 26 27 28 29 30 31 32 33 34 35 36

(e)

Fig. 3.3 Five different human GLP-1 forms. (**a**) GLP-1(1–37), (**b**) GLP-1(1–36)-NH_2, (**c**) GLP-1 (7–37), (**d**) GLP-1(7–36)-NH_2, (**e**) GLP-1(9–36)-NH_2

In recent years, GLP-1 receptor agonists are becoming a therapeutic option in the treatment of diabetes to improve glycemic control. All of the GLP-1 receptor agonists are based on the amino acid sequence of GLP-1 with some modifications to make them remain active in the presence of the enzyme DPP-4 with prolonged half-life [41–51].

3.2 *The History of GLP-1 and GLP-1 Receptor Agonists*

In 1964, Elrick and Mclntyre demonstrated that the insulin secretion in response to oral glucose exceeded that evoked by the intravenous infusion of a similar glucose amount. This observation gave rise to suspect gut-derived substances to be involved in the regulation of insulin secretion after a meal [52].

In 1971, Brown and Dryburgh isolated a peptide from the intestinal mucosa and deduced its amino acid sequence. The peptide was then named "glucose-dependent insulinotropic peptide" (GIP), which was the first incretin isolated. In the 1980s, scientists isolated another incretin, namely glucagon-like peptide-1 (GLP-1) [50]. And in the twenty-first century, a series of GLP-1 receptor agonists have been approved in the market.

3.3 The Activities of GLP-1

GLP-1 can reduce blood glucose mainly by promoting the secretion of insulin in islet β cells and inhibiting the secretion of glucagon in islet α cells. And GLP-1 can also promote the proliferation and differentiation of islet β cells and inhibit their apoptosis. In addition, GLP-1 can inhibit postprandial gastric emptying, reduce appetite and food intake, reduce postprandial blood glucose, and achieve the effect of weight loss. GLP-1 may also increase the uptake and utilization of glucose by peripheral cells, such as adipose tissue and cells, liver cells, and skeletal muscle cells.

Apart from lowering blood glucose, GLP-1 has many other biological activities. For example, GLP-1 may reduce blood pressure and blood lipid and have protective effect on cardiovascular system. GLP-1 also takes part in the regulation of bone turnover, may increase bone density, and may promote bone formation. GLP-1 may also play a role in the formation of water and electrolyte homeostasis in the kidney and improve diabetic nephropathy [15, 49, 52–56].

3.4 The Categories of GLP-1 Receptor Agonists and the Representative Drugs

Because the half-life of GLP-1 is too short, the clinical application is limited. Recently, the structure of GLP-1 is modified to develop GLP-1 receptor agonists, which are also GLP-1 analogues. The GLP-1 receptor agonists on the market or under development are divided into short-acting receptor agonists and long-acting receptor agonists. The short-acting drugs include exenatide (Byetta) and lixisenatide, while long-acting drugs include exenatide once weekly (Bydureon), liraglutide, albiglutide, and dulaglutide (Table 3.2) [47, 50, 56].

Exenatide Exenatide (in Fig. 3.4a) is the first GLP-1 receptor agonist approved for clinical use in 2005. It is a synthetic version of exendin-4, a GLP-1-like peptide isolated from the saliva of the gila monster lizard in 1992. The peptide consists of 39 amino acids and shares 53% homology with human GLP-1. It is resistant to DPP-4 proteolysis and has a longer half-life of 2.4 h. Exenatide (Byetta), produced by AstraZeneca, is administered by subcutaneous injection twice daily [15, 46, 51, 57–60].

Lixisenatide Lixisenatide (in Fig. 3.4b) is a 44-amino acid peptide. Compared to the structure of exendin-4, there is the deletion of a proline residue and addition of

Table 3.2 Introduction of GLP-1 receptor agonists

Type	Generic name	Brand name	Manufacturer	Approval year	Half-life
Short-acting receptor agonists	Exenatide	Byetta	AstraZeneca	2005	2.4 h
	Lixisenatide	Lyxumia	Sanofi	2013	3 h
Long-acting receptor agonists	Exenatide once weekly	Bydureon	AstraZeneca	2012	2.4 h
	Liraglutide	Victoza	Novo Nordisk	2010	13 h
	Albiglutide	Tanzeum	GlaxoSmithKline	2014	5 d
	Dulaglutide	Trulicity	Eli Lilly	2014	4–5 d

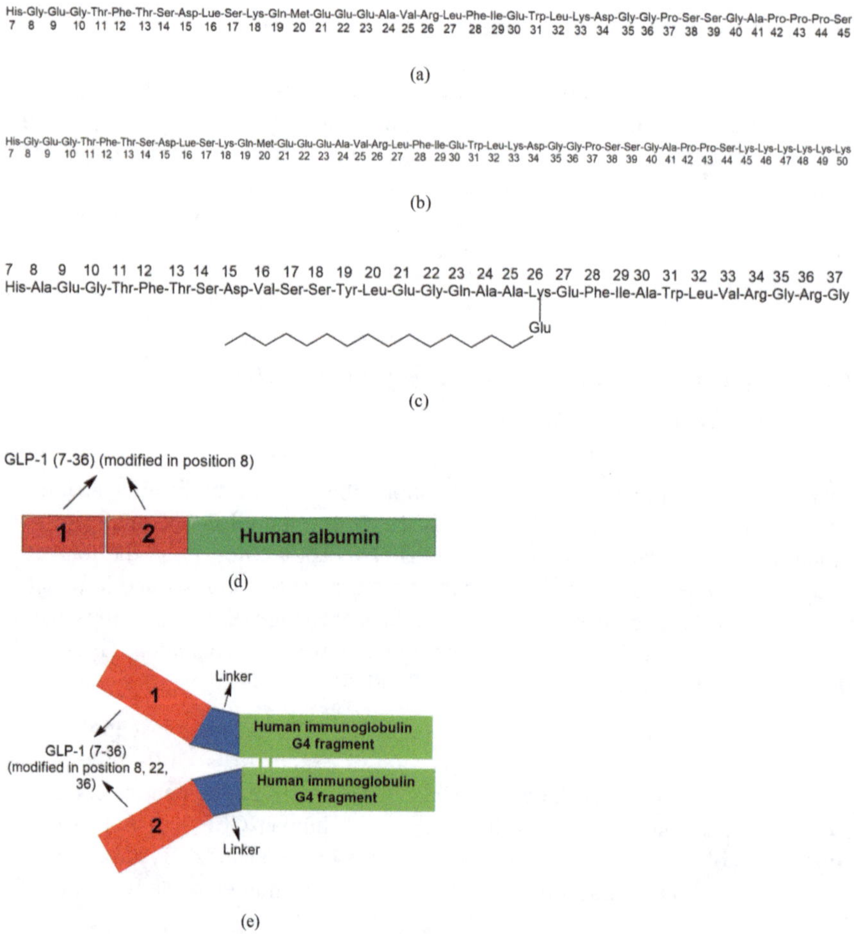

Fig. 3.4 The amino acid sequences or structures of different GLP-1 receptor agonists. (**a**) Exenatide (**b**) Lixisenatide (**c**) Liraglutide (**d**) Albiglutide (**e**) Dulaglutide

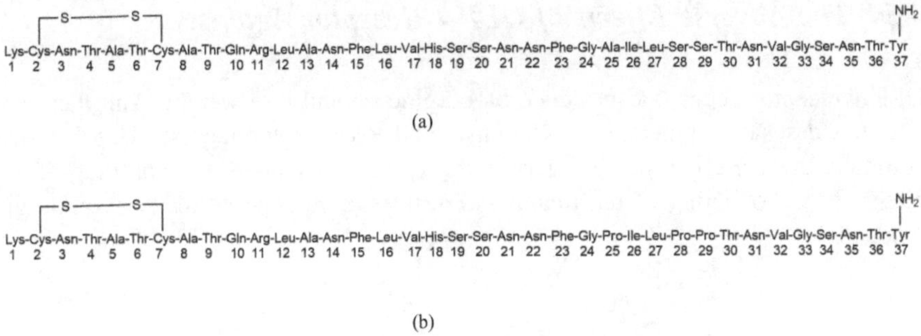

Fig. 3.5 The amino acid sequences of (**a**) human IAPP and (**b**) its analogue, pramlintide

6 lysine residues at the C-terminal end in lixisenatide. Its half-life is about 3 h, and it is injected once daily for the treatment of T2DM [15, 46, 51, 57, 61–63].

Exenatide Once Weekly Exenatide once weekly (Bydureon) is a long-acting version of exenatide, which is suitable for once-weekly administration. The encapsulation of exenatide in the biodegradable poly (lactic-co-glycolic acid) (PLGA) microspheres results in a slow release of exenatide [15, 46, 57, 60, 64–66].

Liraglutide Liraglutide (in Fig. 3.4c) is approved in 2010 and has a longer half-life of 13 h with once-daily administration. The structural modifications in liraglutide include addition of a C16 free fatty acid to Lys26 and lysine substitution with arginine at position 34, resulting in slower subcutaneous absorption, noncovalent albumin binding, and resistance to DPP-4 proteolysis. Liraglutide shares approximately 97% homology with human GLP-1 [46, 51, 57, 67, 68].

Albiglutide Albiglutide (in Fig. 3.4d), developed by GlaxoSmithKline (GSK), was approved by the FDA in 2014. It is composed of a dimer of modified human GLP-1 (a substitution of alanine at position 8 with glycine), which is fused to the N-terminus of recombinant human albumin. Albiglutide has an extended half-life of 5 days allowing for once-weekly dosing [15, 51, 57, 63, 69, 70].

Dulaglutide Dulaglutide (in Fig. 3.4e) is a long-acting GLP-1 receptor agonist for the treatment of T2DM with once-weekly dosing by subcutaneous injection. It is developed by Eli Lilly and approved the by FDA in 2014. Dulaglutide is a dimer of modified human GLP-1 with each monomer fused to a modified human immunoglobulin G4 heavy chain via a small peptide linker. Each chain contains a GLP-1 (7–37) analogue sequence with some substitutions, including alanine at position 8 to glycine, glycine at position 22 to glutamic acid, and arginine at position 36 to glycine. The half-life is about 4–5 days [15, 51, 57, 71, 72].

3.5 The Adverse Effects of GLP-1 Receptor Agonists

GLP-1 receptor agonists can reduce blood glucose and lose weight. And there are very few cases of hypoglycemia when using GLP-1 receptor agonists. But the most common adverse effect of GLP-1 receptor agonists is nausea and vomiting, especially at the beginning of the treatment. Besides, some people may have allergic reactions.

4 Islet Amyloid Polypeptide (IAPP) Analogues

Islet amyloid polypeptide (IAPP), also known as amylin, is a peptide hormone that is co-secreted with insulin from pancreatic β cells after meals [73]. The secretion ratio of IAPP and insulin in the islet of diabetic patients is approximately 1:100 [74]. It was discovered in humans in 1987 by Cooper et al. from the amyloid deposits of diabetic patients [74]. IAPP can maintain glucose homeostasis by slowing gastric emptying and suppressing glucagon secretion [75].

4.1 The Structure and Mechanisms of IAPP

Human IAPP is a 37-amino acid peptide (in Fig. 3.5a), which contains an amidated C-terminus and a disulfide bond between Cys-2 and Cys-7 [76]. It can suppress postprandial glucagon secretion, slow down the rate of gastric emptying, suppress the food intake, and induce satiety [77].

But this peptide forms amyloid deposits in the islets of Langerhans in diabetic patients, and amyloid deposits are believed to contribute to diabetes [76]. In the autopsy of diabetic patients, it was found that about 95% of the patients had amyloid deposits in the pancreatic islet tissue, which may induce the apoptosis of islet β cells and aggravate the condition of diabetes [78]. Due to the inherent tendency of native human amylin to self-aggregate and the insolubility of amyloid deposits, IAPP cannot be used as a potential therapeutic agent for diabetic patients [79].

So it is important to develop IAPP analogues with improved stability and decreased potential for aggregation to reduce blood glucose level [77].

4.2 The Representative IAPP Analogues: Pramlintide

Pramlintide (in Fig. 3.5b) is the first drug approved by the FDA in 2005 as the adjunctive treatment in patients with T1DM or T2DM who fail to achieve desired glycemic control [80]. The amino acids on positions 25 (alanine), 28 (serine), and

29 (serine) are all substituted by proline. It is a synthetic analogue of human amylin with soluble, stable, and nonaggregating features [80]. Pramlintide is administered by subcutaneous injection with a bioavailability of approximately 38–40%. It can regulate the blood glucose level by suppressing the postprandial glucagon secretion, delaying the gastric emptying time, decreasing food intake, and enhancing satiety. It can also reduce the weight of diabetic patients.

Pramlintide is an adjunctive drug, which is often used with insulin for the treatment of diabetes. Apart from diabetes, pramlintide is also used in some other diseases, such as Alzheimer's disease [81]. Pramlintide itself does not lead to hypoglycemia in diabetic patients. But when it is used in combination with insulin, there is also a risk of hypoglycemia. The main side effect of pramlintide, like GLP-1 receptor agonists, is gastrointestinal reactions, which results in severe nausea and vomiting, especially in the early treatment [82].

5 Other Hypoglycemic Biological Drugs

In recent years, some studies have found that inflammatory response and inflammatory cytokines are associated with diabetes and its complications. The cytokines involved include tumour necrosis factor (TNF), interleukin (IL), interferon (IFN), and some others [83, 84]. IL-1β can impair the function of β cells, induce β cells apoptosis, and inhibit the insulin secretion, which contributes to the development of diabetes [85–87]. Therefore, interleukin-1 receptor antagonist (IL-1Ra), a naturally occurring competitive inhibitor of IL-1β, and IL-1β antibody are used as new biological drugs for diabetic patients [85–87]. Gevokizumab (XOMA 052) and Canakinumab (ACZ885) are combinant human-engineered monoclonal IL-1β antibody [84, 88, 89]. Anakinra is a recombinant human IL-1Ra in the treatment of diabetes [85].

There are also some other biological drugs, which are potential drugs to treat diabetes, such as axokine, IC7Fc, and humanin [90–92].

6 Conclusion

Now there are some kinds of biological drugs used for the treatment of diabetes, including insulin and its analogues, GLP-1 receptor agonists, IAPP analogues, and some others. Insulin and its analogues, GLP-1 receptor agonists, and IAPP analogues are all peptide drugs. They are often used as injection, especially subcutaneous injection. But long-term subcutaneous injection may result in subcutaneous fat atrophy, local skin redness, and swelling, bring a lot of pain and inconvenience to diabetic patients, and even lead to local tissue necrosis or infection. So we need to develop some new dosage forms of these peptide drugs. The new dosage forms of insulin have been developed a lot, such as oral insulin, inhaled insulin, nasal insulin, transbuccal insulin, transdermal insulin, and artificial pancreas. The oral dosage of GLP-1 receptor agonists also has been developed, such as oral semaglutide [93–95]. But these new dosage forms of peptide drugs need to improve their bioavailability and safety.

With the deepening research on the pathogenesis of diabetes and the mechanisms of hypoglycemic drugs, the combination therapy of hypoglycemic drugs has attracted more and more attention. The combination of different hypoglycemic drugs can not only improve the curative effect but also reduce the incidence of side effects, which has obvious advantages over the monotherapy of hypoglycemic drugs. The combination of hypoglycemic drugs includes chemical hypoglycemic drugs and insulin, a chemical hypoglycemic drug and another chemical hypoglycemic drug, and chemical hypoglycemic drugs and Chinese patent medicines. Insulin can be used with glucosidase inhibitors like acarbose, metformin, sulfonylureas, DPP-4 enzyme inhibitor, or SGLT-2 inhibitor to control the blood glucose level [96]. Besides, insulin can also be used with hypoglycemic biological drugs, GLP-1 receptor agonists, or IAPP analogue pramlintide [82, 97, 98].

With the deepening understanding of diabetes, new pathogenic mechanisms have been discovered. The drugs for the treatment of diabetes include not only small-molecule chemical drugs like sulfonylureas and biguanides but also some biological drugs such as insulin and its analogues, GLP-1 receptor agonists, and IAPP analogue. So far, biological drugs have made a lot of achievements in the treatment of cancer, autoimmune diseases, and other fields. In recent years, the global investment in the development of hypoglycemic biological drugs keeps increasing. And some progress has been made in the control of blood glucose level, the prevention and treatment of complications, and the improvement of the life quality of diabetic patients. In the future, scientists will continue to optimize the bioavailability, effectiveness, and safety of hypoglycemic biological drugs in order to improve the life quality of patients with diabetes.

References

1. Association AD (2013) Standards of medical care in diabetes-2013. Diabetes Care 36(Suppl 1):11–66
2. American Diabetes Association (2014) Diagnosis and classification of diabetes mellitus. Diabetes Care 37(Suppl 1):81–90
3. Skyler JS, Bakris GL, Bonifacio E, Darsow T, Eckel RH, Groop L, Groop PH, Handelsman Y, Insel RA, Mathieu C, McElvaine AT, Palmer JP, Pugliese A, Schatz DA, Sosenko JM, Wilding JP, Ratner RE (2017) Differentiation of diabetes by pathophysiology, natural history, and prognosis. Diabetes 66(2):241–255
4. Choudhury H, Pandey M, Hua CK, Mun CS, Jing JK, Kong L, Ern LY, Ashraf NA, Kit SW, Yee TS, Pichika MR, Gorain B, Kesharwani P (2018) An update on natural compounds in the remedy of diabetes mellitus: a systematic review. J Tradit Complement Med 8(3):361–376
5. Bai L, Li X, He L, Zheng Y, Lu H, Li J, Zhong L, Tong R, Jiang Z, Shi J, Li J (2019) Antidiabetic potential of flavonoids from traditional Chinese medicine: a review. Am J Chin Med 47(5):933–957

6. Prabhakar PK, Doble M (2011) Mechanism of action of natural products used in the treatment of diabetes mellitus. Chin J Integr Med 17(8):563–574
7. Yadav MK, Tripathi JS (2015) Ethno medicinal plants with antidiabetic activity: a review. Int J Pharm Sci Res 6(12):4570–4578
8. Bai Y, Zang X, Ma J, Xu G (2016) Anti-diabetic effect of Portulaca oleracea L. Polysaccharideandits mechanism in diabetic rats. Int J Mol Sci 17(8):1–14
9. Kooti W, Farokhipour M, Asadzadeh Z, Ashtary-Larky D, Asadi-Samani M (2016) The role of medicinal plants in the treatment of diabetes: a systematic review. Electron Physician 8 (1):1832–1842
10. Wang S, Lu A, Zhang L, Shen M, Xu T, Zhan W, Jin H, Zhang Y, Wang W (2017) Extraction and purification of pumpkin polysaccharides and their hypoglycemic effect. Int J Biol Macromol 98:182–187
11. Xu L, Li Y, Dai Y, Peng J (2018) Natural products for the treatment of type 2 diabetes mellitus: pharmacology and mechanisms. Pharmacol Res 130:451–465
12. Yu A, Adelson D, Mills D (2018) Chinese herbal medicine versus other interventions in the treatment of type 2 Diabetes: a systematic review of randomized controlled trials. J Evid Based Integr Med 23:1–10
13. Tang X, Chen S (2014) Research progress of biopharmaceuticals for type 2 diabetes mellitus. Chin Pharm J 49:907–912
14. Tibaldi JM (2012) Evolution of insulin development: focus on key parameters. Adv Ther 29 (7):590–619. https://doi.org/10.1007/s12325-012-0034-8
15. Gilroy CA, Luginbuhl KM, Chilkoti A (2016) Controlled release of biologics for the treatment of type 2 diabetes. J Control Release 240:151–164. https://doi.org/10.1016/j.jconrel.2015.12.002
16. Heinemann L, Hompesch M (2014) Biosimilar insulins: basic considerations. J Diabetes Sci Technol 8(1):6–13. https://doi.org/10.1177/1932296813516958
17. Lisi GP, Png CY, Wilcox DE (2014) Thermodynamic contributions to the stability of the insulin hexamer. Biochemistry 53(22):3576–3584. https://doi.org/10.1021/bi401678n
18. Bethel MA, Feinglos MN (2005) Basal insulin therapy in type 2 diabetes. J Am Board Fam Pract 18:199–204
19. Young JR, McAdam-Marx C (2010) Treatment of type 1 and type 2 diabetes mellitus with insulin detemir, a long-acting insulin analog. Clin Med Insight Endocrinol Diabet 3:65–80. https://doi.org/10.4137/CMED.S5330
20. Cichocka E, Wietchy A, Nabrdalik K, Gumprecht J (2016) Insulin therapy - new directions of research. Endokrynol Pol 67(3):314–324. https://doi.org/10.5603/EP.2016.0044
21. Yaturu S (2013) Insulin therapies current and future trends at dawn. World J Diabetes 4:1–7
22. Zeng J, Fang Y (2017) Current status and research progress of diabetic drug therapy. Clin Med J 15:1–5. https://doi.org/10.3969/j.issn.1672-3384.2017.10.001
23. Nasrallah SN, Reynolds LR (2012) Insulin Degludec, the new generation basal insulin or just another basal insulin? Clin Med Insights Endocrinol Diabetes 5:31–37. https://doi.org/10.4137/CMED.S9494
24. Atkin S, Javed Z, Fulcher G (2015) Insulin degludec and insulin aspart: novel insulins for the management of diabetes mellitus. Ther Adv Chronic Dis 6(6):375–388. https://doi.org/10.1177/2040622315608646
25. Giugliano D, Esposito K (2012) Efficacy and safety of insulin lispro protamine suspension as basal supplementation in patients with type 2 diabetes. Ther Adv Endocrinol Metab 3 (3):99–108. https://doi.org/10.1177/2042018812442949
26. Haahr H, Heise T (2020) Fast-acting insulin aspart: a review of its pharmacokinetic and pharmacodynamic properties and the clinical consequences. Clin Pharmacokinet 59 (2):155–172. https://doi.org/10.1007/s40262-019-00834-5
27. Evans M, Wilkinson M, Giannpolou A (2019) Fast-acting insulin aspart: the rationale for a new mealtime insulin. Diabetes Ther 10(5):1793–1800. https://doi.org/10.1007/s13300-019-00685-0

28. Hermansen K, Bohl M, Schioldan AG (2016) Insulin aspart in the management of diabetes mellitus: 15 years of clinical experience. Drugs 76(1):41–74. https://doi.org/10.1007/s40265-015-0500-0
29. Helms KL, Kelley KW (2009) Insulin glulisine: an evaluation of its pharmacodynamic properties and clinical application. Ann Pharmacother 43(4):658–668. https://doi.org/10.1345/aph.1E662
30. Becker RHA, Frick AD, Teichert L, Nosek L, Heinemann L, Rave K (2009) Dose-response relationship of insulin glulisine in subjects with type 1 diabetes. Diabetes Obes Metab 11 (1):60–68. https://doi.org/10.1111/j.1463-1326.2008.01013.x
31. Satoru Y (2009) Insulin glulisine in the management of diabetes. Diabetes Metabol Synd Obesity Targets Ther 2:111–115
32. Bolli GB, Luzio S, Marzotti S, Porcellati F, Sert-Langeron C, Charbonnel B, Zair Y, Owens DR (2011) Comparative pharmacodynamic and pharmacokinetic characteristics of subcutaneous insulin glulisine and insulin aspart prior to a standard meal in obese subjects with type 2 diabetes. Diabetes Obes Metab 13:251–257
33. Lih A, Hibbert E, Wong T, Girgis CM, Garg N, Carter JN (2010) The role of insulin glulisine to improve glycemic control in children with diabetes mellitus. Diabetes Metab Syndr Obes 3:403–412. https://doi.org/10.2147/DMSOTT.S5116
34. Owens DR (2011) Insulin preparations with prolonged effect. Diabetes Technol Ther 13(Suppl 1):5–14. https://doi.org/10.1089/dia.2011.0068
35. Veroniki AA, Straus SE, Ashoor HM, Hamid JS, Yu C, Tricco AC (2015) Safety and effectiveness of long-acting versus intermediate-acting insulin for patients with type 1 diabetes: protocol for a systematic review and individual patient data network meta-analysis. BMJ Open 5(12):1–7. https://doi.org/10.1136/bmjopen-2015-010160
36. Andrea CT, Huda MA, Charlene S, Brenda H, David M, Brian H, Catherine HY, Sumit RM, Sharon ES (2013) Safety, effectiveness, and cost of long-acting versus intermediate-acting insulin for type 1 diabetes protocol for a systematic review and network meta-analysis. Syst Rev 2:1–5
37. Nabarro JDN, Stowers JM (1953) The insulin zinc suspensions. Br Med J 7:1027–1030
38. Hedrington MS, Pulliam L, Davis SN (2011) Basal insulin treatment in type 2 Diabetes. Diabetes Technol Ther 13(1):33–42. https://doi.org/10.1089/dia.2011.0062
39. Wang L, Zhan X, Chen S, Mao Z (2015) Progress of novel drug delivery systems of insulin. Chin J Pharm 46:1236–1244. https://doi.org/10.16522/j.cnki.cjph.2015.11.022
40. Han Y, Pan Q (2011) Research and development of non-injectable insulin delivery pathway. Chin Sci Technol Inf 15:162–163. https://doi.org/10.3969/j.issn.1001-8972.2011.15.108
41. Mapelli C, Natarajan SI, Meyer J-P, Bastos MM, Bernatowicz MS, Lee VG, Pluscec J, Riexinger DJ, Sieber-McMaster ES, Constantine KL, Smith-Monroy CA, Golla R, Ma Z, Longhi DA, Shi D, Xin L, Taylor JR, Koplowitz B, Chi CL, Khanna A, Robinson GW, Seethala R, Antal-Zimanyi IA, Stoffel RH, Han S, Whaley JM, Huang CS, Krupinski J, Ewing WR (2009) Eleven amino acid glucagon-like Peptide-1 receptor agonists with Antidiabetic activity. J Med Chem 52(23):7788–7799. https://doi.org/10.1021/jm900752a
42. Marre M, Penfornis A (2011) GLP-1 receptor agonists today. Diabetes Res Clin Pract 93 (3):317–327. https://doi.org/10.1016/j.diabres.2011.01.004
43. Unger JR, Parkin CG (2011) Glucagon-like peptide-1 (GLP-1) receptor agonists: differentiating the new medications. Diab Ther 2(1):29–39. https://doi.org/10.1007/s13300-010-0013-5
44. Pritchett CE, Hajnal A (2012) Glucagon-like peptide-1 regulation of carbohydrate intake is differentially affected by obesogenic diets. Obesity (Silver Spring) 20(2):313–317. https://doi.org/10.1038/oby.2011.342
45. Jeff U (2013) Rationale use of GLP-1 receptor agonists in patients with type 1 Diabetes. Curr Diab Rep 13:663–668. https://doi.org/10.1007/s11892-013-0404-x
46. Bolli GB, Owens DR (2014) Lixisenatide, a novel GLP-1 receptor agonist: efficacy, safety and clinical implications for type 2 diabetes mellitus. Diabetes Obes Metab 16(7):588–601. https://doi.org/10.1111/dom.12253

47. Manandhar B, Ahn J-M (2014) Glucagon-like Peptide-1 (GLP-1) analogs: recent advances, new possibilities, and therapeutic implications. J Med Chem 58(3):1020–1037. https://doi.org/10.1021/jm500810s
48. Hoang HN, Song K, Hill TA, Derksen DR, Edmonds DJ, Kok WM, Limberakis C, Liras S, Loria PM, Mascitti V, Mathiowetz AM, Mitchell JM, Piotrowski DW, Price DA, Stanton RV, Suen JY, Withka JM, Griffith DA, Fairlie DP (2015) Short hydrophobic peptides with cyclic constraints are potent glucagon-like Peptide-1 receptor (GLP-1R) agonists. J Med Chem 58 (9):4080–4085. https://doi.org/10.1021/acs.jmedchem.5b00166
49. Cantini G, Mannucci E, Luconi M (2016) Perspectives in GLP-1 research: new targets, new receptors. Trends Endocrinol Metabol 27(6):427–438. https://doi.org/10.1016/j.tem.2016.03.017
50. Tian L, Jin T (2016) The incretin hormone GLP-1 and mechanisms underlying its secretion. J Diabetes 8(6):753–765. https://doi.org/10.1111/1753-0407.12439
51. Alavi SE, Cabot PJ, Moyle PM (2019) Glucagon-like Peptide-1 receptor agonists and strategies to improve their efficiency. Mol Pharm 16(6):2278–2295. https://doi.org/10.1021/acs.molpharmaceut.9b00308
52. Meier JJ, Nauck MA (2005) Glucagon-like peptide 1(GLP-1) in biology and pathology. Diabetes Metab Res Rev 21(2):91–117. https://doi.org/10.1002/dmrr.538
53. Wen J (2015) The mechanism of human glucagon-like peptide-1 receptor agonists. Chin J Gerontol 35:5981–5983
54. Ren H (2016) The clinical research progress of glucagon-like peptide-1 receptor agonists. Strait Pharm J 28:5–7
55. Muller TD, Finan B, Bloom SR, D'Alessio D, Drucker DJ, Flatt PR, Fritsche A, Gribble F, Grill HJ, Habener JF, Holst JJ, Langhans W, Meier JJ, Nauck MA, Perez-Tilve D, Pocai A, Reimann F, Sandoval DA, Schwartz TW, Seeley RJ, Stemmer K, Tang-Christensen M, Woods SC, DiMarchi RD, Tschop MH (2019) Glucagon-like peptide 1 (GLP-1). Mol Metab 30:72–130. https://doi.org/10.1016/j.molmet.2019.09.010
56. Zhu H, Li G, Fan T, Wu L (2019) The research progress in the treatment of type 2 diabetes with glucagon-like peptide −1 receptor agonists. J Xiangnan Univ (Med Sci) 21:76–79
57. Aroda VR (2018) A review of GLP-1 receptor agonists: evolution and advancement, through the lens of randomised controlled trials. Diabetes Obes Metab 20(Suppl 1):22–33. https://doi.org/10.1111/dom.13162
58. Gentilella R, Bianchi C, Rossi A, Rotella CM (2009) Exenatide: a review from pharmacology to clinical practice. Diabetes Obes Metab 11(6):544–556. https://doi.org/10.1111/j.1463-1326.2008.01018.x
59. Norris SL, Lee N, Thakurta S, Chan BK (2009) Exenatide efficacy and safety: a systematic review. Diabet Med 26(9):837–846. https://doi.org/10.1111/j.1464-5491.2009.02790.x
60. Cirincione B, Mager DE (2017) Population pharmacokinetics of exenatide. Br J Clin Pharmacol 83(3):517–526. https://doi.org/10.1111/bcp.13135
61. Wilkins JJ, Dubar M, Sebastien B, Laveille C (2014) A drug and disease model for lixisenatide, a GLP-1 receptor agonist in type 2 diabetes. J Clin Pharmacol 54(3):267–278. https://doi.org/10.1002/jcph.192
62. Trujillo JM, Goldman J (2017) Lixisenatide, a once-daily prandial glucagon-like Peptide-1 receptor agonist for the treatment of adults with type 2 Diabetes. Pharmacotherapy 37 (8):927–943. https://doi.org/10.1002/phar.1962
63. Jason S, John D (2011) Novel GLP-1 mimetics in diabetes lixisenatide and albiglutide. Futur Prescriber 12:20–22
64. Chitrabhanu B, Stephen G (2012) Bydureon long-acting exenatide for once-weekly injection. Prescriber 1:30–33
65. Pencek R, Blickensderfer A, Li Y, Brunell SC, Chen S (2012) Exenatide once weekly for the treatment of type 2 diabetes: effectiveness and tolerability in patient subpopulations. Int J Clin Pract 66(11):1021–1032. https://doi.org/10.1111/j.1742-1241.2012.03006.x

66. Onishi Y, Koshiyama H, Imaoka T, Haber H, Scism-Bacon J, Boardman MK (2013) Safety of exenatide once weekly for 52 weeks in Japanese patients with type 2 diabetes mellitus. J Diab Invest 4(2):182–189. https://doi.org/10.1111/jdi.12000
67. Gough SCL (2012) Liraglutide from clinical trials to clinical practice. Diabetes Obes Metab 14:33–40
68. Ryan GJ, Hardy Y (2011) Liraglutide: once-daily GLP-1 agonist for the treatment of type 2 diabetes. J Clin Pharm Ther 36(3):260–274. https://doi.org/10.1111/j.1365-2710.2010.01180.x
69. Steve C (2016) Role of albiglutide in the treatment of type 2 diabetes. Prescriber 6:34–35
70. Bush MA, Matthews JE, De Boever EH, Dobbins RL, Hodge RJ, Walker SE, Holland MC, Gutierrez M, Stewart MW (2009) Safety, tolerability, pharmacodynamics and pharmacokinetics of albiglutide, a long-acting glucagon-like peptide-1 mimetic, in healthy subjects. Diabetes Obes Metab 11(5):498–505. https://doi.org/10.1111/j.1463-1326.2008.00992.x
71. Chaplin S, Bain S (2015) Dulaglutide: a once-weekly GLP-1 for type 2 diabetes. Prescriber:28–30
72. Jendle J, Grunberger G, Blevins T, Giorgino F, Hietpas RT, Botros FT (2016) Efficacy and safety of dulaglutide in the treatment of type 2 diabetes: a comprehensive review of the dulaglutide clinical data focusing on the AWARD phase 3 clinical trial program. Diabetes Metab Res Rev 32(8):776–790. https://doi.org/10.1002/dmrr.2810
73. Ole S, Birgitte B, Jorgen R (2004) Amylin agonists: a novel approach in the treatment of diabetes. Diabetes 53:233–238
74. Ernest A, Huba K (2011) Amylin analogues in the treatment of Diabetes mellitus medicinal chemistry and structural basis of its function. Open Med Chem J 5:78–81
75. Andisheh A, Daniel PR (2011) Islet amyloid polypeptide. In: Protein Misfolding diseases: current and emerging principles and therapies. Wiley, pp 517–541
76. Wang H, Abedini A, Ruzsicska B, Raleigh DP (2014) Rationally designed, nontoxic, nonamyloidogenic analogues of human islet amyloid polypeptide with improved solubility. Biochemistry 53(37):5876–5884. https://doi.org/10.1021/bi500592p
77. Roth JD, Erickson MR, Chen S, Parkes DG (2012) GLP-1R and amylin agonism in metabolic disease: complementary mechanisms and future opportunities. Br J Pharmacol 166(1):121–136. https://doi.org/10.1111/j.1476-5381.2011.01537.x
78. Zhang X, Cheng B, Huang K (2010) Recent progress of islet amyloid polypeptide. Chin Bull Life Sci 22:567–574. https://doi.org/10.13376/j.cbls/2010.06.003
79. Richard AP, Keith A, Kevin B, Laurie SLG, Joy EK, Candace XM, Judith R, Timothy JR (1994) Molecular physiology of amylin. J Cell Biochem 55:19–28
80. Anna N, David QP (2006) Adjunctive therapy with Pramlintide in patients with type 1 or type 2 diabetes mellitus. Pharmacotherapy 26:1626–1640
81. John G, Rachel C, Spencer S, Gemma C (2019) Amylin signaling in Diabetes and Alzheimer's disease therapy or pathology. J Neurol Neuromed 4:12–16
82. Yuan R (2005) The research progress in the treatment of diabetes with amylin analogue pramlintide. Foreign Med Sci (Sect Int Med) 32:507–510
83. Zhang L (2016) The research progress of inflammatory cytokines in the pathogenesis of type 2 diabetes mellitus. Chongqing Med 45:1113–1116
84. Tsalamandris S, Antonopoulos AS, Oikonomou E, Papamikroulis GA, Vogiatzi G, Papaioannou S, Deftereos S, Tousoulis D (2019) The role of inflammation in Diabetes: current concepts and future perspectives. Eur Cardiol 14(1):50–59. https://doi.org/10.15420/ecr.2018.33.1
85. Donath MP (2008) The use of interleukin-1-receptor antagonists in the treatment of diabetes mellitus. Nat Clin Pract Endocrinol Metabol 4:1–6
86. Larsen CM, Faulenbach M, Vaag A, Ehses JA, Donath MY, Mandrup-Poulsen T (2009) Sustained effects of Interleukin-1 receptor antagonist treatment in type 2 Diabetes. Diabetes Care 32(9):1663–1668. https://doi.org/10.2337/dc09-0533

87. Ballak DB, Stienstra R, Tack CJ, Dinarello CA, van Diepen JA (2015) IL-1 family members in the pathogenesis and treatment of metabolic disease: focus on adipose tissue inflammation and insulin resistance. Cytokine 75(2):280–290. https://doi.org/10.1016/j.cyto.2015.05.005
88. Cavelti-Weder C, Babians-Brunner A, Keller C, Stahel MA, Kurz-Levin M, Zayed H, Solinger AM, Mandrup-Poulsen T, Dinarello CA, Donath MY (2012) Effects of gevokizumab on glycemia and inflammatory markers in type 2 diabetes. Diabetes Care 35(8):1654–1662. https://doi.org/10.2337/dc11-2219
89. Owyang AM, Issafras H, Corbin J, Ahluwalia K, Larsen P, Pongo E, Handa M, Horwitz AH, Roell MK, Haak-Frendscho M, Masat L (2014) XOMA 052, a potent, high-affinity monoclonal antibody for the treatment of IL-1β-mediated diseases. MAbs 3(1):49–60. https://doi.org/10.4161/mabs.3.1.13989
90. Muzumdar RH, Huffman DM, Atzmon G, Buettner C, Cobb LJ, Fishman S, Budagov T, Cui L, Einstein FH, Poduval A, Hwang D, Barzilai N, Cohen P (2009) Humanin: a novel central regulator of peripheral insulin action. PLoS One 4(7):1–11. https://doi.org/10.1371/journal.pone.0006334
91. Voigt A, Jelinek HF (2016) Humanin: a mitochondrial signaling peptide as a biomarker for impaired fasting glucose-related oxidative stress. Physiol Rep 4(9):1–5. https://doi.org/10.14814/phy2.12796
92. Findeisen M, Allen TL, Henstridge DC, Kammoun H, Brandon AE, Baggio LL, Watt KI, Pal M, Cron L, Estevez E, Yang C, Kowalski GM, O'Reilly L, Egan C, Sun E, Thai LM, Krippner G, Adams TE, Lee RS, Grotzinger J, Garbers C, Risis S, Kraakman MJ, Mellet NA, Sligar J, Kimber ET, Young RL, Cowley MA, Bruce CR, Meikle PJ, Baldock PA, Gregorevic P, Biden TJ, Cooney GJ, Keating DJ, Drucker DJ, Rose-John S, Febbraio MA (2019) Treatment of type 2 diabetes with the designer cytokine IC7Fc. Nature 574 (7776):63–68. https://doi.org/10.1038/s41586-019-1601-9
93. Melanie D, Thomas RP, Marie-Louise HN, Oluf KHH, Serge J, Julio R (2017) Effect of oral semaglutide compared with placebo and subcutaneous semaglutide on glycemic control in patients with type 2 diabetes_ a randomized clinical trial. JAMA 318:1460–1470
94. Granhall C, Sondergaard FL, Thomsen M, Anderson TW (2018) Pharmacokinetics, safety and tolerability of oral semaglutide in subjects with renal impairment. Clin Pharmacokinet 57 (12):1571–1580. https://doi.org/10.1007/s40262-018-0649-2
95. Shi FH, Li H, Cui M, Zhang ZL, Gu ZC, Liu XY (2018) Efficacy and safety of once-weekly semaglutide for the treatment of type 2 diabetes: a systematic review and meta-analysis of randomized controlled trials. Front Pharmacol 9:1–13. https://doi.org/10.3389/fphar.2018.00576
96. Zhang H, Zhang L (2019) The clinical research progress on the combination therapies of diabetic drugs. Tianjin Pharm 31:75–78
97. Balena R, Hensley IE, Miller S, Barnett AH (2013) Combination therapy with GLP-1 receptor agonists and basal insulin: a systematic review of the literature. Diabetes Obes Metab 15:485–502
98. Nuffer W, Guesnier A, Trujillo JM (2018) A review of the new GLP-1 receptor agonist/basal insulin fixed-ratio combination products. Ther Adv Endocrinol Metab 9(3):69–79. https://doi.org/10.1177/2042018817752315

Chapter 4
An Overview of Hypoglycemic Traditional Drugs

Haixia Chen and Nannan Li

Abstract Folk or traditional systems of medicines always played an indispensable role in the global healthcare system. Traditional medicine as abundant natural resources has unique advantages in the treatment of diabetes, hyperlipidemia, and other diseases due to its focus on the regulation of body functions as a whole, multi-layered mechanism and approach, which make traditional drugs widely used in all over the world. There is a long history of using plants to treat diabetes in Asia, Africa, and Americas, and some plants and herbal preparations are effective to diabetes mellitus.

Keywords Diabetes mellitus · Traditional drugs · Plants · Herbal preparation

1 Introduction

Diabetes mellitus (DM) is a chronic secretory disease characterized by a series of metabolic syndrome [1]. At present, many modern chemical drugs are used to treat diabetes in clinic [2]. Although they can obviously reduce blood glucose, their disadvantages should not be ignored. At present, the development direction of diabetes treatment is multi-targets comprehensive treatment [3–5]. The combined intervention of multi-targets drugs can not only reduce the dosage of drugs and save medical resources, but also better control blood glucose and reduce complications caused by diabetes.

In fact, diabetes has existed for a long time and humans have spent thousands of years to treat it. Before the advent of modern medicine and biopharmaceuticals, some countries and regions had used some traditional medicines to treat and relieve some symptoms of diabetes. Traditional medicine has unique advantages in the treatment of diabetes, hyperlipidemia, and other diseases due to its focus on the

H. Chen (✉) · N. Li
Tianjin Key Laboratory for Modern Drug Delivery and High-Efficiency, School of Pharmaceutical Science and Technology, Tianjin University, Tianjin, China
e-mail: chenhx@tju.edu.cn

H. Chen, M. Zhang (eds.), *Structure and Health Effects of Natural Products on Diabetes Mellitus*, https://doi.org/10.1007/978-981-15-8791-7_4

regulation of body functions as a whole, multi-layered mechanism and approach and multi-targets.

Folk or traditional systems of medicines have long played an indispensable role in the global healthcare system. There is a long history of using plants to treat diabetes in Asia, Africa, and Americas. In recent years, with the development of medical technology, European and American countries have begun to screen the pharmacological activities of plants in order to find more effective drugs from natural plants. Therefore, traditional medicine plays a great guiding role in the application of various diseases and the screening of effective active substances. Generally, due to the complexity of the causes and symptoms of diabetes, traditional medicines have broader space in development and will never be replaced by modern medicines.

2 Asian Traditional Medicine

Asia is the largest and most populous of the seven continents, comprising 48 countries and about 1,000 ethnic groups. Cultural diversity makes the application of traditional medicine in Asia diverse. The two most typical forms of traditional medicine in Asia are traditional Chinese medicine and traditional Indian medicine, and these two systems also have profound implications for the treatment of diabetes.

2.1 Traditional Chinese Medicine

The traditional theories and thoughts of Chinese medicine are mainly the theory of Yin-Yang, the theory of five elements, the theory of essence and qi, and visceral manifestation in traditional Chinese medicine.

In the traditional Chinese medicine theory, diabetes belongs to the category of Xiaoke, which is a kind of disease characterized by excessive drinking, eating, urination, fatigue, emaciation, or sweetness of urine [6]. Human body is considered to be an organic whole centering on the five-zang organs in traditional Chinese medicine. The constituents contain heart, lung, liver, spleen, and kidney system. It is pointed out that the weakness of the five viscera is an important factor to cause Xiaoke.

Chinese medicine has a relatively deep research on diabetes about 2000 years. After a long time of clinical research, a variety of Chinese herbs and their preparation forms have been applied to treat various complications. These herbs and prescriptions with effects in clearing heat, dredging the bowels and purging turbid, clearing depression and dispersing knots, nourishing Yin and quenching thirst, invigorating spleen and draining dampness, supplementing Qi and tonifying deficiency, nourishing Yin and tonifying kidney are used to treat diabetes. The advantages of traditional Chinese medicine the treatment of diabetes and its complications of them have been identified.

Traditional Chinese medicine has complex components and can treat diabetes in a multi-target and multi-angle and achieves the comprehensive treatment of diabetes. In traditional Chinese medicine, the most frequently used herbs are *Ophiopogon japonicus (Linn. f.) Ker-Gawl, Glycyrrhiza uralensis Fisch, Panax ginseng C. A. Mey, Astragalus membranaceus (Fisch.) Bunge, Coptis chinensis Franch, Pueraria lobata (Willd)Ohw, Schisandra chinensis (Turcz.) Baill, Scutellaria baicalensis Georgi, Cornus officinalis Sieb, Zucc, Alisma plantago-aquatica Linn, Rehmannia glutinosa Libosc*h, and so on. Modern pharmaceutical chemistry and pharmacology have found that some traditional Chinese medicines do have a good hypoglycemic effect. Some important herbs are introduced in this paper, and others commonly used are shown in Table 4.1.

2.1.1 Some Traditional Chinese Herbal Medicine

2.1.1.1. *Ophiopogon japonicus (Linn. f.) Ker-Gawl*

Ophiopogon japonicus (Maidong in Chinese) belongs to the family Liliaceae, the tuberous roots of *Ophiopogon japonicus* with polysaccharide as the main active ingredient is a traditional Chinese herb. And the polysaccharides in Maidong is popularly used as a functional food additive in China. In addition, Maidong is an effective traditional Chinese medicine used to treat cardiovascular and chronic inflammatory diseases for thousands of years. Some studies have showed the polysaccharide could maintain the antioxidant enzyme levels and improve cardiovascular functions in diabetic rat [7].

Besides, polysaccharide as an important component of plants is effective for diabetes. According to the studies, a variety of polysaccharide components had the function of increasing glucose consumption [8]. A homogeneous polysaccharide fraction *Ophiopogon japonicus* from Sichuan (Chuan-Maidong) significantly ameliorated high-fat diet (HFD)-induced insulin resistance and glucose tolerance. Meanwhile, it regulated the gut microbiota dysbiosis in HFD-fed mice, as indicated by increasing antinobacteria and bifidobacterium, decreasing proteobacteria and type-2 diabetes-enriched taxa (e.g. Desulfovibrionaceae, Dorea, and Ruminococcaceae). In addition, it could improve the metabolic disorder of short-chain fatty acids (SCFAs) in HFD-fed mice [9]. Therefore, this kind of polysaccharide could be developed as a hypoglycemic functional food.

2.1.1.2. Ginseng (The Roots of *Panax ginseng C. A. Mey.*)

Ginseng (Renshen in Chinese) belongs to the family Araliaceae, which is the root of perennial herbaceous plant. Ginseng is regarded as the king of all herbs in China. It is a very precious traditional Chinese medicine and is a medicinal plant contributed for the controlling of many disorders. There is emerging evidence from both animal and clinical studies that ginseng, ginseng constituents including ginsenosides [10, 11], and ginseng-containing formulations can produce beneficial effects in terms of normalization of blood glucose levels and attenuation of cardiovascular complications through a multiplicity of mechanisms. Although more research is required, ginseng may offer a useful therapy for the treatment of diabetes as well as its complications [12]. Korean red ginseng, which is processed from Ginseng, contained substances that promote the secretion of insulin and inhibited renal inflammation, injury, and fibrosis by blocking TGF-beta 1 activation [13].

2.1.1.3 *Schisandra chinensis*

Schisandra chinensis (Wuweizi in Chinese) belongs to the family Schisandraceae. Wuweizi as a famous traditional Chinese medicine is the fruit of *Schisandra chinensis*, containing Schisandrin (Schisandrin $C_{23}H_{32}O_6$) and vitamin C, resin, tannin and a small amount of sugar, it has the effect of relieving cough, nourishing and strengthening, preventing diarrhea and preventing perspiration [14]. Schisandra has both medicinal and food properties, Schisandra polysaccharide can increase glucose consumption [15], have great antioxidant effect [16], hypoglycemic effect [17].

2.1.1.4. *Pueraria lobata* (The Root of Kudzu Vine)

Pueraria (Gegen in Chinese) is the root of *Pueraria lobata (Willd.) Ohwi*. In traditional Chinese medicine, Pueraria is considered to be cool and has therapeutic effects such as relieving muscle and reducing heat, generating fluid to quench thirst and preventing diarrhea [18]. Puerarin, a natural flavonoid in Pueraria root, as one of the components may directly benefit DM by decreasing blood glucose levels, improving insulin resistance, protecting islets, inhibiting inflammation, decreasing oxidative stress and advanced glycation end products (AGEs) formation [19], relieved diabetic nephropathy [20]. Taken together, puerarin might be a potential agent for the treatment of DM and DM complications in future.

2.1.1.5. Fruit of Chinese Wolfberry (*Lycium Chinense Mill.*)

Fruit of Chinese wolfberry (Gouqi in Chinese) is the Fruit of *Lycium chinense Mill.* Chinese wolfberry has effects on nourishing liver and kidney, improving eyesight. The effect of lowering blood glucose is good, especially for the elevation of glycosuria lipid and poor eyesight. Recent research has found that Lycium barbarum polysaccharides (glycocojugates) (LBP), containing several monosaccharides and 17 amino acids, were major bioactive constituents of hypoglycemic effect, the hypoglycemic effect of LBP is multi-target [21–23].

2.1.2 Traditional Chinese Medicine Formula

Of course, in the clinical application of diabetes treatment, most of the herbs are used together to make decoctions, pills, powder, and some other formulas. Compatibility medicine is one of the characteristics of traditional Chinese medicine and compatibility theory presented by "Monarch-Minister-Assistant-Guide" (Jun-Chen-Zuo-Shi in Chinese) is the essence of traditional Chinese medicine theory [24].

Combinations of herbs could produce a variety of effects, some can enhance the original effects, some can cancel or weaken the original effects, some can reduce or eliminate toxic side effects, and some can produce toxic side effects.

Traditional Chinese medicine formula is an important means to prevent and cure diseases and the essence of traditional Chinese medicine. Based on thousands of years of clinical experience, some prescriptions have also been preserved, and at the same time, some other prescriptions have also been found to have hypoglycemic activity. Modern medical research has shown those prescriptions had good results in treating diabetes and complications.

2.1.2.1. Liuwei Dihuang Decoction (LWDHT)

Liuwei Dihaung decoction (LWDHT) is a well-known classic traditional Chinese medicine formula, consists of six herbs including *Rehmannia glutinosa Libosch.* (family: Scrophulariaceae), *Cornus officinalis Sieb.* (family: Cornaceae), *Dioscorea opposite Thunb.* (family: Dioscoreaceae), *Alisma orientale (G. Samuelsson) Juz* (family: Alismataceae), *Poria cocos (Schw.) Wolf* (family: Polyporaceae), and *Paeonia suffruticosa Andrews* (family: Paeoniaceae). It has been used in the treatment of many types of diseases with signs of deficiency of Yin in the kidneys in China clinically. LWDHT could intervene insulin resistance of type 2 diabetes mellitus (T2DM), in part, through regulation of canonical PI3K/Akt signaling pathway of T2DM rats in liver [25]. And LWDHT has neuroprotective effect on memory and cognition deficits in streptozotocin (STZ)-induced diabetic encephalopathy (DE) rats [26]. LW-AFC, a new prescription derived from LWDHT, has been proved a promising drug for Alzheimer's disease [27].

2.1.2.2. Jinqi Recipe

Jinqi Recipe is a commonly prescribed recipe for T2DM in traditional Chinese medicine, including three ingredients such as *Flos ionicera*, *Copits chinensis Franch*, and Astragalus Root. Previous studies have shown that Jinqi recipe had significantly antidiabetic activity [28]. According to the traditional formula, Jinqi Jiangtang tablets as a traditional Chinese patent medicine were used in clinic in China. And clinical studies have shown that the combination of traditional treatment and treatment of T2DM had a good effect on regulating glucose metabolism and improving insulin resistance [29]. Another pathway to treat diabetes was to regulate the gut microbiota and promote the production of SCFAs [30].

2.1.2.3. Huanglian Decoction

Huanglian decoction is a traditional Chinese medicine formula used in the treatment of diabetes for thousands years [31], which consisted of *Coptis chinensis Franch.* (Chinese name: Huanglian), *Zingiber acuminatum Valeton* (Chinese name: Ganjiang), *Panax ginseng C. A. Mey.* (Chinese name: Renshen), *Glycyrrhiza uralensis Fisch* (Chinese name: Gancao), *Pinellia ternata (Thunb.) Makino* (Chinese name: Banxia), *Cinnamomum cassia (L.) J.Presl* (Chinese name: Guizhi), and *Ziziphus jujuba Mill.* (Chinese name: Dazao) [32]. It was proved by metabolomics and network pharmacology that Huanglian decoction work on multi-target and multi-pathway.

Besides the above traditional Chinese herbs, there are some other traditional Chinese herbs used to treat diabetes (Table 4.1).

Table 4.1 Some other Traditional Chinese herbs used to treat diabetes

Family	Names of Plants	Name in Chinese
Alismataceae	*Alisma plantago-aquatica Linn*	Zexie in Chinese
Apiaceae	*Angelica sinensis*	Danggui in Chinese
Cornaceae	*Cornus officinalis*	Shanzhuyu in Chinese
Cucurbitaceae	*Trichosanthes*	Tianhuafen in Chinese, the root of Trichosanthes kirilowii Maxim.
	Momordica charantia L.	Kugua in Chinese
	Gynostemma pentaphyllum (Thunb.) Makino	Jiaogulan in Chinese
Cordycipitaceae	*Ophiocordyceps sinensis*	Dongchongxiacao in Chinese
Lauraceae	*Cinnamomum cassia Presl*	Rougui in Chinese
Labiatae	*Scutellaria baicalensis Georgi*	Huangqin in Chinese
Liliaceae	*Anemarrhena asphodeloides Bunge*	Zhimu in Chinese
	Polygonatum sibiricum	Huangjing in Chinese
Leguminous	*Glycyrrhiza uralensis*	Gancao in Chinese
	Astragalus membranaceus	Huangqi in Chinese
Moraceae	*Morus alba L.*	Sang in Chinese
Polygonaceae	*Rheum palmatum L.*	Dahuang in Chinese
	Wolfiporia cocos	Fuling in Chinese
Ranunculaceae	*Cynanchum otophyllum*	Baishao in Chinese, the root of *Cynanchum otophyllum*
	Coptis chinensis Franch.	Huanglian in Chinese
Solanaceae	*Cortex Lycii*	Digupi in Chinese, the root skin of *Lycium Chinese mill.*
Scrophulariaceae	*Rehmannia*	Dihuang Chinese
Mineral	Gypsum fibrosum (CaSO4)	Shigao in Chinese
	Silkworm cocoon	Canjian in Chinese

2.2 *Traditional Indian Medicine*

Indian civilization as one of the oldest and splendid ancient civilization created in the basin of the Ganges river is comprehensive having multifaceted cultural aspects. At the same time, Indian traditional medicinal system is also one of the oldest traditional medicinal systems in the world. India has the unique characteristic of well-acknowledged traditional systems of medicine, including Ayurveda, Siddha, Unani, Yoga, naturopathy, and homeopathy [33].

Ayurveda is a complete medical care system including physical, psychological, philosophical, ethical, and spiritual consideration for the well-being of mankind; and it describes the causes, treatment, management, and prevention of almost every disease.

DM is most likely one of the well-described disorders in ancient India as "Madhumeha kshaudrameha," which means too much urination with honey like sweet taste. The Ayurvedic approach in the management of diabetes includes a lifestyle modification, exercise, dietary interventions, and different herb and herbal formulation related to the predominant dose, though cleansing measures consider exclusive to the Ayurvedic approach [34].

Yoga was considered as an ancient, traditional, physical, spiritual practice, which is part of traditional Indian culture. And several case studies have reported the Yoga has positive effect on diabetes mellitus and its complication. Yoga could improve glycemic control, improve insulin level and sensitivity, decrease insulin resistant, reduce depression and anxiety, decline the weight of obese people, improve lipid profile and oxidative stress, improve nerve function in diabetic neuropathic condition, and also reduce the risk of cardiovascular complications in diabetes [35, 36].

Traditional Indian medicines include herbs, herbal preparations, herbal materials, and finished herbal products, which contain as active ingredients parts of plants or other plant materials or combinations thereof [37].

In India, where some traditional folk medicines may not be documented, but surveys have shown that Leguminosae family was a commonly used family and the part of most of the plants used was leaf [38]. The most common and effective antidiabetic medicinal plants in Indian were Babul (*Acacia arabica*), garlic (*Allium sativum*), ghrita kumara (*Aloe vera*), ash gourd (*Benincasa hispida*), neem (*Azadirachta indica*), ivy gourd (*Coccinia indica*), methi (*Trigonella foenum-graecum*), and so on. All these plants are a rich source of phytochemicals [39]. Some important traditional Indian drugs against T2DM are described in this article.

2.2.1 *Gymnema sylvestre*

Gymnema sylvestre (gurmar) belongs to the family Asclepiadaceae. It is a herb native to the tropical forests of India and Sri Lanka. It is a potent antidiabetic plant used in Ayurvedic preparations. Several studies have proved its antidiabetic potential in animal model [40, 41]. *Gymnema sylvestre* can help prevent adrenal hormones from stimulating the liver to produce glucose in mice, thereby reducing blood sugar levels [42].

2.2.2 *Ficus religiosa*

Ficus religiosa, commonly known as peepal in India, belongs to family Moraceae. Ficus religiosa has been reported to be used for the treatment of diabetes [43]. Research showed the aqueous extract of Ficus religiosa bark possesses significant antidiabetic activity in streptozotocin-induced diabetic rats [44].

2.2.3 *Ocimum sanctum*

Ocimum sanctum L. (holy basil or tulsi) belongs to the family Lamiaceae. Every part of the plant is used as a therapeutic agent against several diseases. Ethanol extracts of *Ocimum sanctum* leaves improved glucose homeostasis in T2DM rats by enhancing circulating insulin and delaying carbohydrate digestion [45]. Another study showed the level of oxidative stress cytotoxicity and serum blood glucose was significantly reduced in diabetic rats treated with oral administration of aqueous suspension of Ocimum sanctum and Allium sativum daily [46]. It can be used as a dietary adjunct for the management of type 2 diabetes and its complications.

2.2.4 *Trigonella foenum-graecum*

Trigonella foenum-graecum (fenugreek, methi) belongs to the family Fabaceae. Seeds and leaves are the most frequently used parts of the plant. It is used both as a vegetable and as a spice in India. Dietary supplementation of 10 g Fenugreek/day in prediabetes subjects was associated with lower conversion to diabetes with no adverse effects and beneficial possibly due to its decreased insulin resistance [47]. It had protective effect on diabetes-induced oxidative DNA damage in rats [48]. The extract of fenugreek, grape seed, Indian gooseberry, and turmeric was made into capsules, which was clinically safe and effective in combination with other oral hypoglycemic drugs [49].

In addition to India and China, other Asian countries have their own traditional medicines to treat diabetes. For example, a total of 38 species belonging to 37 genera in 28 families were used as herbs for treating diabetes in Thailand, the family mostly used as herbal medicine was Rubiaceae [50]. The seeds of *Phoenix dactylifera L.* (Arecaceae) are used in Arab [51], *Ficus deltoidea var. deltoidea Jack* (FD), a well-known plant is used in Malay folklore medicine [52]. *Andrographis paniculata, Ageratum conyzoides, Swertia chirata, Terminalia arjuna*, and *Azadirachta indica* are used in Bangladesh [53]. *Punica granatum, Rosa damascene, Plantago psyllium, Glycyrrhiza glabra, Coriandrum sativum, Portulaca oleracea,* and *Rumex patientia* are used in Ira [54]. *Ferula persica*, *Paronychia argentea*, and *Pistacia atlantica* are three of the plants widely used in Jordan [55] (Table 4.2).

3 African Traditional Medicine

Africa is also a multicultural region, which is considered to be one of the cradles of the world's ancient humans and civilizations. Traditional African medicine is another amazing kind of traditional medicine which believes that supernatural forces were the cause of diseases. So, medicinal plants and animals are used to treat diseases diagnosed by divination. In Africa, many countries and ethnic groups also have their own traditional medicine to treat diabetes. Different countries and regions also have their own characteristics in using traditional drugs to treat diabetes.

3.1 Congo

Congo is a country in Central Africa, in DR Congo and other developing countries, access to modern healthcare systems remains quite limited, but the prevalence of diabetes is still high. So, an important proportion of patients mainly access traditional knowledge and healing resources to relieve symptoms [87]. Treatments are mostly based on herbal medicines, according to a survey, the leaves, root, and barks of *Albizia adianthifolia*, *Azanza garckeana* (leaves; Katuba), *Cassia occidentalis* (root; Kipushi), *Cassia sieberiana* (leaves; Kashama), *Erythrina abyssinica* were commonly used by the Congolese people [88].

3.2 Algeria

In Algeria, the incidence of diabetes is also increasing year by year. Diabetic patients treat themselves both the usual medicinal drugs and relevant parallel medicinal plants. Diabetic patients and herbalists believe that natural plant treatment is very effective in treating diabetes. This use of treatment with plants remains often undiscussed, limited to diabetic subjects and herbalists because medical doctors are most often against this use or only authorize it while patients stay on drug treatment. According to the survey, the most cited families for the treatment of diabetes in Algeria were Lamiaceae, Asteraceae such as Artemisia arborescens L. [89], Apiaceae, and Fabaceae with a dominance of Asteraceae [90, 91]. In the Northwest Algeria, Sidi Bel Abbes region, Asteraceae, Fabaceae, and Lauraceae were also commonly used families, some medicinal plants commonly used were *Trigonella foenum-graecum*, *Olea europaea*, *Cinamomum cassia*, *Artemisia herba-alba*, *Lupinus albus,* and so on [92]. According to the finding in western Algeria,

Table 4.2 Some other herbs and their pharmacological activities

Family	Names of Plants	Reported antiabetic activity	Country or region
Acanthaceae	*Rhinacanthus nasutus*	Inhibition of oxidative stress and inflammation, potential medicinal or nutritional supplement for the prevention of diabetic nephropathy	Thailand [56]
Annonaceae	*Annona cherimola*	Alpha-glucosidase inhibitor	Mexico [57]
Asteraceae	*Artemisia herba-alba Asso*	Decreased serum insulin concentrations, reduced insulin resistance	Algeria [58]
	Artemisia ludoviciana	Some of the active compounds seem to be acting synergistically on different molecular targets which involved glucose absorption and insulin liberation	Mexico [59]
	Cichorium intybus seeds	Significant reduction in serum glucose and triglycerides levels and decreased the oxidative burden in high-fat-diet-induced diabetic rats	India [60]
Costaceae	*Costus spiralis*	Inhibitory activity for α-glycosidase	Brazil [61]
Cucurbitaceae	*Momordica charantia Linn*	Both insulin secretagogue and insulinomimetic activity of the fruit. Oral administration of the fruit juice or seed powder causes a reduction in fasting blood glucose and improves glucose tolerance in normal and diabetic animals and in humans	India [62]
	Momordica dioica seeds	Antihyperglycemic, antioxidant, and anti-lipid peroxidative activity and thus mitigate STZ-induced oxidative damage	India [63]
	Siraitia grosvenorii	Antihyperglycemic and anti-hyperlipidemic effects of low-polar Siraitia grosvenorii glycosides on T2DM rats, regulating insulin secretion in T2DM rats by increasing GLP-1 levels	China [64]
Dilleniaceae	*Dillenia indica*	Treatment of diabetes-associated complications including diabetic neuropathy and diabetic nephropathy	India [65]
Gentianaceae	*Centaurium erythraea Rafn*	Decreased serum insulin concentrations, reduced insulin resistance	Algeria [58]
Labiatae	*Rotheca myricoides (Hochst.) Steane and Mabb*	Significant antihyperglycemic and antidyslipidemic effects on type 2 diabetes rat model	Africa, Kenya [66]
	Salvia miltiorrhiza	A potential of improving glucose and lipid metabolism in various diabetic animal models	China [67]

(continued)

Table 4.2 (continued)

Family	Names of Plants	Reported antiabetic activity	Country or region
Leguminosae	*Abrus precatorius L.*	The extract of leaf had glucose lowering and pancreato-protective effects	Africa, Ghana [68]
	Bauhinia forficata Link		Brazil [69]
	Pithecellobium dulce Benth. Seed	Significantly inhibition of blood glucose level in sucrose tolerance test by inhibiting enzymes responsible for hydrolysis of sucrose	India [70]
	Cassiae Semen (the dried seed of Cassia obtusifolia L)	Antidiabetic and renoprotective effects in diabetic rats	China [71]
Lilliaceae	*Asparagus gonoclados Baker*	Antidiabetic activity of root tubers in streptozotocin-induced diabetic rats	India [72]
Malvaceae	*Hibiscus rosa-sinensis L*	The flower is beneficial to pregnant rats with diabetes and their offspring, and the leaves could inhibit carbohydrate absorption and increase insulin secretion	Brazil [73, 74]
Melastomataceae	*Leandra lacunose*	Hypoglycemic activity of the hydroalcoholic extract of L. lacunosa aerial parts in alloxan-induced diabetic rats	Brazil [75]
Myrtaceae	*Syzygium jambolanum DC (Black Plum)*	It improved glucose utilization and maintains glucose homeostasis, activates PPARs, inhibitsα-glucosidases, and ameliorates dyslipidemia	India [76]
	Psidium guajava L.	The extract had alpha-glucosidase inhibitory activity and alpha-amylase inhibitory activity, which could improve glucose uptake in muscle cells	Africa, Madagascar [77]
Orchidaceae	*Dendrobium officinale Kimura and Migo*	Polysaccharide ameliorates diabetic hepatic glucose metabolism via glucagon-mediated signaling pathways and modifying liver- glycogen structure	China [78]
	Prosthechea karwinskii	P. karwinskii leaf extract inhibited reactive oxygen species and exerted an anti-inflammatory effect. It can be employed to treat other pathological conditions associated with oxidative stress	Mexico [79]
Passifloraceae	*Passiflora nitida Kunth*	A decrease of total cholesterol, a hypoglycemic effect, and an antioxidant activity with alloxan-diabetic mice	Brazil [80]
	Passiflora suberosa L.	Inhibition of intestinal absorption of (79%) glucose	Sri Lanka [81]

(continued)

Table 4.2 (continued)

Family	Names of Plants	Reported antiabetic activity	Country or region
Rhizophoraceae	*Rhizophora mucronata*	Inhibition of α-amylase and α-glucosidase enzymes, reducing diabetic complications	India [82]
Rosaceae	*Prunus Africana*	Ability to reduce the dipeptidyl peptidase-4 (DPP-4) enzyme, ability to decrease the serum blood glucose level	Africa [83]
	Potentilla discolor Bunge	21 antidiabetic compounds in Potentilla discolor Bunge regulate 33 diabetes-related proteins in 28 signal pathways and involve 21 kinds of diabetes-related diseases	China [84]
Rutaceae	*Aegle marmelos fruit*	Lowering of insulin resistance	India [85]
Sterculiaceae	*Cola nitida*	Significant depletion of blood glucose, serum triglycerides, LDL-cholesterol, fructosamine, ALT, and uric acids in type 2 diabetic rats	Africa [86]

non-insulin dependent patients used more medicinal plants than insulin-dependent patients [93].

3.3 Nigeria

An ethnobotanical survey of antidiabetic plants was conducted in Northwest Nigeria, the researchers found 54 species of traditional medicinal plants and treatments for diabetes, belonging to 33 families, with Cassia sieberiana being cited the most and ranked first, Azadirachta indica, Ficus exasperata, and Schwenckia americana ranked second [94]. The most preferred method of preparation by the healers is concoction and decoction of fresh leaves, stem bark, and roots [95].

3.4 Uganda

Traditional medicines were widely accepted within community in Uganda due to the difficulty of accessing medical care for diabetes and the shortage of diabetes medicines [96]. In Uganda, there is also some documentation on traditional medicines to treat diabetes. A total of 18 names of medicinal plants recorded by a survey, *Aloe vera var*, *Solanum indicum,* and *Vernonia amygydalina* were the most commonly mentioned plants and thus they are widely regarded as effective herbs for the treatment of diabetes. The main parts that were used to prepare the herbal medicine were leaves and the preparations were made in water. In all the cases, only the oral route was used in traditional herbs or medicines in the treatment at diabetes [97].

3.5 Sudan

Sudan folklore medicine is characterized by a unique combination of Islamic, Arabic, and African cultures. In poor communities, traditional medicine has remained as the most reasonable source of treatment of several diseases and microbial infections. Although the traditional medicine is accepted in Sudan, to date there is no updated review available, which focuses on most effective and frequently used Sudanese medicinal plant [98]. *Z. spinachristi* is commonly used in ethnomedicine for the treatment of many illnesses. Its leaves have a good hypoglycemic effect.

3.6 South Africa

Hypoxis hemerocallidea (AP) is a commonly used traditional medicine treating DM and one of the important herbal medicines developed in South Africa. The mechanism of AP regulating blood glucose effect is in the following three aspects, regulating blood glucose level, potential hypoglycemic mechanism, and its safety [99, 100]. *Rumex acetosa Linn* belongs to Polygonaceae, the species is distributed worldwide (African, Asian, American, and European countries). *Rumex acetosa* is used traditionally as vegetables and for its medicinal uses. Its diverse uses in traditional and cultural applications have geared much research towards its phytochemical and pharmacological activities. Many cultures around the world use the leaves and aerial parts as vegetables, other parts of this medicinal plant are employed in the management of a number of ailments such as constipation, diarrhea, jaundice, mild diabetes and as an analgesic, antihypertensive, against gallbladder, liver and skin disorders, and inflammation [101]. *Morinda Lucida* is a folk plant in Africa that treats frequent urination in adults, which is not necessarily associated with diabetes but is associated with weakness and rapid wasting. Modern research shows that *Morinda Lucida* mitigated against derangements in the measured renal and hepatic function parameters as well as oxidative stress induced by alloxan-induced hyperglycemia [102].

4 Traditional American Medicine

In Central America, out of 535 identified species used to manage diabetes and its complications, 104 species are used to manage diabetes and 16 of the 20 species reported were found in vitro and in vivo preclinical experimental evidence of hypoglycemic effect. According to the research reports in recent years, it is found that the following medicinal plants have higher frequency of occurrence and more studies on hypoglycemic effect: *Momordica charantia L., Neurolaena lobata (L.) R. Br. ex Cass., Tecoma stans (L.) Juss. ex Kunth, Persea americana Mill., Psidium guajava L., Anacardium occidentale L.,* and *Hamelia patens Jacq.* Several of the species that are used to manage diabetes in Central America are also used to treat conditions that may arise as its consequence such as kidney disease, urinary problems, and skin conditions [103]. In the Americas, countries such as Belize, Costa Rica, El Salvador, Guatemala, Honduras, Nicaragua, and Panama also use plants to treat diabetes and related symptoms.

In fact, Mexico is an ancient civilization of Central America, and Mexico also has some traditional medicines for the treatment of diabetes and complications. Brazil, the largest country in South America, was inhabited by Indian in ancient times, and some plants were used to treat diabetes (Table 4.2).

Mexico is a megadiverse country that has 3600 to 4000 species of medicinal plants, of which approximately 800 are used to treat conditions related to diabetes mellitus [104]. Here are some traditional Mexican medicines and traditional Brazilian medicines.

4.1 Cecropia obtusifolia Bertol *(Cecropiaceae)*

Cecropia obtusifolia Bertol belongs to the family Cecropiaceae, it is used for the treatment of DM in Central and South America, and chlorogenic acid and isoorientin were the main hypoglycemic active components [105]. *Cecropia obtusifolia* is widely used in Mexican traditional medicine due to its reputed hypoglycemic effect. Recent research has found that the aqueous extracts of *Cecropia obtusifolia* have a significant hypoglycemic effect with no adverse effects and that the mechanism of action is not brought about by stimulating the insulin secretion [106].

4.2 Calea ternifolia

Calea ternifolia belongs to the family Ranunculaceae, which is used in the treatment of diabetes, and it is commercialized as a dietary supplement in several countries. However, some results show renal and hepatic toxicity; therefore, more profound research on the toxicity of this plant is needed [107].

4.3 *Xoconostle (*Opuntia joconostle*)*

Xoconostle fruit is the acidic cactus pear fruit of *Opuntia joconostle* of the Cactaceae family. Xoconostle is rich in nutrients and health functions, with a variety of biological activities. In traditional Mexican medicine, this fruit was believed to have various properties, such as ashypoglycemic, hypocholesterolemic, anti-inflammatory, antiulcerogenic, and immunostimulant, among others. Modern research has found that Xoconostle possesses a glucose- and lipid-lowering effect in both healthy and diabetes-induced rats [108]. Another research found it had the inhibitory effects on α-amylase and α-glucosidase on simulated intestinal conditions [109].

4.4 Achillea millefolium L. *(Asteraceae)*

Achillea millefolium L. (Asteraceae) is a perennial herb used in Mexican folk medicine for treatment of several pathologies, including inflammatory and spasmodic gastrointestinal disorders, hepatobiliary complaints, overactive cardiovascular, respiratory ailments and diabetes. The hydro-alcoholic extract of Achillea millefolium L could protect pancreatic cells [110]. In vivo experiments, it worked through multi-target modes of action that involve antihyperglycemic (α-glucosidases inhibition), hypoglycemic (insulin secretion) and potential insulin sensitizer (PPAR gamma/GLUT4 overexpression) actions [111].

4.5 Cucurbita ficifolia

Cucurbita ficifolia belongs to the family Cucurbitaceae, it is used in Mexican traditional medicine as an anti-diabetic and anti-inflammatory agent and its actions can be mediated by antioxidant mechanisms. An aqueous extract of *Cucurbita ficifolia* with hypoglycemic action could improve glutathione (GSH) redox state, increasing glutathione pool, GSH, GSH/GSSG ratio, The mechanism is related to its antioxidant properties, which supports its use as an alternative treatment for the control of DM, and prevent the induction of complications by oxidative stress [112]. Other studies have found that it could increase insulin secretion in RINm5F cells through an influx of $Ca^{2+}$from the endoplasmic reticulum [113].

5 Conclusion

A large number of plants have been used to experimentally treat the symptoms of this disease, including a large number of valuable medicinal plants from marine algae and fungi to advanced plants. The large and widely distributed families are Fabaceae, Asteraceae, Lamiaceae, Liliaceae, and Euphorbiaceae. In fact, different regions may use the same plants to treat diabetes, which shows the reliability of some traditional medicines.

The large number of taxa reported to have been used traditionally or experimentally for the treatment of diabetes may be coincidental. However, traditional drugs still have great value in treating diabetes and its symptoms. The drug theory system of each country and nation can also provide better suggestions for modern medicine. In addition to spirit and some natural therapies, the use and compatibility of various ethnic groups also have great enlightenment on the development of drugs.

With the development of modern extraction and separation technology, more and more effective components have been extracted from these plants, including flavonoids, alkaloids, polysaccharides, volatile oils, quinines, terpenoids, lignin, coumarins, saponins, cardiac glycosides, phenolic acids, amino acids, and enzymes. For example, Dimethylbiguanide (metformin), a derivative of a natural product, was got from Galega officinalis Linn., a plant native to Europe. It is an alkaloid with good therapeutic effect for type 2 diabetes mellitus. It has been on the market for more than 60 years since 1957 and is still widely used in clinic as a first-line drug. It is an economic, safe, and effective classic drug [114].

The modern medicine can't replace the traditional medicine, traditional medicine is also gradually developed, and combined with modern medicine to play a better therapeutic effect. For example, Xiaoke Pill is one of the most popular proprietary Chinese medicines (PCM) used in China for treating mild, moderate DM in recent decades. Xiaoke Pill was developed from two ancient Traditional Chinese Medicine formulae, namely *Yuquan Powder* and *Xiaoke Formula*. It is a composite drug composed of 7 Chinese materia medica, namely Puerariae Radix, Rehmanniae Radix, Astragali Radix, Trichosanthis Radix, Stylus Zeae Maydis, Schisandrae Sphenantherae Fructus and Dioscoreae Rhizoma and glibenclamide, which is a second-generation sulphonylureas drug [115]. A number of antidiabetic herbal formulations have been granted patent recently.

Etiology of diabetes is multifactorial and multi-targeted herbal drugs would be comparatively safer than modern drugs and may be used in management of prediabetes and prevention of progression to diabetes. With the interpenetration and intersection of multiple disciplines, especially the combination of natural product research and molecular biology, more valuable drug resources will be got from nature, it will be the new direction for human to develop hypoglycemic drugs.

References

1. Simmons RK, Alberti KGMM, Gale EAM, Colagiuri S, Tuomilehto J, Qiao Q, Ramachandran A, Tajima N, Mirchov IB, Ben-Nakhi A, Reaven G, Sambo BH, Mendis S, Roglic G (2010) The metabolic syndrome: useful concept or clinical tool? Report of a WHO Expert Consultation. Diabetologia 53(4):600–605. https://doi.org/10.1007/s00125-009-1620-4
2. Tang YZ, Wang G, Jiang ZH, Yan TT, Chen YJ, Yang M, Meng LL, Zhu YJ, Li CG, Li Z, Yu P, Ni CL (2015) Efficacy and safety of vildagliptin, sitagliptin, and linagliptin as add-on therapy in Chinese patients with T2DM inadequately controlled with dual combination of insulin and traditional oral hypoglycemic agent. Diabetol Metab Syndr 7:9. https://doi.org/10.1186/s13098-015-0087-3
3. Xiao SW, Liu C, Chen MJ, Zou JF, Zhang ZM, Cui X, Jiang S, Shang EX, Qian DW, Duan JN (2020) Scutellariae radix and coptidis rhizoma ameliorate glycolipid metabolism of type 2 diabetic rats by modulating gut microbiota and its metabolites. Appl Microbiol Biot 104 (1):303–317. https://doi.org/10.1007/s00253-019-10174-w
4. Bannister M, Berlanga J (2016) Effective utilization of oral hypoglycemic agents to achieve individualized HbA1c targets in patients with type 2 diabetes mellitus. Diab Therapy 7 (3):387–399. https://doi.org/10.1007/s13300-016-0188-5
5. Mishra GP, Sharma R (2016) Identification of potential PPAR gamma agonists as hypoglycemic agents: molecular docking approach. Interdiscip Sci 8(3):220–228. https://doi.org/10.1007/s12539-015-0126-7
6. Sun XL, Zhang B, Wang SH, Liu SY, Zhou QY (2020) Analysis of the rule of TCM compatibility in TCM prescriptions containing Ginseng Radix ET Rhizoma in ancient books for Xiaoke Bing. Evid-Based Compl Alt 2020:7. https://doi.org/10.1155/2020/9472304
7. Zhang J, Fan S, Mao Y, Ji Y, Jin L, Lu J, Chen X (2016) Cardiovascular protective effect of polysaccharide from Ophiopogon japonicus in diabetic rats. Int J Biol Macromol 82:505–513. https://doi.org/10.1016/j.ijbiomac.2015.09.069
8. Gong YJ, Zhang J, Gao F, Zhou JW, Xiang ZN, Zhou CG, Wan LS, Chen JC (2017) Structure features and in vitro hypoglycemic activities of polysaccharides from different species of Maidong. Carbohydr Polym 173:215–222. https://doi.org/10.1016/j.carbpol.2017.05.076
9. Wang H-Y, Guo L-X, Hu W-H, Peng Z-T, Wang C, Chen Z-C, Liu EYL, Dong TTX, Wang T-J, Tsim KWK (2019) Polysaccharide from tuberous roots of Ophiopogon japonicus regulates gut microbiota and its metabolites during alleviation of high-fat diet-induced type-2 diabetes in mice. J Funct Foods 63. https://doi.org/10.1016/j.jff.2019.103593
10. Zhu YY, Zhu CH, Yang HX, Deng JJ, Fan DD (2020) Protective effect of ginsenoside Rg5 against kidney injury via inhibition of NLRP3 inflammasome activation and the MAPK signaling pathway in high-fat diet/streptozotocin-induced diabetic mice. Pharmacol Res 155:11. https://doi.org/10.1016/j.phrs.2020.104746
11. Wei Y, Yang H, Zhu C, Deng J, Fan D (2020) Hypoglycemic effect of ginsenoside Rg5 mediated partly by modulating gut microbiota dysbiosis in diabetic db/db mice. J Agr Food Chem 68(18):5107–5117. https://doi.org/10.1021/acs.jafc.0c00605
12. Karmazyn M, Gan XT (2019) Ginseng for the treatment of diabetes and diabetes-related cardiovascular complications: a discussion of the evidence. Can J Physiol Pharmacol 97 (4):265–276. https://doi.org/10.1139/cjpp-2018-0440
13. Karunasagara S, Hong GL, Park SR, Lee NH, Jung DY, Kim TW, Jung JY (2020) Korean red ginseng attenuates hyperglycemia-induced renal inflammation and fibrosis via accelerated autophagy and protects against diabetic kidney disease. J Ethnopharmacol 254:10. https://doi.org/10.1016/j.jep.2020.112693
14. Liu J, Xu J, Guo J (2019) Review of active constituents and pharmacological activities of Schisandrae Chinensis Fructus (五味子活性成分及药理作用研究进展). Chinese J Exp Tradit Med Formulae 25(11):206–215

15. Jin D, Zhao T, Feng WW, Mao GH, Zou Y, Wang W, Li Q, Chen Y, Wang XT, Yang LQ, Wu XY (2016) Schisandra polysaccharide increased glucose consumption by up-regulating the expression of GLUT-4. Int J Biol Macromol 87:555–562. https://doi.org/10.1016/j.ijbiomac.2016.03.028
16. Niu J, Xu G, Jiang S, Li H, Yuan G (2017) In Vitro Antioxidant activities and anti-diabetic effect of a polysaccharide from Schisandra sphenanthera in rats with type 2 diabetes. Int J Biol Macromol 94(Pt A):154–160. https://doi.org/10.1016/j.ijbiomac.2016.10.015
17. Du XX, Tao X, Liang S, Che JY, Yang S, Li H, Chen JG, Wang CM (2019) Hypoglycemic effect of acidic polysaccharide from Schisandra chinensis on T2D rats induced by high-fat diet combined with STZ. Biol Pharm Bull 42(8):1275–1281. https://doi.org/10.1248/bpb.b18-00915
18. Yang L, Chen J, Lu H, Lai J, He Y, Liu S, Guo X (2019) Pueraria lobata for Diabetes Mellitus: past, present and future. Am J Chin Med 47(7):1419–1444. https://doi.org/10.1142/S0192415X19500733
19. Chen X, Yu J, Shi J (2018) Management of diabetes mellitus with puerarin, a natural isoflavone from Pueraria lobata. Am J Chin Med 46(8):1771–1789. https://doi.org/10.1142/S0192415X18500891
20. Li XL, Zhu QQ, Zheng R, Yan JY, Wei MG, Fan YC, Deng YY, Zhong YF (2020) Puerarin attenuates diabetic nephropathy by promoting autophagy in podocytes. Front Physiol 11:11. https://doi.org/10.3389/fphys.2020.00073
21. Luo Q, Cai Y, Yan J, Sun M, Corke H (2004) Hypoglycemic and hypolipidemic effects and antioxidant activity of fruit extracts from Lycium barbarum. Life Sci 76(2):137–149. https://doi.org/10.1016/j.lfs.2004.04.056
22. Yang S, Si L, Fan L, Jian W, Pei H, Lin R (2018) Polysaccharide IV from Lycium barbarum L. Improves lipid profiles of gestational diabetes mellitus of pregnancy by upregulating ABCA1 and downregulating sterol regulatory element-binding transcription 1 via miR-33. Front Endocrinol (Lausanne) 9:49. https://doi.org/10.3389/fendo.2018.00049
23. Xia H, Tang H, Wang F, Yang X, Wang Z, Liu H, Pan D, Yang C, Wang S, Sun G (2019) An untargeted metabolomics approach reveals further insights of Lycium barbarum polysaccharides in high fat diet and streptozotocin-induced diabetic rats. Food Res Int 116:20–29. https://doi.org/10.1016/j.foodres.2018.12.043
24. Luan X, Zhang LJ, Li XQ, Rahman K, Zhang H, Chen HZ, Zhang WD (2020) Compound-based Chinese medicine formula: From discovery to compatibility mechanism. J Ethnopharmacol 254:11. https://doi.org/10.1016/j.jep.2020.112687
25. Dai B, Wu Q, Zeng C, Zhang J, Cao L, Xiao Z, Yang M (2016) The effect of Liuwei Dihuang decoction on PI3K/Akt signaling pathway in liver of type 2 diabetes mellitus (T2DM) rats with insulin resistance. J Ethnopharmacol 192:382–389. https://doi.org/10.1016/j.jep.2016.07.024
26. Liu JP, Feng L, Zhang MH, Ma DY, Wang SY, Gu J, Fu Q, Qu R, Ma SP (2013) Neuroprotective effect of Liuwei Dihuang decoction on cognition deficits of diabetic encephalopathy in streptozotocin-induced diabetic rat. J Ethnopharmacol 150(1):371–381. https://doi.org/10.1016/j.jep.2013.09.003
27. Cheng X, Huang Y, Zhang Y, Zhou W (2020) LW-AFC, a new formula from the traditional Chinese medicine Liuwei Dihuang decoction, as a promising therapy for Alzheimer's disease: pharmacological effects and mechanisms. Adv Pharmacol (San Diego, Calif) 87:159–177. https://doi.org/10.1016/bs.apha.2019.10.005
28. Han C (2009) Comparison of anti-hyperglycemic effect of inorganic constituents and organic in traditional Chinese medicine, Jinqi compound recipe. Biol Trace Elem Res 131(1):55–61. https://doi.org/10.1007/s12011-009-8344-7
29. Gao HJ, Yang YX, Deng JQ, Liang JQ, Zhang WH, Feng XZ (2019) A systematic review and meta-analysis on the efficacy and safety of traditional Chinese patent medicine Jinqi Jiangtang Tablet in the treatment of type 2 diabetes. Complement Ther Med 47:11. https://doi.org/10.1016/j.ctim.2019.01.016

30. Cao Y, Yao GW, Sheng YY, Yang L, Wang ZX, Yang Z, Zhuang PW, Zhang YJ (2019) JinQi Jiangtang tablet regulates gut microbiota and improve insulin sensitivity in type 2 diabetes mice J Diabetes Res 12. https://doi.org/10.1155/2019/1872134
31. Pang B, Ni Q (2019) Application of classical formula in treatment of diabetes. Zhongguo Zhong yao za zhi = Zhongguo zhongyao zazhi = China J Chinese Materia Medica 44 (18):3895–3898. https://doi.org/10.19540/j.cnki.cjcmm.20190416.504
32. Pan L, Li Z, Wang Y, Zhang B, Liu G, Liu J (2020) Network pharmacology and metabolomics study on the intervention of traditional Chinese medicine Huanglian decoction in rats with type 2 diabetes mellitus. J Ethnopharmacol 258:112842. https://doi.org/10.1016/j.jep.2020.112842
33. Shrivastava SR, Shrivastava PS, Ramasamy J (2015) Mainstreaming of Ayurveda, Yoga, Naturopathy, Unani, Siddha, and Homeopathy with the health care delivery system in India. J Tradit Complement Med 5(2):116–118. https://doi.org/10.1016/j.jtcme.2014.11.002
34. Nakanekar A, Kohli K, Tatke P (2019) Ayurvedic polyherbal combination (PDBT) for prediabetes: a randomized double blind placebo controlled study. J Ayurveda Integr Med 10 (4):284–289. https://doi.org/10.1016/j.jaim.2018.05.004
35. Sharma R, Shahi VK, Khanduri S, Goyal A, Chaudhary S, Rana RK, Singhal R, Srikanth N, Dhiman KS (2019) Effect of Ayurveda intervention, lifestyle modification and Yoga in prediabetic and type 2 diabetes under the National Programme for Prevention and Control of Cancer, Diabetes, Cardiovascular Diseases and Stroke (NPCDCS)-AYUSH integration project. Ayu 40(1):8–15. https://doi.org/10.4103/ayu.AYU_105_19
36. Jyotsna VP (2014) Prediabetes and type 2 diabetes mellitus: evidence for effect of yoga. Indian J Endocrinol Metab 18(6):745–749. https://doi.org/10.4103/2230-8210.141318
37. Sen S, Chakraborty R, De B (2016) Management of Diabetes Mellitus. Diabetes Mellitus in 21st Century, 153–174.
38. Dixit S, Tiwari S (2020) Investigation of anti-diabetic plants used among the ethnic communities of Kanpur division, India. J Ethnopharmacol 253:11. https://doi.org/10.1016/j.jep.2020.112639
39. Rizvi SI, Mishra N (2013) Traditional Indian medicines used for the management of diabetes mellitus. J Diabetes Res 2013:712092. https://doi.org/10.1155/2013/712092
40. Kumar GS, Shetty AK, Salimath PV (2005) Modulatory effect of fenugreek seed mucilage and spent turmeric on intestinal and renal disaccharidases in streptozotocin induced diabetic rats. Plant Foods Hum Nutr 60(2):87–91. https://doi.org/10.1007/s11130-005-5104-5
41. Kamble B, Gupta A, Moothedath I, Khatal L, Janrao S, Jadhav A, Duraiswamy B (2016) Effects of Gymnema sylvestre extract on the pharmacokinetics and pharmacodynamics of glimepiride in streptozotocin induced diabetic rats. Chem Biol Interact 245:30–38. https://doi.org/10.1016/j.cbi.2015.12.008
42. Gholap S, Kar A (2003) Effects of Inula racemosa root and Gymnema sylvestre leaf extracts in the regulation of corticosteroid induced diabetes mellitus: involvement of thyroid hormones. Pharmazie 58(6):413–415
43. Singh D, Singh B, Goel RK (2011) Traditional uses, phytochemistry and pharmacology of Ficus religiosa: a review. J Ethnopharmacol 134(3):565–583. https://doi.org/10.1016/j.jep.2011.01.046
44. Pandit R, Phadke A, Jagtap A (2010) Antidiabetic effect of Ficus religiosa extract in streptozotocin-induced diabetic rats. J Ethnopharmacol 128(2):462–466. https://doi.org/10.1016/j.jep.2010.01.025
45. Hannan J, Ojo O, Rokeya L, Khaleque J, Akhter M, Flatt P, Abdel-Wahab Y (2015) Actions underlying antidiabetic effects of Ocimum sanctum leaf extracts in animal models of type 1 and type 2 diabetes. Eur J Med Plants 5(1):1–12. https://doi.org/10.9734/ejmp/2015/11840
46. Jayant SK, Srivastava N (2019) Effects of Ocimum sanctum and Allium sativum extracts against diabetes and determination of DNA damage and cytotoxicity in alloxan induced diabetic rats. Biosci Biotechnol Res Commun 12(4):1100–1109. https://doi.org/10.21786/bbrc/12.4/33

47. Gaddam A, Galla C, Thummisetti S, Marikanty RK, Palanisamy UD, Rao PV (2015) Role of Fenugreek in the prevention of type 2 diabetes mellitus in prediabetes. J Diabetes Metab Disord 14:74. https://doi.org/10.1186/s40200-015-0208-4
48. Sharma S, Mishra V, Srivastava N (2020) Protective effect of Trigonella foenum-graecum and Cinnamomum zeylanicum against diabetes induced oxidative DNA damage in rats. Indian J Biochem Biophys 57(1):15–26
49. Banerji S, Banerjee S (2016) A formulation of grape seed, Indian gooseberry, turmeric and fenugreek helps controlling type 2 diabetes mellitus in advanced-stage patients. Eur J Integr Med 8(5):645–653. https://doi.org/10.1016/j.eujim.2016.06.012
50. Neamsuvan O, Madeebing N, Mah L, Lateh W (2015) A survey of medicinal plants for diabetes treating from Chana and Nathawee district, Songkhla province, Thailand. J Ethnopharmacol 174:82–90. https://doi.org/10.1016/j.jep.2015.07.050
51. Abdelaziz DH, Ali SA, Mostafa MM (2015) Phoenix dactylifera seeds ameliorate early diabetic complications in streptozotocin-induced diabetic rats. Pharm Biol 53(6):792–799. https://doi.org/10.3109/13880209.2014.942790
52. Abdel-Rahman RF, Ezzat SM, Ogaly HA, Abd-Elsalam RM, Hessin AF, Fekry MI, Mansour DF, Mohamed SO (2020) Ficus deltoidea extract down-regulates protein tyrosine phosphatase 1B expression in a rat model of type 2 diabetes mellitus: a new insight into its antidiabetic mechanism. J Nutr Sci 9:e2. https://doi.org/10.1017/jns.2019.40
53. Rafe MR (2017) A review of five traditionally used anti-diabetic plants of Bangladesh and their pharmacological activities. Asian Pac J Trop Med 10(10):933–939. https://doi.org/10.1016/j.apjtm.2017.09.002
54. Rashidi AA, Mirhashemi SM, Taghizadeh M, Sarkhail P (2013) Iranian medicinal plants for diabetes mellitus: a systematic review. Pakistan J Biol Sci: PJBS 16(9):401–411
55. Hamdan II, Afifi FU (2004) Studies on the in vitro and in vivo hypoglycemic activities of some medicinal plants used in treatment of diabetes in Jordanian traditional medicine. J Ethnopharmacol 93(1):117–121. https://doi.org/10.1016/j.jep.2004.03.033
56. Zhao LL, Makinde EA, Shah MA, Olatunji OJ, Panichayupakaranant P (2019) Rhinacanthins-rich extract and rhinacanthin C ameliorate oxidative stress and inflammation in streptozotocin-nicotinamide-induced diabetic nephropathy. J Food Biochem 43(4):e12812. https://doi.org/10.1111/jfbc.12812
57. Calzada F, Solares-Pascasio JI, Ordonez-Razo RM, Velazquez C, Barbosa E, Garcia-Hernandez N, Mendez-Luna D, Correa-Basurto J (2017) Antihyperglycemic activity of the leaves from Annona cherimola Miller and Rutin on Alloxan-induced diabetic rats. Pharmacogn Res 9(1):1–6. https://doi.org/10.4103/0974-8490.199781
58. Hamza N, Berke B, Cheze C, Le Garrec R, Lassalle R, Agli AN, Robinson P, Gin H, Moore N (2011) Treatment of high fat diet induced type 2 diabetes in C57BL/6J mice by two medicinal plants used in traditional treatment of diabetes in the east of Algeria. J Ethnopharmacol 133 (2):931–933. https://doi.org/10.1016/j.jep.2010.11.019
59. Anaya-Eugenio GD, Rivero-Cruz I, Rivera-Chavez J, Mata R (2014) Hypoglycemic properties of some preparations and compounds from Artemisia ludoviciana Nutt. J Ethnopharmacol 155(1):416–425. https://doi.org/10.1016/j.jep.2014.05.051
60. Chester K, Zahiruddin S, Ahmad A, Khan W, Paliwal S, Ahmad S (2019) Bioautography-based identification of antioxidant metabolites of Solanum nigrum L. and exploration its hepatoprotective potential against D-galactosamine-induced hepatic fibrosis in rats. Pharmacogn Mag 15(62):104–110. https://doi.org/10.4103/pm.pm_359_18
61. de Oliveira AP, Coppede JS, Bertoni BW, Crotti AEM, Franca SC, Pereira AMS, Taleb-Contini SH (2018) Costus spiralis (Jacq.) Roscoe: a novel source of flavones with alpha-glycosidase inhibitory activity. Chem Biodivers 15(1). https://doi.org/10.1002/cbdv.201700421
62. Raman A, Lau C (1996) Anti-diabetic properties and phytochemistry of Momordica charantia L. (Cucurbitaceae). Phytomed Int J Phytother Phytopharmacol 2(4):349–362. https://doi.org/10.1016/s0944-7113(96)80080-8

63. Rao PS, Mohan GK (2017) In vitro alpha-amylase inhibition and in vivo antioxidant potential of Momordica dioica seeds in streptozotocin-induced oxidative stress in diabetic rats. Saudi J Biol Sci 24(6):1262–1267. https://doi.org/10.1016/j.sjbs.2016.01.010
64. Zhang Y, Zhou G, Peng Y, Wang M, Li X (2020) Anti-hyperglycemic and anti-hyperlipidemic effects of a special fraction of Luohanguo extract on obese T2DM rats. J Ethnopharmacol 247:112273. https://doi.org/10.1016/j.jep.2019.112273
65. Kamboj P, Talukdar NC, Banerjee SK (2019) Therapeutic benefit of Dillenia indica in diabetes and its associated complications. J Diabetes Res 2019:4632491. https://doi.org/10.1155/2019/4632491
66. Chege BM, Waweru MP, Frederick B, Nyaga NM (2019) The freeze-dried extracts of Rotheca myricoides (Hochst.) Steane & Mabb possess hypoglycemic, hypolipidemic and hypoinsulinemic on type 2 diabetes rat model. J Ethnopharmacol 244:112077. https://doi.org/10.1016/j.jep.2019.112077
67. Jia Q, Zhu R, Tian Y, Chen B, Li R, Li L, Wang L, Che Y, Zhao D, Mo F, Gao S, Zhang D (2019) Salvia miltiorrhiza in diabetes: a review of its pharmacology, phytochemistry, and safety. Phytomedicine 58:152871. https://doi.org/10.1016/j.phymed.2019.152871
68. Boye A, Acheampong DO, Gyamerah EO, Asiamah EA, Addo JK, Mensah DA, Brah AS, Ayiku PJ (2020) Glucose lowering and pancreato-protective effects of Abrus Precatorius (L.) leaf extract in normoglycemic and STZ/Nicotinamide - induced diabetic rats. J Ethnopharmacol:112918. https://doi.org/10.1016/j.jep.2020.112918
69. Cechinel-Zanchett CC, da Silva R, Tenfen A, Siebert DA, Micke G, Vitali L, Cechinel V, de Andrade SF, de Souza P (2019) Bauhinia forficata link, a Brazilian medicinal plant traditionally used to treat cardiovascular disorders, exerts endothelium-dependent and independent vasorelaxation in thoracic aorta of normotensive and hypertensive rats. J Ethnopharmacol 243:8. https://doi.org/10.1016/j.jep.2019.112118
70. Kumar M, Govindrajan J, Nyola NK (2017) Antihyperglycemic Potential of Saponin-enriched Fraction from Pithecellobium dulce Benth. Seed extract. Pharmacogn Res 9(Suppl 1):S23–S26. https://doi.org/10.4103/pr.pr_18_17
71. Wang Q, Zhou J, Xiang Z, Tong Q, Pan J, Wan L, Chen J (2019) Anti-diabetic and renoprotective effects of Cassiae Semen extract in the streptozotocin-induced diabetic rats. J Ethnopharmacol 239:111904. https://doi.org/10.1016/j.jep.2019.111904
72. Rajasekhar A, Peddanna K, Vedasree N, Munirajeswari P, Nagaraju N, Badri KR, Chippada AR (2019) Antidiabetic activity of root tubers of Asparagus gonoclados Baker in streptozotocin induced diabetic rats. J Ethnopharmacol 242:112027. https://doi.org/10.1016/j.jep.2019.112027
73. Afiune LAF, Leal-Silva T, Sinzato YK, Moraes-Souza RQ, Soares TS, Campos KE, Fujiwara RT, Herrera E, Damasceno DC, Volpato GT (2017) Beneficial effects of Hibiscus rosa-sinensis L. flower aqueous extract in pregnant rats with diabetes. PLoS One 12(6): e0179785. https://doi.org/10.1371/journal.pone.0179785
74. Ansari P, Azam S, Hannan JMA, Flatt PR, Wahab Y (2020) Anti-hyperglycaemic activity of H. rosa-sinensis leaves is partly mediated by inhibition of carbohydrate digestion and absorption, and enhancement of insulin secretion. J Ethnopharmacol 253:10. https://doi.org/10.1016/j.jep.2020.112647
75. Cunha WR, Arantes GM, Ferreira DS, Lucarini R, Silva ML, Furtado NA, da Silva Filho AA, Crotti AE, Araujo AR (2008) Hypoglicemic effect of Leandra lacunosa in normal and alloxan-induced diabetic rats. Fitoterapia 79(5):356–360. https://doi.org/10.1016/j.fitote.2008.04.002
76. Baliga MS, Fernandes S, Thilakchand KR, D'Souza P, Rao S (2013) Scientific validation of the antidiabetic effects of Syzygium jambolanum DC (black plum), a traditional medicinal plant of India. J Altern Complement Med 19(3):191–197. https://doi.org/10.1089/acm.2011.0752
77. Beidokhti MN, Eid HM, Villavicencio MLS, Jager AK, Lobbens ES, Rasoanaivo PR, McNair LM, Haddad PS, Staerk D (2020) Evaluation of the antidiabetic potential of Psidium guajava L. (Myrtaceae) using assays for alpha-glucosidase, alpha-amylase, muscle glucose uptake,

liver glucose production, and triglyceride accumulation in adipocytes. J Ethnopharmacol 257:112877. https://doi.org/10.1016/j.jep.2020.112877
78. Liu Y, Yang L, Zhang Y, Liu X, Wu Z, Gilbert RG, Deng B, Wang K (2020) Dendrobium officinale polysaccharide ameliorates diabetic hepatic glucose metabolism via glucagon-mediated signaling pathways and modifying liver-glycogen structure. J Ethnopharmacol 248:112308. https://doi.org/10.1016/j.jep.2019.112308
79. Barragan-Zarate GS, Lagunez-Rivera L, Solano R, Pineda-Pena EA, Landa-Juarez AY, Chavez-Pina AE, Carranza-Alvarez C, Hernandez-Benavides DM (2020) Prosthechea karwinskii, an orchid used as traditional medicine, exerts anti-inflammatory activity and inhibits ROS. J Ethnopharmacol 253:112632. https://doi.org/10.1016/j.jep.2020.112632
80. Montefusco-Pereira CV, de Carvalho MJ, de Araujo Boleti AP, Teixeira LS, Matos HR, Lima ES (2013) Antioxidant, anti-inflammatory, and hypoglycemic effects of the leaf extract from Passiflora nitida Kunth. Appl Biochem Biotechnol 170(6):1367–1378. https://doi.org/10.1007/s12010-013-0271-6
81. Sudasinghe HP, Peiris DC (2018) Hypoglycemic and hypolipidemic activity of aqueous leaf extract of Passiflora suberosa L. PeerJ 6:17. https://doi.org/10.7717/peerj.4389
82. Adhikari A, Ray M, Das AK, Sur TK (2016) Antidiabetic and antioxidant activity of Rhizophora mucronata leaves (Indian sundarban mangrove): an in vitro and in vivo study. Ayu 37(1):76–81. https://doi.org/10.4103/ayu.AYU_182_15
83. Komakech R, Kang Y (2019) Ethnopharmacological potential of African cherry [Prunus africana]. J Herb Med:17–18. https://doi.org/10.1016/j.hermed.2019.100283
84. Wang N, Zhu F, Shen M, Qiu L, Tang M, Xia H, Chen L, Yuan Y, Ma S, Chen K (2019) Network pharmacology-based analysis on bioactive anti-diabetic compounds in Potentilla discolor bunge. J Ethnopharmacol 241:111905. https://doi.org/10.1016/j.jep.2019.111905
85. Mudi SR, Akhter M, Biswas SK, Muttalib MA, Choudhury S, Rokeya B, Ali L (2017) Effect of aqueous extract of Aegle marmelos fruit and leaf on glycemic, insulinemic and lipidemic status of type 2 diabetic model rats. J Complement Integr Med 14(2). https://doi.org/10.1515/jcim-2016-0111
86. Erukainure OL, Sanni O, Ijomone OM, Ibeji CU, Chukwuma CI, Islam MS (2019) The antidiabetic properties of the hot water extract of kola nut (Cola nitida (Vent.) Schott & Endl.) in type 2 diabetic rats. J Ethnopharmacol 242:112033. https://doi.org/10.1016/j.jep.2019.112033
87. Deaton AS, Tortora R (2015) People in sub-Saharan Africa rate their health and health care among the lowest in the world. Health Aff (Millwood) 34(3):519–527. https://doi.org/10.1377/hlthaff.2014.0798
88. Amuri B, Maseho M, Simbi L, Okusa P, Duez P, Byanga K (2017) Hypoglycemic and antihyperglycemic activities of nine medicinal herbs used as antidiabetic in the region of Lubumbashi (DR Congo). Phytother Res 31(7):1029–1033. https://doi.org/10.1002/ptr.5814
89. Abderrahim A, Belhamel K, Chalard P, Figueredo G (2019) Chemotypes and radical scavenging activity of the essential oils from Artemisia arborescens L. growing in three areas of Bejaia (Algeria). J Food Meas Charact 13(3):2491–2499. https://doi.org/10.1007/s11694-019-00169-6
90. Hamza N, Berke B, Umar A, Cheze C, Gin H, Moore N (2019) A review of Algerian medicinal plants used in the treatment of diabetes. J Ethnopharmacol 238:111841. https://doi.org/10.1016/j.jep.2019.111841
91. Telli A, Esnault MA, Khelil AOE (2016) An ethnopharmacological survey of plants used in traditional diabetes treatment in south-eastern Algeria (Ouargla province). J Arid Environ 127:82–92. https://doi.org/10.1016/j.jaridenv.2015.11.005
92. Aissa L, Zohra M, El Houda HN, Zohra DF (2019) Ethnobotanical study of medicinal plants used for the treatment of Diabetes mellitus in Sidi Bel Abbes region (North-west Algeria). Bol Latinoam Caribe Plantas M 18(4):392–410. https://doi.org/10.35588/blacpma.19.18.4.25
93. Allali H, Benmehdi H, Dib MA, Tabti B, Ghalem S, Benabadji N (2008) Phytotherapy of diabetes in West Algeria. Asian J Chem 20(4):2701–2710

94. Salihu Shinkafi T, Bello L, Wara Hassan S, Ali S (2015) An ethnobotanical survey of antidiabetic plants used by Hausa-Fulani tribes in Sokoto, Northwest Nigeria. J Ethnopharmacol 172:91–99. https://doi.org/10.1016/j.jep.2015.06.014
95. Akharaiyi FC, Akinyemi AJ, Isitua CC, Ogunmefun OT, Opakunle SO, Fasae JK (2017) Some antidiabetic medicinal plants used by traditional healers in Ado Ekiti, Nigeria. Bratisl Med J 118(8):504–505. https://doi.org/10.4149/bll_2017_097
96. Rutebemberwa E, Lubega M, Katureebe SK, Oundo A, Kiweewa F, Mukanga D (2013) Use of traditional medicine for the treatment of diabetes in Eastern Uganda: a qualitative exploration of reasons for choice. BMC Int Health Hum Rights 13:7. https://doi.org/10.1186/1472-698x-13-1
97. Ssenyange CW, Namulindwa A, Oyik B, Ssebuliba J (2015) Plants used to manage type II diabetes mellitus in selected districts of central Uganda. Afr Health Sci 15(2):496–502. https://doi.org/10.4314/ahs.v15i2.24
98. Karar MGE, Kuhnert N (2017) Herbal drugs from Sudan: traditional uses and phytoconstituents. Pharmacogn Rev 11(22):83–103. https://doi.org/10.4103/phrev.phrev_15_15
99. Oguntibeju OO, Meyer S, Aboua YG, Goboza M (2016) Hypoxis hemerocallidea significantly reduced hyperglycaemia and hyperglycaemic-induced oxidative stress in the liver and kidney tissues of streptozotocin-induced diabetic male wistar rats. Evid-Based Compl Alt:10. https://doi.org/10.1155/2016/8934362
100. Muthaphuli N, Tshisikhawe MP (2016) Ethnomedicinal survey of plant species used by the Vhavenda people in the management of diabetes mellitus in Muraga Village, Vhembe District, Limpopo Province. S Afr J Bot 103:337–337. https://doi.org/10.1016/j.sajb.2016.02.127
101. Bello OM, Fasinu PS, Bello OE, Ogbesejana AB, Adetunji CO, Dada AO, Ibitoye OS, Aloko S, Oguntoye OS (2019) Wild vegetable Rumex acetosa Linn.: its ethnobotany, pharmacology and phytochemistry – a review. S Afr J Bot 125:149–160. https://doi.org/10.1016/j.sajb.2019.04.018
102. Adeneye AA, Olagunju AJ, Babatunde O, Oluyemi O, Awosope O, Oyekunle G, Ayuba H, Omoregie O, Omole O (2019) Morinda lucida Aqueous Stem Bark Extract Ameliorates Hepato-Renal Dysfunctions in experimental diabetes model. Niger J Physiol Sci 34(1):33–42
103. Giovannini P, Howes MJ, Edwards SE (2016) Medicinal plants used in the traditional management of diabetes and its sequelae in Central America: a review. J Ethnopharmacol 184:58–71. https://doi.org/10.1016/j.jep.2016.02.034
104. Laguna-Hernandez G, Rio-Zamorano CA, Meneses-Ochoa IG, Brechu-Franco AE (2017) Histochemistry and immunolocalisation of glucokinin in antidiabetic plants used in traditional Mexican medicine. Eur J Histochem 61(2):2782. https://doi.org/10.4081/ejh.2017.2782
105. Cadena-Zamudio JD, Klicasio-Torres MD, Guerrero-Analco JA, Ibarra-Laclette E (2019) Ethnopharmacological studies of Cecropia obtusifolia (Urticaceae) and its importance in the treatment of type 2 diabetes mellitus: a mini-review. Acta Bot Mex 126:12. https://doi.org/10.21829/abm126.2019.1361
106. Revilla-Monsalve MC, Andrade-Cetto A, Palomino-Garibay MA, Wiedenfeld H, Islas-Andrade S (2007) Hypoglycemic effect of Cecropia obtusifolia Bertol aqueous extracts on type 2 diabetic patients. J Ethnopharmacol 111(3):636–640. https://doi.org/10.1016/j.jep.2007.01.014
107. Gonzalez-Yanez MGE, Rivas-Morales C, Oranday-Cardenas MA, Verde-Star MJ, Nunez-Gonzalez MA, Sanchez E, Leos-Rivas C (2019) Safety of Aqueous extract of Calea ternifolia used in mexican traditional medicine. Evid Based Complement Alternat Med 2019:7478152. https://doi.org/10.1155/2019/7478152
108. Paiz RC, Juarez-Flores BI, Rivera JRA, Ortega NCC, Aguero JAR, Chavez EG, Fuentes GA (2010) Glucose-lowering effect of xoconostle (Opuntia joconostle A. Web., Cactaceae) in diabetic rats. J Med Plants Res 4(22):2326–2333
109. Medina-Pérez G, Zaldívar-Ortega AK, Cenobio-Galindo AJ, Afanador-Barajas LN, Vieyra-Alberto R, Estefes-Duarte JA, Campos-Montiel RG (2019) Antidiabetic activity of cactus acid

fruit extracts: simulated intestinal conditions of the inhibitory effects on α-amylase and α-glucosidase. Appl Sci 9(19). https://doi.org/10.3390/app9194066

110. Zolghadri Y, Fazeli M, Kooshki M, Shomali T, Karimaghayee N, Dehghani M (2014) Achillea Millefolium L. Hydro-Alcoholic extract protects pancreatic cells by down regulating IL-1beta and iNOS gene expression in diabetic rats. Int J Mol Cell Med 3(4):255–262
111. Chavez-Silva F, Ceron-Romero L, Arias-Duran L, Navarrete-Vazquez G, Almanza-Perez J, Roman-Ramos R, Ramirez-Avila G, Perea-Arango I, Villalobos-Molina R, Estrada-Soto S (2018) Antidiabetic effect of Achillea millefollium through multitarget interactions: alpha-glucosidases inhibition, insulin sensitization and insulin secretagogue activities. J Ethnopharmacol 212:1–7. https://doi.org/10.1016/j.jep.2017.10.005
112. Diaz-Flores M, Angeles-Mejia S, Baiza-Gutman LA, Medina-Navarro R, Hernandez-Saavedra D, Ortega-Camarillo C, Roman-Ramos R, Cruz M, Alarcon-Aguilar FJ (2012) Effect of an aqueous extract of Cucurbita ficifolia Bouche on the glutathione redox cycle in mice with STZ-induced diabetes. J Ethnopharmacol 144(1):101–108. https://doi.org/10.1016/j.jep.2012.08.036
113. Miranda-Perez ME, Ortega-Camarillo C, Del Carmen Escobar-Villanueva M, Blancas-Flores-G, Alarcon-Aguilar FJ (2016) Cucurbita ficifolia Bouche increases insulin secretion in RINm5F cells through an influx of Ca(2+) from the endoplasmic reticulum. J Ethnopharmacol 188:159–166. https://doi.org/10.1016/j.jep.2016.04.061
114. Alusik S, Paluch Z (2015) Metformin: the past, presence, and future. Minerva Med 106(4):233–238
115. Chan CC, Zhang HW, Chan K, Lin ZX (2016) Xiaoke Pill and anti-diabetic drugs: a review on clinical evidence of possible herb-drug interactions. Chin J Integr Med. https://doi.org/10.1007/s11655-015-2106-5

Chapter 5
Glycosides from Natural Sources in the Treatment of Diabetes Mellitus

Kaveri M. Adki and Yogesh A. Kulkarni

Abstract Diabetes is one of the principal causes of death in developed and developing countries. Many synthetic drugs are being used for the treatment of diabetes. But these drugs have many adverse effects. Hence there is an immediate requirement of new therapies that can be useful for better management of diabetes. From ancient times, herbal drugs are well accepted for their therapeutic values in different disease conditions. Natural products obtained from medicinal plants can be one of the best options for the treatment of various diseases including diabetes. Plants synthesize various secondary metabolites like terpenoids, saponins, tannins, flavonoids, anthraquinones, alkaloids, and glycosides. Glycosides consist of sugar (glycone) moiety joined to a non-sugar moiety (aglycone) via a glycosidic bond. Many plants synthesize glycosides, which can be hydrolyzed to give glycone and aglycone part by enzyme hydrolysis. Various glycosides as well as aglycones are reported to have many biological activities. Glycosides like rutin, puerarin, gymnemic acid I, and stevioside have been reported for significant antidiabetic activity. Aglycones like securigenin, strictinin, and christinin-A have been reported for their antidiabetic activity. The mechanism of their antidiabetic activity involves stimulation of insulin secretion, inhibition of α-amylase, α-glucosidase, and tyrosine phosphatase 1B enzymes involved in glycemic control. The present book chapter focuses on the effect of various plant derived glycosides and aglycones in diabetes.

Keywords Antidiabetic activity · Glycosylamine · O-glycoside · C-glycoside · Thioglycoside · Rutin · Puerarin · Gymnemic acid I · Securigenin

---

K. M. Adki · Y. A. Kulkarni (✉)
Shobhaben Pratapbhai Patel School of Pharmacy & Technology Management, SVKM's NMIMS, Mumbai, India

H. Chen, M. Zhang (eds.), *Structure and Health Effects of Natural Products on Diabetes Mellitus*, https://doi.org/10.1007/978-981-15-8791-7_5

1 Introduction

Diabetes mellitus (DM) comprises a group of metabolic disorders characterized by hyperglycemia. DM may arise due to defects in insulin secretion and its action, or both, which alters carbohydrate, fat, and protein metabolism. Diabetic patients are at increased risk of cardiovascular, cerebrovascular, nonalcoholic fatty liver diseases, cataract, and erectile dysfunction. Diabetics are also at increased risk of infectious diseases like tuberculosis. Excessive thirst, increased appetite, polyuria, and weight loss are characteristic symptoms of diabetes. Ketoacidosis or a nonketotic hyperosmolar state is the most severe clinical manifestations which lead to dehydration, coma, and death. According to the World Health Organization (WHO), diabetes is classified into following types—Type 1 diabetes, Type 2 diabetes, hybrid types of diabetes, other specific types of diabetes, and unclassified diabetes. Type 1 and type 2 diabetes have been differentiated based on age at onset, amount of loss of β cell function, level of insulin resistance, amount of diabetes-associated autoantibodies, and required amount of insulin treatment for survival [1]. However, now diabetes cannot be categorized into type 1 and type 2 only, due to various reasons like less distinct phenotypes, increased prevalence in young population, and growing number of diabetes subtypes due to knowledge of molecular genetics. The updated knowledge of pathophysiology of diabetes has opened an area of precision medicine development for diabetes [2]. The detail classification of diabetes has been shown in Fig. 5.1.

The key underlying characteristic cause of all forms of diabetes is the destruction or dysfunction of pancreatic β-cells. Other associated causes include genetic abnormalities and predisposition, epigenetic, auto-immunity, insulin resistance, inflammation, and environment related factors [3]. Currently, four diagnostic tests and standard blood glucose levels for diabetes with signs and symptoms of diabetes are recommended as mentioned in Table 5.1 [1].

Diabetes is one of the fastest growing global health problems of the twenty-first century. Diabetes is reported in all populations and regions in the world, including undeveloped parts of low- and middle-income countries. The number of diabetic patients is rising steadily. International Diabetes Federation (IDF) has provided the current statistics of diabetes in its atlas (ninth edition). In 2019, 463 million people are suffering from diabetes and this number is proposed to reach 578 million by 2030 and 700 million by 2045. Hyperglycemia causes almost four million deaths per year. Diabetes affects not only the individual but also has a great impact to affect their families. It has broad socio-economic consequences and may affect national productivity and economies [4, 5].

From ancient times, plants are accepted as a medication for the treatment and management of various disorders. Plants synthesize an array of secondary metabolites, which have significant physiological effects. Important secondary metabolites synthesized by plants are terpenoids, saponins, tannins, alkaloids, glycosides, anthraquinones, coumarins, flavonoids, and sulfur-containing compounds [6].

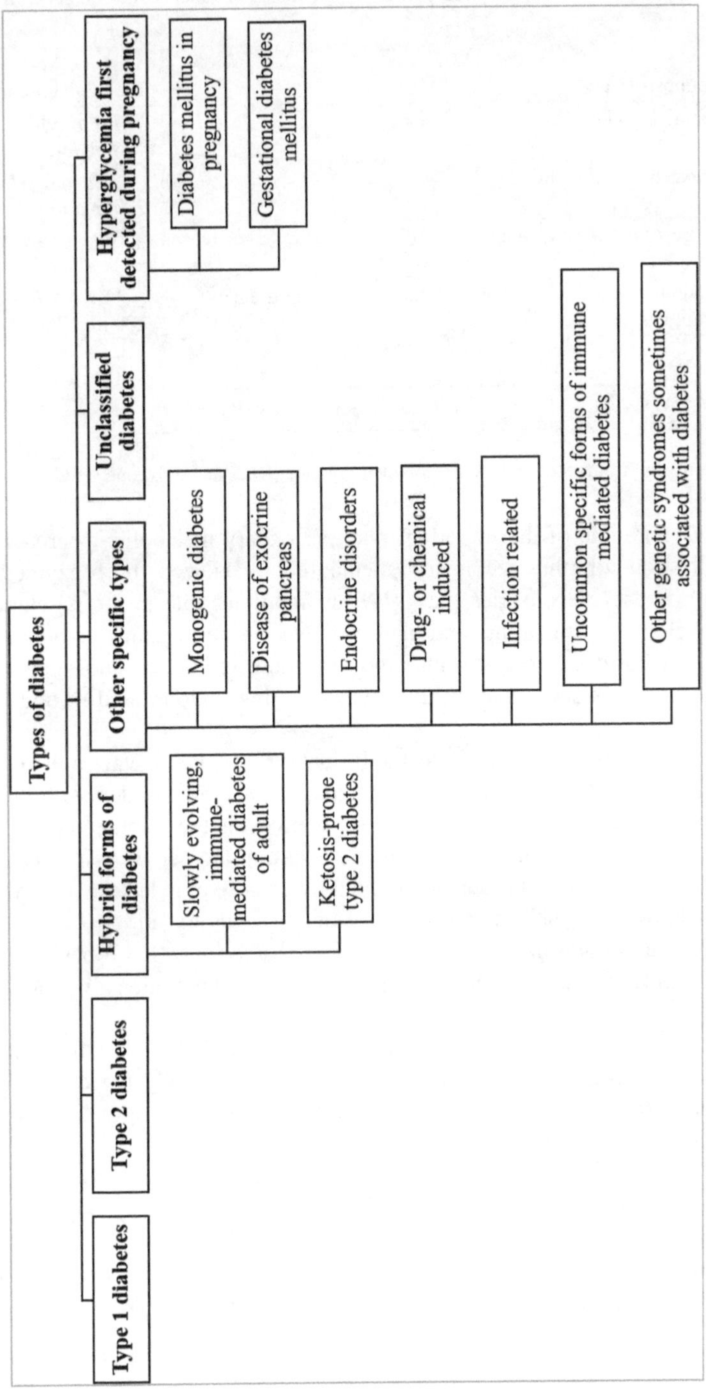

Fig. 5.1 Recent classification of diabetes mellitus [4]

Table 5.1 Diagnostic tests and standard blood glucose levels in patients with diabetic signs and symptoms

Sr. No.	Diagnosis tests	Blood glucose levels
1.	Fasting blood glucose	≥ 7.0 mM/L or 126 mg/dL
2.	Glycated hemoglobin	≥ 11.1 mM/L or 200 mg/dL
3.	2-h post-load plasma glucose after a 75 g oral glucose tolerance test	≥ 6.5% or 48 mM/mol
4.	Random blood glucose levels (in patients with signs and symptoms of diabetes)	≥ 11.1 mM/L or 200 mg/dL

$$R_1\text{-C-OH} + R_2\text{-C-OH} \longleftrightarrow R_1\text{-}\mathbf{C\text{-}O\text{-}C}\text{-}R_2 + H_2O$$

Fig. 5.2 Formation of glycosidic linkage between glycone (R_1C) and aglycone (R_2C)

Glycosides are one of the important plant secondary metabolites, derived from post-modification of the secondary metabolites catalyzed by enzymes like glycosyltransferases. Glycosides may play an important role in the regulation of growth, development, signaling, and plant's defense system against herbivores and pathogens. Glycosides frequently undergo acylation, oxidation, and degradation. Glycosides are produced mainly by plants during abiotic and biotic stress conditions [7].

Glycosides are formed during acetal formation, i.e. hemiketal or hemiacetal hydroxyl group (-OH) of a monosaccharide undergoes condensation with the -OH of a second molecule to form a glycosidic bond with the elimination of water molecule. The sugars which provide hemiacetal group are named as "glycosides." For example, if a hemiacetal group is provided by glucose or galactose, the resultant molecule is known as glucoside and galactoside, respectively [8].

Glycosides are made up of glycone (R_1C) and aglycone (R_2C) part, which is chemically and functionally independent part linked together with a glycosidic bond (Fig. 5.2).

Depending upon types of glycosidic linkages glycosides are classified into four main groups as follows: O-glycosides (the most abundant glycosides in plants), C-glycosides (resistant to hydrolysis), S-glycosides (thioglycosides), and N-glycosides (present in nucleosides), respectively (Fig. 5.3).

Plant glycosides can also be classified based on the aglycone moiety as phenolic, coumarins, flavonoid, anthraquinone, saponin, cardiac, cyanogenic, and thioglycosides (Table 5.2).

The therapeutic activity of glycosides mainly depends on the aglycone while water solubility; pharmacokinetic properties depend on the saccharide part. A

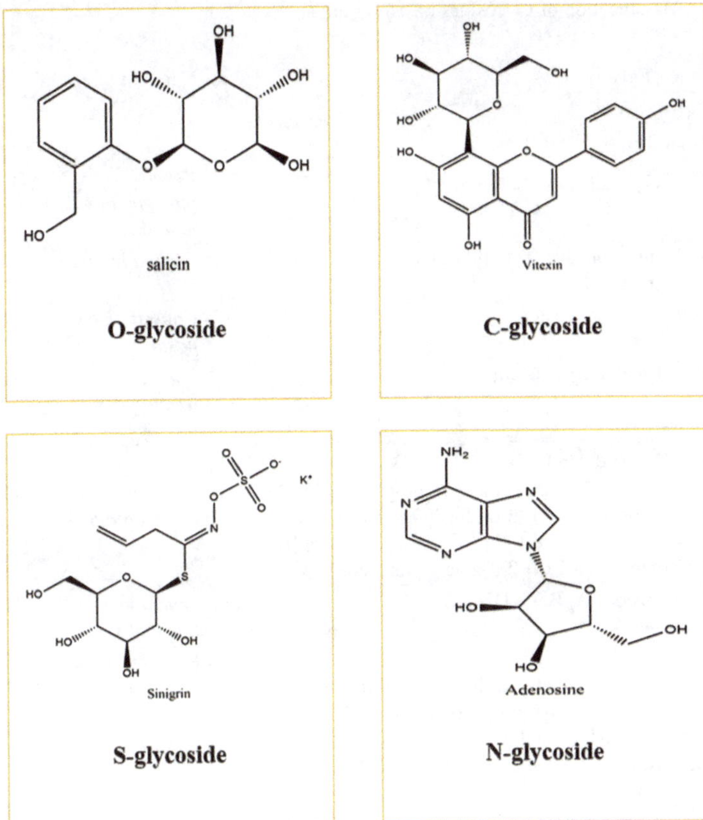

Fig. 5.3 Types of glycosides depending on the type of glycosidic linkage

β-linked glycosidic bond is more active compared to other linkages in plants. These glycosides are resistant to human digestive systems and hence poorly absorbed in the gastrointestinal tract. These glycosides further travel to the ileum and large bowel. These glycosides are broken to aglycone with the help of microbial flora and are rapidly absorbed into the blood stream [39].

2 Glycosides in Diabetes

Literature has shown various glycosides, and aglycones have shown significant effects in treatment of diabetes. These plant glycosides act via different mechanisms [40].

Table 5.2 Classification of glycosides based on the aglycone part

Sr. No.	Name of glycoside	Name of plant	Reference
Phenolic glycosides			
1	Sorcomisides A and B	*Sorbus commixta*	[9]
2	Platyphylloside	*Betula platyphylla var. japonica*	[10]
3	7S, 8R-urolignoside-9'-O-β-D-glucoside (1) and scrophenoside G	*Ginkgo biloba L*	[11]
4	Sesquilignan glycoside	*Capsella bursa-pastoris*	[12]
5	Salicortin and Tremulacin	*Populus tremuloides Michx.*	[13]
Coumarin and chromone glycosides			
6	Umbelliferone 7-O-α-d-glucopyranosyl(1 → 3)-[β-d-apiofuranosyl(1 → 6)]-β-d-glucopyranoside and Umbelliferone 7-O-α-d-glucopyranosyl(1 → 4)-[β-d-apiofuranosyl(1 → 6)]-β-d-glucopyranoside	*Hydrangea paniculata Sieb*	[14]
7	Officinalisides A, B, C, D and E	*Scindapsus officinalis*	[15]
8	(3'S)-3'-O-β-d-apiofuranosyl-(1 → 6)-β-d-glucopyranosylhamaudol; (2'S)-4'-O-β-d-apiofuranosyl-(1 → 6)-β-d-glucopyranosylvisamminol, 3'-O-glucopyranosylhamaudol, 4'-O-β-d-glucopyranosylvisamminol, and 4'-O-β-d-glucopyranosyl-5-O-methylvisamminol	*Saposhnikovia divaricata*	[16]
9	Genglycoside A–F	*Gendarussa vulgaris*	[17]
10	Moriramulosid A and Moriramulosid B	*Morus alba L*	[18]
Flavonoid glycosides			
11	Bulbiferumoside	*Sedum bulbiferum*	[19]
12	Barringosides A–F	*Barringtonia acutangula*	[20]
13	Nelumbosides A–D	*Nelumbo Nucifera*	[21]
14	Chrysoeriol-7-O-β-D-xyloside, luteolin-7-O-β-D-apiofuranosyl-(1 → 2)-β-D-xylopyranoside, chrysoeriol-7-O-β-D-apiofuranosyl-(1 → 2)-β-D-xylopyranoside, chrysoeriol-7-O-α-L-rhamnopyranosyl-(1 → 6)-β-D-(4-″-hydrogeno sulfate) glucopyranoside and isorhamnetin-3-O-α-L-rhamnopyranosyl-(1 → 6)-β-D-glucopyranoside	*Graptophyllum grandulosum*	[22]
15	3′-hydroxyl epimedoside A, neo-sagittasine A, Dihydrofuran-baohuoside I, furan-baohuoside I	*Epimedium brevicornum*	[23]

(continued)

Table 5.2 (continued)

Sr. No.	Name of glycoside	Name of plant	Reference
Anthraquinone glycosides			
16	Emodin-8-O-β-d-glucoside, aloe-emodin-8-O-β-d-glucoside, and chrysophanol-8-O-β-d-glucoside	*Rheum palmatum*	[24]
17	Rumpictuside A	*Rumex pictus*	[25]
18	Chrysophanol 1-O-β-d-glucoside, chrysophanol 8-O-β-d-glucoside, and physion 8-O-β-d-glucoside	*Rheum tanguticum*	[26]
Cardiac glycosides			
19	Peruvoside	*Cascabela thevetia*	[27]
20	[2'-O-acetylacoschimperoside P (1) and oleandrigenin-3-O-α-l-2'-O-acetylvallaropyranoside	*Vallaris glabra*	[28]
21	19-nor-10-hydroxymethylstrebloside	*Streblus asper*	[29]
22	Convallatoxin	*Digitalis purpurea*	[30]
23	Strophanthidin	*Strophanthus kombe*	[31]
Cyanogenic glycosides			
24	Amygdalin	*Prunus armeniaca*	[32]
25	Linamarin	*Hibiscus sabdariffa*	[33]
26	Dhurrin	*Sorghum forage*	[34]
Thioglycosides			
27	6-Methylsulfinylhexyl isothiocyanate	*Wasabia japonica*	[35]
28	Glucosinolates	*Brassica rapa*	[36]
29	Glucoraphanin	*Brassica oleracea*	[37]
30	Sulforaphane	*Brassica oleracea*	[38]

2.1 Rutin

Rutin (2-(3,4-dihydroxyphenyl)-5,7-dihydroxy-3-[(2S,3R,4S,5S,6R)-3,4,5-trihydroxy-6-[[(2R,3R,4R,5R,6S)-3,4,5-trihydroxy-6-methyloxan-2-yl]oxymethyl]oxan-2-yl]oxychromen-4-one) was discovered in 1842. Rutin (quercetin-3-O-rutinoside) is a flavonol glycoside (Fig. 5.4). It is also known as vitamin P, rutoside, sophorin, and quercetin-3-rutinoside. Chemically rutin comprised of flavonol aglycone (quercetin) along with disaccharide (rutinose). It is found in plants like *Fagopyrum esculentum, Vitis vinifera, Amaranthus cruentus, Guiera senegalensis, Capparis spinosa, Allium cepa, Asparagus officinalis, Malus pumila, and Ruta graveolens* [41]. The name "rutin" came from plant *Ruta graveolens*. The highest concentrations

Fig. 5.4 Chemical structure of Rutin

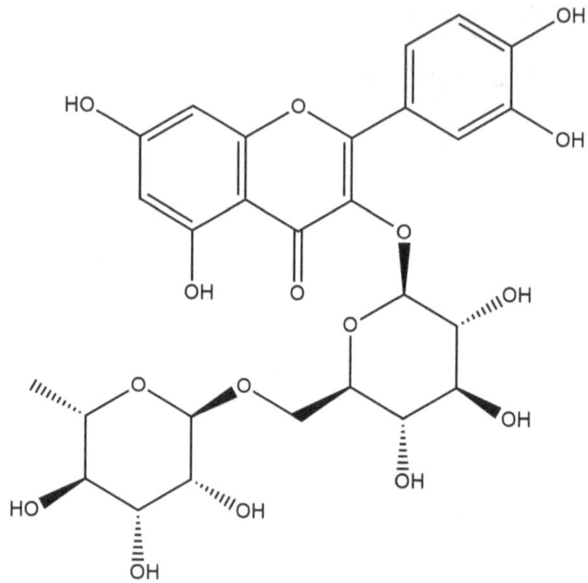

of rutin have been reported in grapes and buckwheat. Rutin is mainly found in plant parts such as leaves, flowers, fruit skins, and roots. It has antioxidant, anti-inflammatory, antidiabetic, anticancer, neuroprotective, and cardio protective activities [42].

Buckwheat is a major source of rutin. It is an edible crop that is rich in flavonoids, carbohydrates, glycosides, and chemical compounds. It is widely distributed in temperate zones in countries like Europe, China, North America, Japan, and Korea. Buckwheat is widely documented for its medicinal property in various diseases like cardiovascular and diabetes due to its high nutritive value. Rutin in buckwheat is responsible for its antidiabetic activity (Fig. 5.5). The rutin content in buckwheat is around 0.8–1.7%. Nowadays many scientists are working on buckwheat for its antidiabetic and other nutraceutical functions [43].

Rutin was extracted from stems, leaves, and flowers of buckwheat. The extraction was done by using a 60% ethanol and 5% ammonia in water solution. Rutin content was estimated from the extract by capillary electrophoresis. Capillary electrophoresis was done by using 50 mM borate (pH 9.3) and 100 mM sodium dodecyl sulfate as a running buffer. Absorbance was recorded at 380 nm. The concentration of rutin was found to be 131–476 ppm in bran fractions and 19–168 ppm in flour fractions. The study revealed rutin is present in plant parts at different concentrations that about 300 ppm (in leaves), 1000 ppm (in stems), and 46,000 ppm (in flowers). The study reported that buckwheat is an important nutritional source of rutin [44].

Kamalakkannan and group have also studied the antihyperglycemic activity of rutin in *Streptozotocin* (STZ)-induced diabetic rats. Diabetic animals were treated

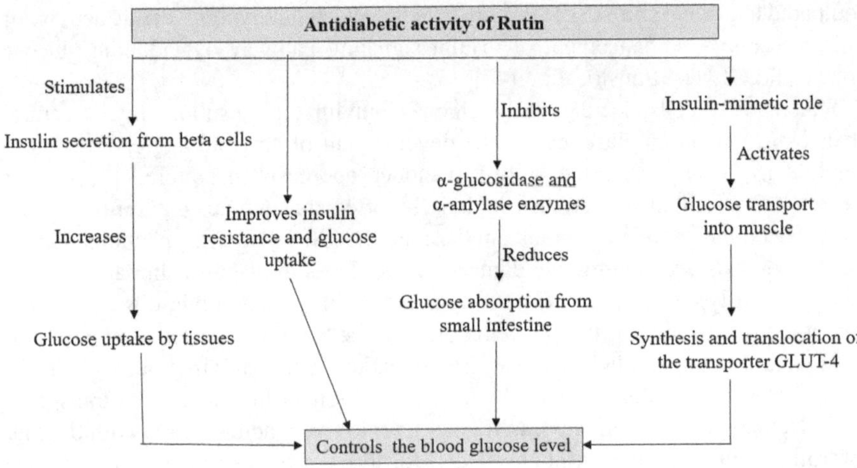

Fig. 5.5 Mechanism of antidiabetic activity of Rutin [49]

with rutin 100 mg/kg, orally for 45 days. The study reported that rutin administration improved the antioxidant status, increased insulin levels, and lowered elevated plasma glucose levels in diabetic rats compared to the normal control group. Antioxidant activity of rutin has also protected kidney, brain, and liver [45].

Abnormalities in glucose metabolism constitute one of the most common problems in DM. The antihyperglycemic activity of rutin was studied in STZ-induced diabetes in male Wistar rats. Rats were treated intraperitoneally with rutin 50 mg/kg for 45 days. The study showed that rutin lowered the blood glucose and lipid profile in diabetic rats. Further rutin has also showed reduced elevated glycogen and triacylglycerol levels in hepatic and cardiac tissue. The study proved that the rutin administration could regulate hyperglycemia by controlling the liver functions [46].

It is documented that rutin has ability to enhance phosphorylation of insulin-dependent receptor kinase and glucose transporter type 4 (GLUT4). The antihyperglycemic mechanism of rutin was studied by insulin-dependent receptor kinase activity. The in vitro and in vivo mechanism of antidiabetic activity of rutin was studied in mouse muscle myoblast cells and insulin resistant mice, respectively. Insulin receptor kinase auto-phosphorylation was triggered by using insulin in differentiated mouse muscle myoblast cells. Mouse muscle myoblast cell (C2C12) and mouse muscle myoblast (L6) cells were incubated with rutin 100 μM for 90 min. These cells were analyzed by western blot, insulin receptor assay, glucose uptake capacity, GLUT4 translocation assay, and glucose uptake assay. Rutin also potentiated the phosphorylation of the insulin receptor kinase. Co-treatment of cells with rutin reduced inhibition of insulin-dependent translocation of GLUT4. C57BL/6 mice were treated with rutin 25 mg/kg orally for 4 days. These insulin resistance mice were used for the estimation of glucose tolerance efficacy. In vivo rutin treatment showed antidiabetic potential via regulating blood glucose levels compared to normal animals. The study concluded that antihyperglycemic activity of rutin was due to enhancement of insulin signaling pathway by inducing glucose uptake and GLUT4 translocation [47].

Pancreatic β-cells secrete the hormones amylin and insulin. Any structural changes in human amylase lead to the development of amyloid aggregates. These amyloid aggregates have the ability to induce apoptosis in cultured β-cells and destroy the β-cell in ex-vivo islets. It is reported that human amylin-transgenic mice selectively express human amylase in their β-cells. This phenomenon of β-cell apoptosis and progressive damage to islet leads to diabetes similar to that in patients with type 2 diabetes. Hence, aggregation of human amylase is considered one of the risk factors for the causation of diabetes. Rutin is reported as an inhibitor of misfolding and also helps in the disaggregation of human amylase. Aitken and group have studied the antidiabetic activity of rutin in human amylin-transgenic mice. Hemizygous human amylase-transgenic mice were administered with 0.5 mg/ml rutin orally via the drinking water for 60 days. Hemizygous male mice (non-transgenic) were considered as a control group. Control group animals were administered with pure water. Quantification of misfolding of human amylase was measured by ion-mobility mass spectrometry and time-dependent thioflavin-T spectroscopy. The study showed that rutin has the potential to inhibit misfolding, disaggregation of human amylase in transgenic mice. The study also revealed that rutin has the ability to change the conformation of misfolded human amylase towards the normal physiological human amylase. Rutin administration lowered elevated blood glucose. The study concluded that rutin administration has the capacity to suppress the human amylase aggregation and increases the lifespan of diabetic human amylase-transgenic mice [48].

2.2 Puerarin

Puerarin (7-hydroxy-3-(4-hydroxyphenyl)-8-[(2S,3R,4R,5S,6R)-3,4,5-trihydroxy-6-hydroxymethyl)oxan-2-yl]chromen-4-one) is the major bioactive compound isolated from the root of the *Pueraria lobata*. Chemically puerarin is 8-C-glucoside of daidzein (Fig. 5.6). The first time puerarin was isolated from Ohwi in the late 1950s. Puerarin is present in plants like *Ziziphus jujuba* and *Pueraria phaseoloides*. It has been reported for its significant effects in cerebrovascular, cardiovascular diseases, Parkinson's disease, Alzheimer's disease, cancer, diabetes, and diabetic complications [50].

Puerarin is also isolated form the Kudzu plant. It is a perennial, semi-woody, and leguminous plant native to South East Asia. It is frequently used in traditional Chinese medicine. Kudzu root has been reported for more than 70 phytoconstituents including isoflavonoides, glycosides, and triterpenoids. Among 70 phytoconstituents, Puerarin is the major phytoconstituent of Kudzu root. Kudzu root has been used for the treatment of fever, diarrhea, acute dysentery, diabetes, and cardiovascular diseases [51].

Fig. 5.6 Chemical structure of Puerarin

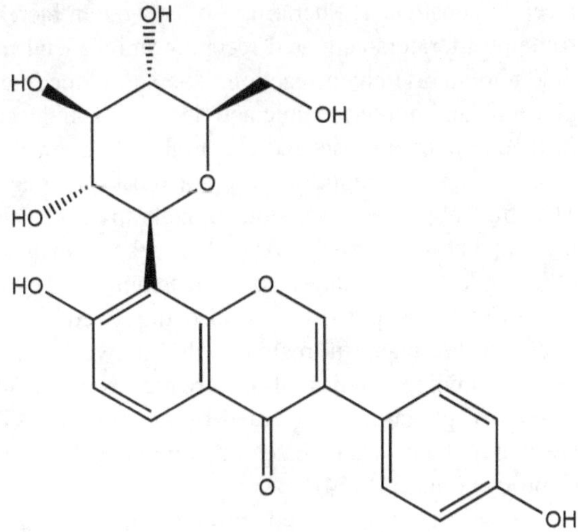

Puerarin

Dried Kudzu roots were subjected to grinding in mill and sieved by using 20–40 mesh. 100 g of the sample was mixed with 1000 ml of n-butanol/water in the 1:1 (v/v) ratio. Sample was kept for extraction at 25 °C for 60 min. Further, the extract was subjected to filtration. The filtrate was kept for complete phase separation. Then the water phase was used to get puerarin. The study documented that puerarin can be separated from the hydrophilic and hydrophobic impurities by double solvent extraction. Puerarin extraction is pH dependent. The first solvent extraction was done in a beaker by using magnetic stirrer at 300 rpm. The aqueous solution of the extract was stirred for 1 h with an equal volume of n-butanol. The pH of the mixture was maintained at 2–8 with sodium hydroxide or hydrochloric acid solution (1 mol/ L). Puerarin was found to be unstable at pH above 8. The mixture was kept for complete phase separation in a funnel. The content of puerarin was estimated in aqueous and n-butanol by using HPLC. The purification of puerarin was done by decreasing the pH. The resultant mixture stirred for 1 h with an equal volume of distilled water. The pure form of puerarin was quantified by using HPLC. HPLC was carried out by using HPLC grade methanol and Agilent system, consisting of a quaternary pump, a Waters column (150 mm × 3.9 mm, 5 m particle size), a column thermostat, a degasser unit, an auto sampler, and a UV-detector [52].

Antihyperglycemic effects of puerarin were studied in STZ-induced diabetic mice. Diabetic mice were treated with puerarin 20, 40, 80 mg/ kg orally for 14 days. After treatment, plasma glucose, insulin levels, and lipid profiles were analyzed. Puerarin treatment decreased blood glucose levels and increased serum insulin levels significantly. It also improved the lipid profile in diabetic mice. Further, histopathological studies of the pancreas revealed that puerarin also protected the pancreas from diabetic lesions. Western blot analysis of pancreatic tissues showed upregulated levels of insulin-like growth factor-1 and insulin receptor substrate-1. Puerarin also showed an increase in the expression of peroxisome proliferators-activated receptor and skeletal muscle insulin receptor in a real time-polymerase chain reaction. The study documented that puerarin exerts hypoglycemic and hypolipidemic activity in STZ-induced diabetic mice via maintaining metabolic homoeostasis and elevating insulin level [53].

Glucagon-like peptide 1 receptor (GLP-1R) plays an important role in diabetes. The effect of puerarin was studied in db/db mice. The db/db mice were subjected to a high fat diet for 12 weeks. After 12 weeks animals were orally treated with puerarin (150 mg/kg) for 55 days. Puerarin treatment showed an increase in body weight, controlled blood glucose, and also improved glucose tolerance in diabetic db/db mice. Further puerarin treatment also showed decreased levels of cell apoptosis in immunostaining studies of pancreatic sections. The mechanism of antidiabetic activity of puerarin was studied by targeting the GLP-1R signaling pathway. The western blot analysis revealed the upregulated expression of GLP-1R and activation of protein kinase B [54].

Puerarin was isolated and purified from *Pueraria lobata.* The antidiabetic activity of puerarin was studied in male Wistar rats. Diabetes was induced by using intravenous injection of STZ 60 mg/kg. Diabetic rats were treated with puerarin at a dose of 20 mg/kg for 3 days. The effect of puerarin was estimated by assessing the plasma glucose levels, intravenous glucose challenge test, glucose uptake into soleus muscle, gene expression analysis by northern and western blot analysis. The results showed that intravenous bolus injection of puerarin lowered the plasma glucose levels at 20 mg/kg in diabetic rats. Puerarin significantly reduced the increased levels of plasma glucose in the glucose challenge test in diabetic rats. Puerarin improved the uptake of radio-labeled glucose in the isolated soleus muscle of diabetic rats in a concentration dependent manner. Gene expression study of mRNA and protein levels showed upregulation of GLUT4 in soleus muscle (Fig. 5.7) [55].

Recently, synergistic antidiabetic activity of puerarin and pumpkin polysaccharides were reported in type 2 diabetic mice. Type 2 diabetes was induced by a high fat diet and intraperitoneal injection of STZ (35 mg/kg) for 3 consecutive days. Diabetic mice were treated orally with pumpkin polysaccharides (400 mg/kg) and (puerarin 200 mg/kg) for 8 weeks. The results showed that pumpkin polysaccharides and puerarin combination lowered elevated plasma glucose level, lowered lipid profile, decreased oxidative stress in diabetic mice. Further hypoglycemic activity of pumpkin polysaccharides and puerarin combination was confirmed due to upregulated expression of nuclear factor E2 related factor 2, phosphoinositide-3-kinase, and heme oxygenase-1 in Nrf2 and PI3K signaling pathway [56].

2.3 *Gymnemic Acid I*

Gymnemic acid I ((2S,3S,4S,5R,6R)-6-[(3S,4R,4aR,6aR,6bS,8S,8aR,9R,10R,12aS,14aR,14bR)-8a-(acetyloxymethyl)-8,9-dihydroxy-4-(hydroxymethyl)-4,6a,6b,11,11,14b-hexamethyl-10-[(E)-2-methylbut-2-enoyl]oxy-1,2,3,4a,5,6,7,8,9,10,12,12a,14,

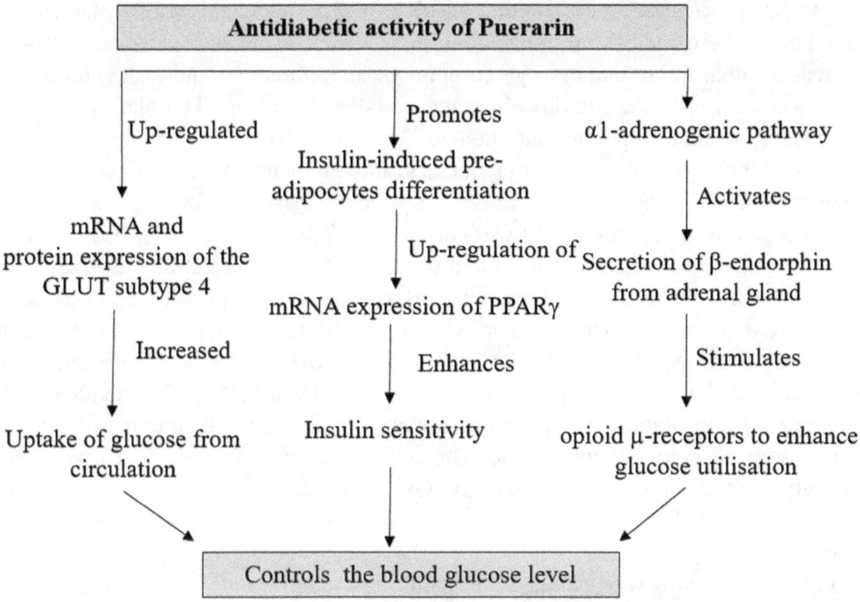

Fig. 5.7 Mechanism of antidiabetic activity of puerarin [53]

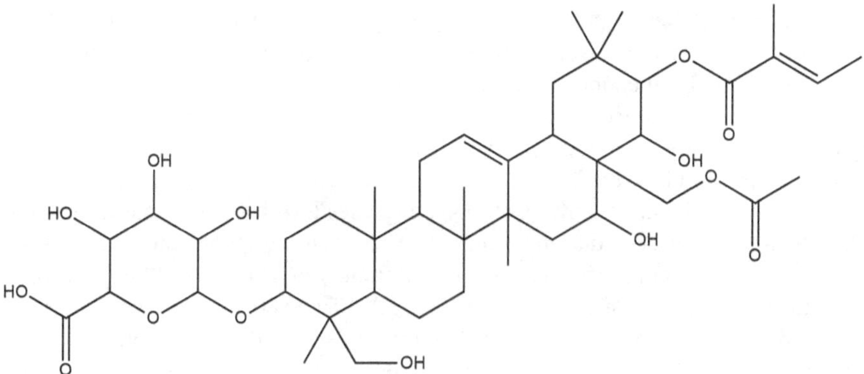

Gymnemic acid I

Fig. 5.8 Chemical structure of Gymnemic acid I

14a-tetradecahydropicen-3-yl]oxy]-3,4,5-trihydroxyoxane-2-carboxylic acid) is a triterpenoid glycoside (Fig. 5.8). Gymnemic acid I majorly found in *Gymnema sylvestre.* Gymnemic acid I also found in a variety of plants including *Hovenia dulcis, Ziziphus jujuba, Gymnema alterniflorum, Stephanotis lutchuensis,* and *Styrax japonicus* plants. *Gymnema sylvestre* herb traditionally used to control obesity and the treatment of diabetes in India and Japan [57].

Gymnema sylvestre is a perennial, slow-growing, medicinal woody plant. It is found in southern and central India and tropical Africa. From ancient times it is used to treat asthma, inflammation, eye complaints, inflammation, and snakebite. Leaf extract is used as a diuretic, laxative, and cough suppressant. The plant is rich in glycosides, saponins, and anthraquinones. The major glycoside gymnemic acid I is present abundantly in leaves. Gymnemic acid I documented as an active constituent responsible for the antidiabetic activity of *Gymnema sylvestre* [58].

Extraction of gymnemic acid I was carried out by Hiroshimin and group. Leaves of *Gymnema sylvestre* were cleaned and dried in a hot air oven at 50 °C. The dried leaves were crushed and passed through a 40 mesh sieve. The powdered sample was again dried at 50 °C. The powder was subjected to extraction in the Soxhlet apparatus in hot condition with 95% ethanol. The extracted ethanol was dried in vacuum drier. The yield of gymnemic acid I obtained was 6.15%. Gymnemic acid I has the ability to suppress sweetness in humans via inhibiting the binding of sweets to the sweet receptors of the tongue. The acyl group of gymnemic acid I have the capacity to reduce the antisweet activity. Gymnemic acid reduces the sweetness of artificial as well as natural sweeteners like aspartame and thaumatin, respectively [59].

Gymnemic acid I was evaluated in insulin resistant human liver carcinoma cell lines (HepG2) and male *Sprague–Dawley* rats. In vitro assay, HepG2 cells were treated with gymnemic acid I at 0.25, 0.5, 0.1, 1.5, 2.0 mg/ml concentration and various parameters like cell apoptosis, microculture tetrazolium assay, glucose uptake analysis, glucose production analysis, glycogen content determination, and reactive oxygen species levels were estimated. The cell apoptosis study was estimated by using a cell apoptosis kit, and microculture tetrazolium assay was carried out by using MTT labeling reagent, glucose uptake analysis was done by using fluorescent D-glucose analog 2-NBDG, glucose production analysis was done by using glucose oxidase method kit, glycogen content determination was done by using a commercial kit, and reactive oxygen species levels were estimated by using 2′,7′-dichlorofluorescein diacetate and fluorescent spectrophotometer. The study showed that gymnemic acid has effectively improved glucose uptake by 15.2%. The oral administration of gymnemic acid I at 22, 40, and 80 mg/kg to type 2 diabetic rats lowered fasting blood glucose levels by 26.7%. Further, the molecular level studies revealed that gymnemic acid I showed improvement in the dilatation and degranulation of the smooth endoplasmic reticulum in HepG2 cells and type 2 diabetic rat livers. Expressional analysis of gymnemic acid I treatment revealed at protein and mRNA level showed the inhibition of ORP150 expression in HepG2 cell lines and type 2 diabetic rat livers. Further, effect of gymnemic acid I on the endoplasmic reticulum stress indicators like PERK, JNK, and eIF2α was studied. Gymnemic acid I treatment lowered the expression of p-c-Jun and simultaneously inhibited the expression of p-PERK and p-eIF2α. The study reported that the antidiabetic activity of gymnemic acid I may be associated with the facilitation of insulin signal transduction and alleviation of endoplasmic reticulum stress in insulin resistant HepG2 and diabetic rats (Fig. 5.9) [60].

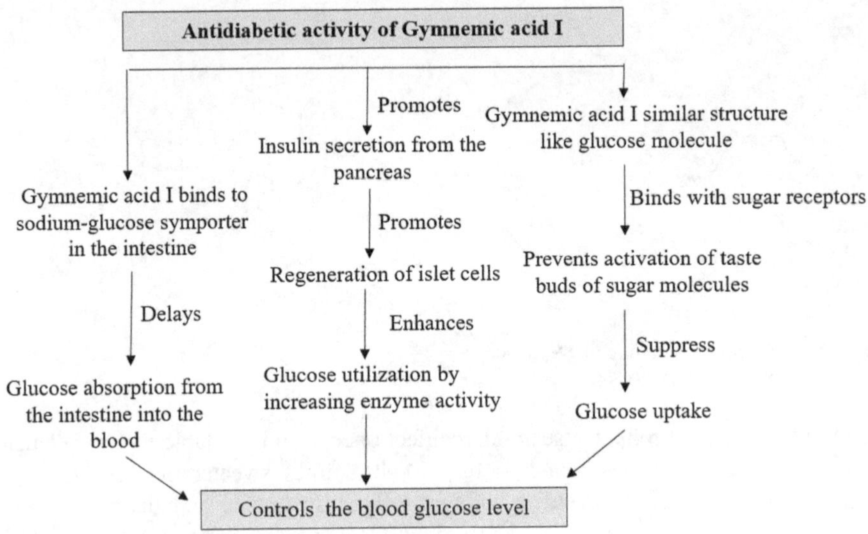

Fig. 5.9 Mechanism of antidiabetic activity of Gymnemic acid I [62]

The molecular mechanism of antihyperglycemic activity of gymnemic acid I was studied by Li and group in male *Sprague–Dawley* rats. Diabetes was induced by feeding the animals with a high fat diet for 4 weeks and subsequent STZ (30 mg/kg) intraperitoneal injection. After confirmation of diabetes, animals were grouped and treated with gymnemic acid I at dose of 40 and 80 mg/kg for 6 weeks. The study revealed that gymnemic acid decreased insulin concentration by 16.1% and fasting blood glucose by 26.7%. Further, western blotting and real-time polymerase chain reaction analysis revealed that gymnemic acid I treatment downregulated the expression of glycogen synthesis kinase-3β and phosphatidylinositol-3-kinase. The study indicated that gymnemic acid I treatment promotes insulin signal transduction via activating AMPK-mediated and PI3K/Akt-signaling pathways in diabetic rats [61].

2.4 *Stevioside*

Stevioside ([(2S,3R,4S,5S,6R)-3,4,5-trihydroxy-6-(hydroxymethyl)oxan-2-yl] (1R,4S,5R,9S,10R,13S)-13-[(2S,3R,4S,5S,6R)-4,5-dihydroxy-6-(hydroxymethyl)-3-[(2S,3R,4S,5S,6R)-3,4,5-trihydroxy-6-(hydroxymethyl)oxan-2-yl]oxyoxan-2-yl]oxy-5,9-dimethyl-14-methylidenetetracyclo[11.2.1.01,10.04,9]hexadecane-5-carboxylate) consists of aglycone-steviol fused with three glucose molecules (Fig. 5.10). Stevioside is majorly found in leaves of *Stevia rebaudiana*. Stevioside was discovered by Bridel and Lavielle in 1931. *Stevia rebaudiana* plant is found in several Asian countries and South America. Stevia is used as a sweetener in food and beverages in Brazil and Paraguay. Stevioside is majorly used in vegetable products,

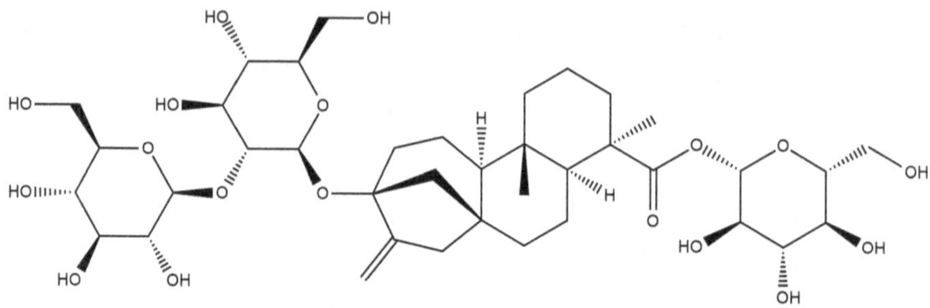

Fig. 5.10 Chemical structure of Stevioside

soft drinks, fruit products, seafood, confectionery, and as table-top sweetener [63, 64]. Stevioside was found to be 150–400 times sweeter than saccharose. Hence it is used as a sweetener in weight loss treatments and heart diseases. Stevioside extraction can be efficiently done by using a green method developed by Lopez. A green method for the extraction of steviol was done by using a factorial design. Five variables, including temperature, time, agitation, sample–solvent ratio, and grinding time, are considered for the efficient extraction of stevioside in Box–Behnken surface design. In this extraction, hot water was used as a solvent. The optimized method for maximum recovery of stevioside from the leaves of *Stevia rebaudiana* was carried out using 20 min extraction time, 200 gm/L leaf water ratio, 75 °C temperature, and intermediate grinding. The maximum yield of stevioside was 188.64 mg/L by using this method. This method was found to be low cost, easy, and environment friendly so it can be used for industrial level extraction of stevioside [65].

Insulinomimetic activity of stevioside was carried out in L6 myotubes and 3T3-L1 adipocytes cell line. Diabetes was induced by using 100 nM of human insulin to the insulin sensitive cell lines. These diabetic cells were incubated with stevioside at a concentration of 1 μM–100 μM for 24 h. The mechanism of action of stevioside was studied at protein and gene level by using anti GLUT4 antibody ELISA kit and quantitative polymerase chain reaction, respectively. Radioactive glucose uptake studies are also done by measuring the rate of glucose absorption by cells. The study revealed that stevioside showed insulinomimetic activity by activating the GLUT4 gene at 100 μM concentration in L6 and 3 T3-L1 cells. This upregulated GLUT4 expressions and glucose uptake [66].

The mechanism behind the antidiabetic activity of stevioside was studied in β-cell INS-1 and normal mouse islet cell lines. Both cells were subjected to sequential perifusion (by using 16.7 mM/L of glucose and stevioside (1 mM/L)). The study showed that stevioside enhanced insulin secretion in a dose-dependent manner in mouse islets in the presence of 16.7 mM/L glucose. The insulinotropic activity of stevioside also conserved in the presence as well as in the absence of extracellular Ca^+ [67].

In vivo antidiabetic activity of stevioside was studied in NMRI Haan mice. Diabetes was induced by using an intraperitoneal injection of alloxan at a dose of 150 mg/kg. Animals were treated with stevioside at a dose of 20 mg/kg, orally before and after diabetes induction. Antidiabetic activity of stevioside was analyzed by oral glucose tolerance test and adrenaline test (pretreatment and post-treatment of stevioside). A significant difference in glucose levels was found in mice pre-treated with saline (23.32 ± 2.14 mM/L) and stevioside (14.70 ± 4.95 mM/L) in alloxan induced diabetes. The study revealed that stevioside prevented a significant increased level of glucose in an oral glucose tolerance test. Very small β cell loss was found in mice pre-treated with stevioside compared to the alloxan treated group. The study concluded that stevioside has potential use in diabetes even at a low dose [68].

Antidiabetic potential of stevioside was also studied in male albino rats in the high fat-low dose STZ model. Animals were kept on a high fat diet (5% vegetable oil and 45% tallow) for 4 weeks and then diabetes was induced using intraperitoneal injection of STZ (35 mg/kg). Diabetic animals were treated orally with 12.5, 25, and 50 mg/kg of stevioside for 21 days. Effect of stevioside was observed by analyzing levels of glucose, lipid profile, and oxidative stress parameters like lipid peroxidation and nitric oxide in the kidney and liver. The mechanistic antidiabetic activity of stevioside was evaluated by using a DNA ladder assay. Stevioside significantly decreased levels of plasma insulin, glucose, dipeptidyl peptidase IV, and oxidative stress. DNA ladder assay also confirmed the antidiabetic activity of stevioside by reducing the DNA fragmentation in the kidney and liver of the diabetic rats. The computational analysis of antidiabetic activity of stevioside was done by using predicting activity spectra for substances-based method. The in silico results have also shown antidiabetic activity which is in line with inhibition of G-protein-coupled receptor kinase and beta-adrenergic receptor kinase. The study concluded that the antidiabetic activity of stevioside might be due to the prevention of DNA fragmentation [69].

2.5 *Securigenin*

Securigenin is the aglycone part of securidaside. Securigenin is majorly present in the seeds of *Securigera securidaca.* The aglycone securigenin has an unsaturated lactone ring and an aldehyde group. Securigenin glycosides are used in Persian folk medicine to decrease blood sugar [70].

Securigera securidaca is an annual herb occurring in Europe, West Asia, and Africa. From ancient times it is considered as Iranian folk medicine. *Securigera securidaca* has been reported for the treatment of disorders like diabetes, epilepsy, hyperlipidemia, hypertension, diuretic, gastric disturbances, and hypokalemic disorders. This plant contains phytochemical compounds like phenols, alkaloids, flavonoids, saponins, and tannins. The ethanolic and aqueous extracts of *Securigera securidaca* have been reported for various pharmacological activities [71].

Securigenin isolation was done by using hydroalcoholic extract by Hadjzadeh and the group. The seeds of *Securigera securidaca* were powdered and macerated using a 70% ethanol solution. The powder was soaked in 3.2 lit of 72 h. The extract was filtered and concentrated at 40–45 °C for 72 h. The resulting extract was purified. Purified extract was further used for the estimation of antihyperglycemic activity in STZ- induced diabetic female Wistar rats. Diabetic animals were treated with purified hydroalcoholic extract at 100 and 200 mg/kg daily in drinking water for 4 weeks. Treatment of *Securigera securidaca* seed extract has shown a significant reduction in the levels of serum glucose and total cholesterol at 200 mg/kg dose compared to diabetic control animals [72].

The study of chloroform and methanol fractions of seeds of *Securigera securidaca* was carried out by Tofighi and the group. Securigenin-3-O-inositol-(1 → 3)-bglucopyranosyl-(1 → 4)-b-xylopyranoside, Securigenin-3-O-b-glucopyranosyl-(1 → 4)-b-xylopyranoside and securigenin-3-O-a-rhamnopyranosyl-(1 → 4)-a-glucopyranoside were isolated as active constituents of *Securigera securidaca.* The diabetes was induced by using STZ *i.p.* injection in male mice. Once diabetes confirmed, the animals were treated with methanol fraction 100 mg/kg and chloroform fraction at a dose of 400 mg/kg. The hypoglycemic effect of methanol fraction (100 mg/kg) and chloroform fraction (400 mg/kg) was comparable with standard antidiabetic drug glibenclamide (3 mg/kg). Methanol fraction at a dose of 400 mg/kg and chloroform fraction at a dose of 600 mg/kg showed an equal hypoglycemic effect to insulin 12.5 IU/kg. The hypoglycemic effect of these extracts was due to an increase in insulin secretion as compared with normal animals [73].

3 Conclusion

Diabetes is the main leading causes of mortality and morbidity throughout the world. Many antidiabetic drugs available in the market have side effects and toxicity. In this scenario, natural product research has gained importance due to its multifunctional and less toxic profiles compared to synthetic medicines. Glycosides and aglycone part of different glycosides are documented for antidiabetic activity. Antidiabetic activity of glycosides may be linked to a reduction in glucose absorption, an increase in insulin secretion and glucose uptake. Many plant extracts having glycosides have been reported for antidiabetic effects but isolation, characterization, and identification of glycosides are lacking. Systematic preclinical studies with special focus on detail molecular level effects of the glycosides or aglycones can be designed. Clinical studies of established glycosides with antidiabetic property are also important area to be explored. Thus, glycosides need a systematic approach focused on preclinical and clinical studies which will help for development of new drugs for diabetes.

References

1. World Health Organization (2019) In: Classif. diabetes. https://apps.who.int/iris/handle/10665/325182. Accessed 5 Sept 2019
2. Leslie RD, Palmer J, Schloot NC, Lernmark A (2016) Diabetes at the crossroads: relevance of disease classification to pathophysiology and treatment. Diabetologia 59:13–20
3. Skyler JS, Bakris GL, Bonifacio E et al (2017) Differentiation of diabetes by pathophysiology, natural history, and prognosis. Diabetes 66:241–255
4. Global Report On Diabetes (2016) The World Health Organization Library. Cataloguing-in-publication data global report on diabetes
5. International Diabetes Federation Diabetes Atlas (2019) 9th edition
6. Guerriero G, Berni R, Muñoz-Sanchez JA et al (2018) Production of plant secondary metabolites: examples, tips and suggestions for biotechnologists. Genes (Basel) 9:309
7. Brunetton J (2008) Pharmacognosy, phytochemistry, medicinal plants (2nd ed. - retirage broch). Lavoisier, Paris
8. Bartnik M, Facey PC (2017) Glycosides. In: Pharmacognosy: fundamentals, applications and strategy. Elsevier Inc., London, pp 101–161
9. Kim CS, Oh J, Subedi L et al (2018) Two new phenolic glycosides from *Sorbus commixta*. Chem Pharm Bull 66:839–842. https://doi.org/10.1248/cpb.c18-00280
10. Huh JY, Lee S, Ma EB et al (2018) The effects of phenolic glycosides from *Betula platyphylla* var. japonica on adipocyte differentiation and mature adipocyte metabolism. J Enzyme Inhib Med Chem 33:1167–1173. https://doi.org/10.1080/14756366.2018.1491846
11. Shan SJ, Zhang PP, Luo J, Kong LY (2018) Two new phenolic glycosides isolated from ginkgo seeds. Chin J Nat Med 16:505–508. https://doi.org/10.1016/S1875-5364(18)30086-4
12. Cha JM, Suh WS, Lee TH et al (2017) Phenolic glycosides from *Capsella bursa*-pastoris (L.) medik and their anti-inflammatory activity. Molecules 22:1023. https://doi.org/10.3390/molecules22061023
13. Lastra RA, Kenkel NC, Daayf F (2017) Phenolic glycosides in *Populus tremuloides* and their effects on long-term ungulate browsing. J Chem Ecol 43:1023–1030. https://doi.org/10.1007/s10886-017-0895-z
14. Ma J, Li CJ, Yang JZ et al (2017) New phenylpropanoid and coumarin glycosides from the stems of *Hydrangea paniculata* Sieb. Molecules 22:133. https://doi.org/10.3390/molecules22010133
15. Yu J, Song X, Wang D et al (2017) Five new chromone glycosides from *Scindapsus officinalis* (Roxb.) Schott. Fitoterapia 122:101–106. https://doi.org/10.1016/j.fitote.2017.09.002
16. Ma SY, Shi LG, Gu ZB et al (2018) Two new chromone glycosides from the roots of *Saposhnikovia divaricata*. Chem Biodivers 15:e1800253. https://doi.org/10.1002/cbdv.201800253
17. Sun Y, Gao M, Chen H et al (2019) Six new Coumarin glycosides from the aerial parts of *Gendarussa vulgaris*. Molecules 24:1456. https://doi.org/10.3390/molecules24081456
18. Yao J, He H, Xue J et al (2019) *Mori Ramulus* (Chin.Ph.)—the dried twigs of *Morus alba* L./Part 1: discovery of two novel coumarin glycosides from the anti-hyperuricemic ethanol extract. Molecules 24:629. https://doi.org/10.3390/molecules24030629
19. Warashina T, Miyase T (2017) Flavonoid glycosides from *Sedum bulbiferum*. Chem Pharm Bull 85:1199–1204. https://doi.org/10.1248/cpb.c17-00678
20. Vien LT, Van QTT, Hanh TTH et al (2017) Flavonoid glycosides from *Barringtonia acutangula*. Bioorganic Med Chem Lett 27:3776–3781. https://doi.org/10.1016/j.bmcl.2017.06.066
21. Jiang XL, Wang L, Wang EJ et al (2018) Flavonoid glycosides and alkaloids from the embryos of *Nelumbo nucifera* seeds and their antioxidant activity. Fitoterapia 125:184–190. https://doi.org/10.1016/j.fitote.2018.01.009

22. Tagousop CN, Tamokou JDD, Ekom SE et al (2018) Antimicrobial activities of flavonoid glycosides from *Graptophyllum grandulosum* and their mechanism of antibacterial action. BMC Complement Altern Med 18:1–10. https://doi.org/10.1186/s12906-018-2321-7
23. Li F, Du BW, Lu DF et al (2017) Flavonoid glycosides isolated from *Epimedium brevicornum* and their estrogen biosynthesis-promoting effects. Sci Rep 7:1–12. https://doi.org/10.1038/s41598-017-08203-7
24. Cheng FR, Cui HX, Fang JL et al (2019) Ameliorative effect and mechanism of the purified anthraquinone-glycoside preparation from *Rheum palmatum* L. on type 2 diabetes mellitus. Molecules 24:1454. https://doi.org/10.3390/molecules24081454
25. El-Kashak WA, Elshamy AI, Mohamed TA et al (2017) Rumpictuside a: unusual 9,10-anthraquinone glucoside from *Rumex pictus* Forssk. Carbohydr Res 448:74–78. https://doi.org/10.1016/j.carres.2017.05.023
26. Chen T, Li H, Zou D et al (2016) Separation of three anthraquinone glycosides including two isomers by preparative high-performance liquid chromatography and high-speed countercurrent chromatography from *Rheum tanguticum* maxim. ex Balf. J Sep Sci 39:3105–3112. https://doi.org/10.1002/jssc.201600487
27. Anandhi D, Pandit VR, Kadhiravan T et al (2019) Cardiac arrhythmias, electrolyte abnormalities and serum cardiac glycoside concentrations in yellow oleander (*Cascabela thevetia*) poisoning–a prospective study. Clin Toxicol 57:104–111. https://doi.org/10.1080/15563650.2018.1499930
28. Kruakaew S, Seeka C, Lhinhatrakool T et al (2017) Cytotoxic cardiac glycoside constituents of *Vallaris glabra* leaves. J Nat Prod 80:2987–2996. https://doi.org/10.1021/acs.jnatprod.7b00554
29. Ren Y, Chen W-L, Lantvit DD et al (2017) Cardiac glycoside constituents of *Streblus asper* with potential antineoplastic activity HHS public access. J Nat Prod 80:648–658. https://doi.org/10.1021/acschembio.5b01018
30. Anderson SE, Barton CE (2017) The cardiac glycoside convallatoxin inhibits the growth of colorectal cancer cells in a p53-independent manner. Mol Genet Metab Rep 13:42–45. https://doi.org/10.1016/j.ymgmr.2017.07.011
31. Knittel DN, Lorenz P, Huber U et al (2016) Characterization of the cardiac glycoside and lipid profiles of *Strophanthus kombé* Oliv. Seeds. Zeitschrift fur Naturforsch Sect C J Biosci 71:55–64. https://doi.org/10.1515/znc-2015-0186
32. Saleem M, Asif J, Asif M, Saleem U (2018) Amygdalin from apricot kernels induces apoptosis and causes cell cycle arrest in cancer cells: an updated review. Anti Cancer Agents Med Chem 18:1650–1655. https://doi.org/10.2174/1871520618666180105161136
33. Misihairabgwi JM, Ishola A, Sulyok M, Krska R (2019) Mycotoxin and cyanogenic glycoside assessment of the traditional leafy vegetables mutete and omboga from Namibia. Food Addit Contam Part B Surveill 12:245–251. https://doi.org/10.1080/19393210.2019.1616829
34. Pandey AK, Madhu P, Bhat BV (2019) Down-regulation of CYP79A1 gene through antisense approach reduced the cyanogenic glycoside Dhurrin in [*Sorghum bicolor* (L.) Moench] to improve fodder quality. Front Nutr 6:122. https://doi.org/10.3389/fnut.2019.00122
35. Kim CS, Oh J, Subedi L et al (2018) Rare thioglycosides from the roots of *Wasabia japonica*. J Nat Prod 81:2129–2133. https://doi.org/10.1021/acs.jnatprod.8b00570
36. Soundararajan P, Kim JS (2018) Anti-carcinogenic glucosinolates in cruciferous vegetables and their antagonistic effects on prevention of cancers. Molecules 23:2983
37. Xu L, Nagata N, Ota T (2018) Glucoraphanin: a broccoli sprout extract that ameliorates obesity-induced inflammation and insulin resistance. Adipocytes 7:218–225
38. Vanduchova A, Anzenbacher P, Anzenbacherova E (2019) Isothiocyanate from broccoli, sulforaphane, and its properties. J Med Food 22:121–126
39. Bone K, Mills S (2012) Principles and practice of phytotherapy: modern herbal medicine. Elsevier Ltd, London
40. Khattak S, Khan H (2016) Phyto-glycosides as therapeutic target in the treatment of diabetes. mini-reviews. Med Chem 18:208–215. https://doi.org/10.2174/1389557516666160909112751

41. Frutos MJ, Rincón-Frutos L, Valero-Cases E (2018) Rutin. In: Nonvitamin and nonmineral nutritional supplements. Elsevier, London, pp 111–117
42. Ganeshpurkar A, Saluja AK (2017) The pharmacological potential of rutin. Saudi Pharm J 25:149–164
43. Lu CL, Zheng Q, Shen Q et al (2017) Uncovering the relationship and mechanisms of Tartary buckwheat (*Fagopyrum tataricum*) and type II diabetes, hypertension, and hyperlipidemia using a network pharmacology approach. Peer J 2017:e4042. https://doi.org/10.7717/peerj.4042
44. Kreft S, Knapp M, Kreft I (1999) Extraction of rutin from buckwheat (*Fagopyrum esculentum* moench) seeds and determination by capillary electrophoresis. J Agric Food Chem 47:4649–4652. https://doi.org/10.1021/jf990186p
45. Kamalakkannan N, Prince PSM (2006) Rutin improves the antioxidant status in streptozotocin-induced diabetic rat tissues. Mol Cell Biochem 293:211–219. https://doi.org/10.1007/s11010-006-9244-1
46. Fernandes AAH, Novelli ELB, Okoshi K et al (2010) Influence of rutin treatment on biochemical alterations in experimental diabetes. Biomed Pharmacother 64:214–219. https://doi.org/10.1016/j.biopha.2009.08.007
47. Hsu CY, Shih HY, Chia YC et al (2014) Rutin potentiates insulin receptor kinase to enhance insulin-dependent glucose transporter 4 translocation. Mol Nutr Food Res 58:1168–1176. https://doi.org/10.1002/mnfr.201300691
48. Aitken JF, Loomes KM, Riba-Garcia I et al (2017) Rutin suppresses human-amylin/hIAPP misfolding and oligomer formation in-vitro, and ameliorates diabetes and its impacts in human-amylin/hIAPP transgenic mice. Biochem Biophys Res Commun 482:625–631. https://doi.org/10.1016/j.bbrc.2016.11.083
49. Ghorbani A (2017) Mechanisms of antidiabetic effects of flavonoid rutin. Biomed Pharmacother 96:305–312
50. Zhou YX, Zhang H, Peng C (2014) Puerarin: a review of pharmacological effects. Phyther Res 28:961–975
51. Wong KH, Li GQ, Li KM et al (2011) Kudzu root: traditional uses and potential medicinal benefits in diabetes and cardiovascular diseases. J Ethnopharmacol 134:584–607
52. Xu HN, He CH (2007) Separation and purification of puerarin with solvent extraction. Sep Purif Technol 56:397–400. https://doi.org/10.1016/j.seppur.2007.06.003
53. Wu K, Liang T, Duan X et al (2013) Anti-diabetic effects of puerarin, isolated from *Pueraria lobata* (Willd.), on streptozotocin-diabetogenic mice through promoting insulin expression and ameliorating metabolic function. Food Chem Toxicol 60:341–347. https://doi.org/10.1016/j.fct.2013.07.077
54. Yang L, Yao D, Yang H et al (2016) Puerarin protects pancreatic β-cells in obese diabetic mice via activation of GLP-1R signaling. Mol Endocrinol 30:361–371. https://doi.org/10.1210/me.2015-1213
55. Hsu FL, Liu IM, Kuo DH et al (2003) Antihyperglycemic effect of puerarin in streptozotocin-induced diabetic rats. J Nat Prod 66:788–792. https://doi.org/10.1021/np0203887
56. Chen X, Qian L, Wang B et al (2019) Synergistic hypoglycemic effects of pumpkin polysaccharides and puerarin on type II diabetes mellitus mice. Molecules 24:955. https://doi.org/10.3390/molecules24050955
57. Izawa K, Amino Y, Kohmura M et al (2010) Human-environment interactions - taste. In: Comprehensive natural products II: chemistry and biology. Elsevier Ltd, London, pp 631–671
58. Pothuraju R, Sharma RK, Chagalamarri J et al (2014) A systematic review of *Gymnema sylvestre* in obesity and diabetes management. J Sci Food Agric 94:834–840
59. Liu HM, Kiuchi T (1992) Isolation and structure elucidation of Gymnemic acids, Antisweet principles of *Gymnema sylvestre*. Chem Pharm Bull (Tokyo) 40:1366–1375. https://doi.org/10.1248/cpb.40.1366

60. Li Y, Sun M, Liu Y et al (2019) Gymnemic acid alleviates type 2 diabetes mellitus and suppresses endoplasmic reticulum stress in vivo and in vitro. J Agric Food Chem 67:3662–3669. https://doi.org/10.1021/acs.jafc.9b00431
61. Li Y, Liu Y, Liang J et al (2019) Gymnemic acid ameliorates Hyperglycemia through PI3K/AKT- and AMPK-mediated signaling pathways in type 2 diabetes mellitus rats. J Agric Food Chem 67:13051–13060. https://doi.org/10.1021/acs.jafc.9b04931
62. Tiwari P, Ahmad K, Hassan Baig M (2016) *Gymnema sylvestre* for diabetes: from traditional herb to future's therapeutic. Curr Pharm Des 23:1667–1676. https://doi.org/10.2174/1381612823666161108162048
63. Jorge K (2003) Soft drinks | chemical composition. In: Encycl food Sci Nutr, 2nd edn. Elsevier Ltd, London, pp 5346–5352. https://doi.org/10.1016/B0-12-227055-X/01101-9
64. Bassoli A, Merlini L (2003) Sweetners | intensive. In: Encyclopedia of food sciences and nutrition. Academic Press, Cambridge, pp 5688–5695
65. Lopez-Carbon V, Sayago A, Gonzalez-Domínguez R, Fernandez-Recamales A (2019) Simple and efficient green extraction of steviol glycosides from *stevia rebaudiana* leaves. Foods 8:402. https://doi.org/10.3390/foods8090402
66. Bhasker S, Madhav H, Chinnamma M (2015) Molecular evidence of insulinomimetic property exhibited by steviol and stevioside in diabetes induced L6 and 3T3L1 cells. Phytomedicine 22:1037–1044. https://doi.org/10.1016/j.phymed.2015.07.007
67. Jeppesen PB, Gregersen S, Poulsen CR, Hermansen K (2000) Stevioside acts directly on pancreatic β cells to secrete insulin: actions independent of cyclic adenosine monophosphate and adenosine triphosphate-sensitive K+-channel activity. Metabolism 49:208–214. https://doi.org/10.1016/S0026-0495(00)91325-8
68. Ilic V, Vukmirovic S, Stilinovic N et al (2017) Insight into anti-diabetic effect of low dose of stevioside. Biomed Pharmacother 90:216–221. https://doi.org/10.1016/j.biopha.2017.03.045
69. Rotimi SO, Rotimi OA, Adelani IB et al (2018) Stevioside modulates oxidative damage in the liver and kidney of high fat/low streptozocin diabetic rats. Heliyon 4:e00640. https://doi.org/10.1016/j.heliyon.2018.e00640
70. Zatula VV, Chernobrovaya NV, Kolesnikov DG (1966) A chemical study of the structure of securigenin and its bioside securidaside. Chem Nat Compd 2:359–360. https://doi.org/10.1007/BF00564226
71. Hosseinzadeh H, Ramezani M, Danaei AR (2002) Antihyperglycaemic effect and acute toxicity of *Securigera securidaca* L. seed extracts in mice. Phyther Res 16:745–747. https://doi.org/10.1002/ptr.1020
72. Hadjzadeh M-A-R, Moradi R, Ghorbani A et al (2015) Antihyperglycemic and antihyperlipidemic effects of hydroalcoholic extract of *Securigera securidaca* seeds in streptozotocin-induced diabetic rats. Adv Biomed Res 4:33. https://doi.org/10.4103/2277-9175.150427
73. Tofighi Z, Moradi-Afrapoli F, Ebrahimi SN et al (2017) Securigenin glycosides as hypoglycemic principles of *Securigera securidaca* seeds. J Nat Med 71:272–280. https://doi.org/10.1007/s11418-016-1060-7

Chapter 6
Isolation and Structure Elucidation of Hypoglycemic Compounds

Haixia Chen and Tingting Zhang

Abstract Due to the mild action and low toxicity of natural hypoglycemic compounds, the research of natural compounds has become a hotspot in the development of new hypoglycemic drugs in the world, so the isolation and structure elucidation of hypoglycemic compounds have become increasingly important. This article reviews the literature and summarizes the isolation and structural elucidation of major hypoglycemic compounds such as polysaccharides, flavonoids, alkaloids, saponins, and terpenoids.

Keywords Hypoglycemic · Crude extraction · Isolation and purification · Structure elucidation · Polysaccharides · Flavonoids · Alkaloids · Saponins · Terpenoids

1 Introduction

Diabetes has been described as the common metabolic disorder of carbohydrate, protein, and fat metabolism, which is due to absolute or relative lack of insulin and is characterized by hyperglycemia [1]. Chinese medicine refers to this as "Xiaoke Disease" [2], whose obvious symptoms mainly are polydipsia, polyphagia, polyuria, weight loss, fatigue, and weakness. Two main types of diabetes based on their clinical manifestations are identified as Type I diabetes known as Juvenile diabetes or insulin-sensitive diabetes and Type II diabetes or non-insulin dependent diabetes mellitus (NIDDM) [3]. Diabetes is becoming the third "killer" of the health of mankind along with cancer, cardiovascular, and cerebrovascular diseases because of its high prevalence, morbidity, and mortality [1]. Its complications are an important factor leading to increased mortality of diabetes. Complications are mainly divided into two categories [2]: 1. microvascular complications, peripheral nerve mutations, retinopathy, diabetic nephropathy, etc.; 2. macrovascular complications,

H. Chen (✉) · T. Zhang
Tianjin Key Laboratory for Modern Drug Delivery and High-Efficiency, School of Pharmaceutical Science and Technology, Tianjin University, Tianjin, China
e-mail: chenhx@tju.edu.cn

H. Chen, M. Zhang (eds.), *Structure and Health Effects of Natural Products on Diabetes Mellitus*, https://doi.org/10.1007/978-981-15-8791-7_6

such as myocardial infarction, stroke, occlusive disease of lower extremity blood vessels, etc. Diabetes is a serious, long-term condition with a major impact on the lives and well-being of individuals, families, and societies worldwide [4].

Diabetes has been conventionally treated with orthodox medicines [1]. Administration of exogenous insulin is the treatment for all type-1 diabetic patients and for some type-2 patients who do not achieve adequate blood glucose control with oral hypoglycemic drugs [5]. And various chemical medicine, including sulfonylureas, thiazolidinedione, α-glucosidase inhibitors, Biguanide, DPP-4 inhibitors, and GLP-1 analogs, have been employed for therapy diabetes [6]. However, these medicines all have limited efficacy, limited tolerability, and/or significant mechanism based side effects [7]. They mainly focus on lowering blood sugar, but their effects on diabetic complications are not good [2]. Therefore, the research on natural compounds with mild effects and small toxic side effects has become a hot spot in the development of new drugs in the world. WHO once recognized traditional medicine as "an accessible, affordable and culturally acceptable form of healthcare trusted by large numbers of people, which stands out as a way of coping with the relentless rise of chronic non-communicable diseases in the midst of soaring healthcare costs and nearly universal austerity" [8].

The types of natural active ingredients mainly include polysaccharides, flavonoids, alkaloids, saponins, terpenoids, etc. [9]. They are mainly extracted from plants and microorganisms [2]. A large number of studies have proved that the active ingredients of hypoglycemic are extracted from many plants. Yam, ginseng, lotus seeds, etc. contain active ingredients that promote insulin secretion [2], α-glucosidase inhibitors are extracted from leaves of *Phoebe bournei* [10]. Microbes are also one of the sources of traditional Chinese medicine for the development of hypoglycemic drugs, especially actinomycetes. Their secondary metabolites have important economic utilization value and various medicinal activity values. The first α-glycosidase inhibitor was found in the culture broth of *Actinoplanes* SE50 [2].

2 Polysaccharides

Polysaccharide is a kind of natural macromolecular polymer, which is usually composed of more than 10 monosaccharides through glycosidic linkages in linear or branched chains, with a molecular weight of tens of thousands or even millions [11]. They widely exist in the plants, microorganism, algae, and animals [12]. Polysaccharides are considered an important class of bioactive natural products, which have been widely studied in the biochemical and medical areas due to specific bioactivities, such as anti-diabetic activities [13]. At present, most polysaccharides have the pharmacological characteristics of multiple pathways, multiple targets, and multi-directions to lower blood sugar [14], and act on diabetes through multiple mechanisms and multiple links. Some polysaccharides have been reported to have anti-diabetic effects through a variety of mechanisms, such as targeting b-cell dysfunction, insulin enhancement, and inhibiting a-amylase and a-glucosidase [15].

2.1 Crude Extraction

The question of how to extract polysaccharides has been the focus of researchers (Table 6.1). In general, polysaccharides are extracted by using solvent extraction method. The most classical and most convenient method is water extraction and alcohol precipitation [16–18], which is widely used in industry. Furthermore, some modern methods are also used to improve the extraction process, such as ultrasonic extraction method [19], microwave extraction method [20], enzyme-assisted extraction method [21], supercritical fluid extraction method [22], and ionic liquid-microwave-assisted extraction method [23, 24]. It is more efficient to extract by these methods. Also during the extraction, in order to maximize the extraction rate, some optimization methods are commonly used to get the best extraction solid-liquid ratio, temperature, time, etc., such as response surface method optimization [23], orthogonal design optimization [25], orthogonal test optimization [26], and uniform design optimization [27].

Table 6.1 Comparison of polysaccharides crude extraction methods [28]

Method	Advantages and disadvantages	Example	Reference
Water extraction and alcohol precipitation	Convenient, longer time, and low efficiency	Sugarcane leaf polysaccharides, *Moringa oleifera* polysaccharides, Corn silk polysaccharides	[16–18]
Ultrasonic extraction method	Higher extraction efficiency, but may destroy the structure of macromolecules and affect the biological activity	Mulberry leaf polysaccharide	[19]
Microwave extraction method	Higher extraction efficiency, low equipment requirements, less reagents used, less pollution, and short production cycle	Hawthorn polysaccharide	[20]
Enzyme-assisted extraction method	Higher extraction efficiency, but owing to the presence of enzymes, it is necessary to consider factors such as temperature, pH, and reaction time	*Moringa oleifera* polysaccharide	[21]
Supercritical fluid extraction method	Higher extraction efficiency, higher extraction rate, higher security, but expensive equipment	Bamboo leaf polysaccharides	[22]
Ionic liquid-microwave-assisted extraction method	Higher extraction efficiency, higher extraction rate, environmental protection, but not widely used in industry	Schisandra polysaccharide	[23, 24]

2.2 *Isolation and Purification*

2.2.1 Removal of Proteins and Pigments

Common methods for removing proteins (Table 6.2) are: polyamide chromatography [29, 30], Sevage method [31], trichloroacetic acid method [31], enzymatic hydrolysis [32], etc. At the same time, some polysaccharides contain a lot of pigments, which also seriously restricts the further purification of polysaccharides. The main methods of decolorization (Table 6.3) are polyamide chromatography [29], activated carbon [33], adsorption resin method [32], H_2O_2 decolorization

Table 6.2 Comparison of protein removal methods [28]

Method	Advantages and disadvantages	Example	Reference
Polyamide chromatography	High removal rate, less loss of polysaccharides, protection of polysaccharide structure, avoiding the large use of organic solvents	Polysaccharides from Inonotus obliquus	[29, 30]
Sevage method	Easy to use, mild action, protection of polysaccharide structure, but requirement for a lot of organic solvents, time-consuming	Mulberry leaf polysaccharide	[31]
Trichloroacetic acid method (TCA)	Good protein removal effect, but easy to degraded	Mulberry leaf polysaccharide	[31]
Enzymatic hydrolysis	High removal rate, mild action, but easy to cause difficulty in filtering sugar	Ginkgo fruit polysaccharide	[32]

Table 6.3 Comparison of decolorization methods [28]

Method	Advantages and disadvantages	Example	Reference
Polyamide chromatography	Good decolorization effect, high polysaccharide retention rate	Polysaccharides from *Inonotus obliquus*	[29]
Activated carbon	Gentle action, not too high decolorization rate, but with the high loss rate of polysaccharides, and activated carbon is difficult to filter	Mulberry leaf polysaccharide	[33]
Adsorption resin method	Good bleaching effect, high polysaccharide retention rate, simple operation	Ginkgo fruit polysaccharide	[32]
H_2O_2 decolorization method	Easy to destroy polysaccharide structure, easy to remain in polysaccharide extract	*Eucommia ulmoides* polysaccharide	[34]
Reverse micelle solution method	Good bleaching effect, expensive, and difficult to recycle	Ginkgo fruit polysaccharide	[32]
Organic solvent repeated washing method	A large use of organic solvents, easy to cause harm to human body and the environment, and not conducive to industrial production	Polysaccharides from *Inonotus obliquus*	[29]

method [34], reverse micelle solution method [32], organic solvent repeated washing method [29], etc.

2.2.2 Further Purification

Choose different methods to purify polysaccharides further according to the acidity, alkalinity, molecular weight, and other properties of the polysaccharide. Commonly used purification methods include membrane dialysis, fractional precipitation, chemical precipitation, and chromatography. Through membrane dialysis [35], the different molecular weights of polysaccharides were got, but the membrane was easily to be jammed and the selection of membrane was hard. The fractional precipitation is the separation process of the polysaccharide according to the molecular weight from large to small precipitation, Li Qian [36] uses fractional precipitation to separate yam polysaccharides with different structures. The chemical precipitation method [37] (Table 6.4) mainly includes the quaternary ammonium salt precipitation and metal complexing method, which is basically suitable for most crude polysaccharides. Chromatography is the most widely used separation and purification method. According to the physical and chemical properties of the target, the most suitable stationary phase and mobile phase are selected. Commonly used are cellulose column chromatography, ion exchange column chromatography, gel column chromatography, affinity column chromatography, of which ion exchange column chromatography and gel column chromatography are the most widely used (Table 6.5).

Table 6.4 Comparison of two chemical precipitation methods [37]

Method	Mechanism	Advantages and disadvantages
Quaternary ammonium salt precipitation	Precipitation was settling out in low ionic strength solution, for complex formation could appear between long chain quaternary ammonium salt and acidic polysaccharide or high molecular weight polysaccharide	Simple and easy, but with bigger loss and lower yield
Metal complexing method	Polysaccharide with all kinds of copper, barium, calcium, and lead ions to form complex and precipitation	Simple but with lower yield

Table 6.5 Comparison of two chromatography methods

Method	Example	Advantages and disadvantages	Example	Reference
Gel column chromatography	Sephadex-gel column chromatography, Sepharose-gel column chromatography	Quick, easy, good separation effect, but with rigorous separation condition	Plantago depressa polysaccharide	[38]
Ion exchange chromatography	DEAE-cellulose column chromatography	Bigger exchange capacity, better separation effect, but expensive	Mulberry leaf polysaccharide	[31]

Hou Xiaotao [16] used DEAE-cellulose column chromatography and gel column chromatography to purify sugarcane polysaccharide, and then used Q-Sepharose Fast Flow (QFF) strong anion exchange chromatography and Sepharcryl S-300 (S300) gel filtration chromatography column to separate more specific components. Wang Fang [17] used diethylaminoethylcellulose, DEAE-52 column chromatography and dextran G-200 gel chromatography column to purify the Moringa oleifera polysaccharide.

2.3 *Structure Elucidation*

Polysaccharides are formed by the condensation and dehydration of multiple monosaccharide molecules, and are a type of saccharides with complex and large molecular structure [39]. The structural unit of polysaccharides is monosaccharides. According to the number of types of monosaccharides, it can be divided into homogeneous polysaccharide and heterogeneous polysaccharide. The structural units are connected by glycosidic bonds. Common glycosidic bonds are α-1,4-, β-1,4- and α-1,6-glycosidic bonds. The structural units can be connected into a straight chain (such as amylose) or can form a branched chain (such as amylopectin).

2.3.1 Primary Structure of Polysaccharides

Polysaccharides are complex biological macromolecules, which can be divided into primary, secondary, tertiary, and quaternary structures similar to proteins [39]. At present, the primary structure analysis technology for polysaccharides is relatively mature. In view of the complexity of polysaccharide structure (especially heteroglycan), it is difficult to resolve its primary structure by one or several methods, and it must be combined with a large number of chemical analysis and modern instrumental analysis methods [40]. Taking "partial degradation-methylation-nuclear magnetic resonance" as the core technology for the primary structure characterization of polysaccharides is a practical and advanced means at this stage, and combined with ESI-MS, Maldi-TofMS, ion mobility-mass spectrometry, and other technologies, the primary structure of polysaccharides can be scientifically and finely analyzed [39] (Table 6.6).

2.3.2 Higher-Level Structure of Polysaccharides

People also have been exploring the higher-level structure of polysaccharides. Use Atomic Force Microscope [48] to take high-speed three-dimensional images of polysaccharides. Using X-ray diffraction to speculate the parameters of the spirochete such as symmetry and pitch of the polysaccharide [41]. The environmental scanning electron microscope (ESEM) can directly observe the three-dimensional

Table 6.6 Methods for the analysis of primary structure of polysaccharides [39]

Detections	Methods
Polysaccharide purity and relative molecular mass	High-performance gel permeation chromatography (HPGPC) [41], dynamic and static light scattering instrument [42], field flow separation technology, etc.
Monosaccharide composition and ratio [43]	Complete acid hydrolysis combined with HPLC, GC, GC-MS, high-performance anion exchange chromatography (HPAEC), etc.
Sugar ring forms of glycosides	IR, NMR
Monosaccharide residue type and glycosidic linkage site	Methylation [44–46] and GC-MS [16], Smith degradation, MSn, NMR, ion mobility-mass spectrometry
Glycosidic linkage sequence	Selective acid hydrolysis method [47], glycosidase sequential hydrolysis, 1D NMR, and 2D NMR
Glycoside-substituted anomeric forms (α- and β-)	Hydrolysis of glycosidic bonds, NMR, IR, laser Raman spectroscopy

structure of the sample surface, and can also directly observe the sample dynamically [49]. And the use of Circular Dichromatography (CD) is effective to understand the spatial structure of polysaccharides and their solution behavior [50].

3 Flavonoids and Their Glycosides

Flavonoids are a ubiquitous group of naturally occurring polyphenolic compounds characterized by the flavan nucleus and represent one of the most prevalent classes of compounds in fruits, vegetables, and plant-derived beverages [51]. They are usually in free form or in combination with sugar to form glycosides. The best described pharmacological property of flavonoids is their capacity to act as potent antioxidant that has been reported to play an important role in the alleviation of diabetes [52].

3.1 *Crude Extraction*

The traditional extraction method of flavonoids is the solvent extraction method. Solvent extraction methods include hot water extraction [53, 54] and organic solvent extraction [55, 56]. Organic solvent extraction is currently the most commonly used method for extracting flavonoids from plant materials. Ethanol is widely used by researchers due to its advantages of light pollution, low cost, and easy recovery. In order to increase the extraction efficiency, compared to polysaccharides, ultrasonic extraction method [57], microwave extraction method [58, 59], microwave-ultrasonic collaborative extraction [60], enzyme-assisted extraction method [61], supercritical fluid extraction method [62] can also be used to improve the extraction process of flavonoids.

Meanwhile, alkaline extraction and acid precipitation method and high-voltage pulsed electric field assisted extraction method can also be used to get higher extraction efficiency. Liu Jinxiang et al. used the alkali-soluble acid precipitation method to extract the total flavonoids in *Ginkgo biloba* leaves. The extraction rate of total flavonoids in *Ginkgo biloba* leaves reached 86.4% [63]. This method had slightly lower extraction rate but without complex operation, pollution of environment and organic solvent. Yin Yongguang et al. [64] extracted total flavonoids from dried pine needles by high-voltage pulsed electric field method, which has the advantages of uniform transmission, short action time, low heat production, and low energy consumption. However, due to its high cost, it has not been widely used.

3.2 *Isolation and Purification*

At present, several methods, such as membrane separation method, solvent extraction method, high-speed countercurrent chromatography method (HSCCC), and column chromatography are available in the literature for the enrichment and separation of active flavonoids components from plant extracts (Table 6.7).

3.3 *Structure Elucidation*

Flavonoids are phenolic substances formed in plants from amino acids including phenylalanine and tyrosine and malonate, with more than 4000 individual compounds known [71]. The basic flavonoid structure contains flavan nucleus, which consists of 15 carbon atoms arranged in three rings (C6–C3–C6). They can be subdivided into different subgroups depending on the carbon of the C ring on which B ring is attached, and the degree of unsaturation and oxidation of the C ring [72]. Flavonoids in which B ring is linked in position 3 of the ring C are called isoflavones; those in which B ring is linked in position 4, neoflavonoids, while those in which the B ring is linked in position 2 can be further subdivided into several subgroups on the basis of the structural features of the C ring. These subgroups are: flavones, flavonols, flavanones, flavanonols, flavanols or catechins, and anthocyanins. And flavonoids with open C ring are called chalcones (Figs. 6.1 and 6.2).

3.3.1 Determine the Type of Compounds

Because of their structural characteristics, flavonoids can form colored complexes with certain reducing agents, such as certain metal ions and reagents, and exhibit color changes [73]. Flavonoids react with reagents such as aluminum trichloride, concentrated hydrochloric acid, and magnesium powder (or zinc powder), sodium hydroxide, and ammonia to produce characteristic colors. These types of color reaction can be used to preliminarily identify their types. The advantage of this

Table 6.7 Comparison of several isolation methods

Method	Advantages and disadvantages	Example	Reference
Membrane separation method	Simple operation, safety and energy saving, high product yield, and strong activity, but it has large fixed asset investment and high equipment requirements. Contain ultrafiltration, microfiltration, nanofiltration, and reverse osmosis	Lotus leaf flavonoids	[65]
Solvent extraction method	Economical, simple, and convenient operation, large processing capacity, and low energy consumption, but low efficiency and with toxicity	Huangqi leaves flavonoids	[66]
High-speed countercurrent chromatography method (HSCCC)	No requirement for a solid carrier, automatic operation, high separation speed, high efficiency and good reproducibility	Mulberry leaves flavonoids	[67]
Macroporous adsorption resin chromatography (AB-8, X-5, HPD-750, etc.)	Widely used in large-scale production, high stability, large adsorption capacity, strong reproducibility, long service life, and low cost	Citrus peel flavonoids	[58]
Silica gel column chromatography	Requirement for proper eluents. Separations of flavonoid aglycones mainly choose chloroform-methanol mixture as eluent; separations of flavonoid glycosides mainly choose ethyl acetate-acetone-water or chloroform-methanol-water as eluent	Mulberry root flavonoids	[68]
Polyamide column chromatography method	Simple process, unique adsorption, low cost, long service life, and low pollution	Ginkgo biloba leaves flavonoids	[69]
Sephadex column chromatography (Sephadex-G and Sephadex-LH20)	Molecular sieve effect, fast elution speed, small loss, and repeated use	Litchi pulp flavonoids	[70]

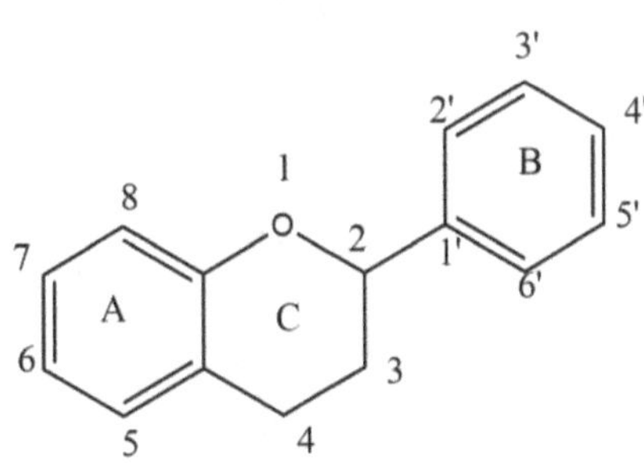

Fig. 6.1 Skeleton of diphenylpropane

Fig. 6.2 Flavonoid subgroups [72]

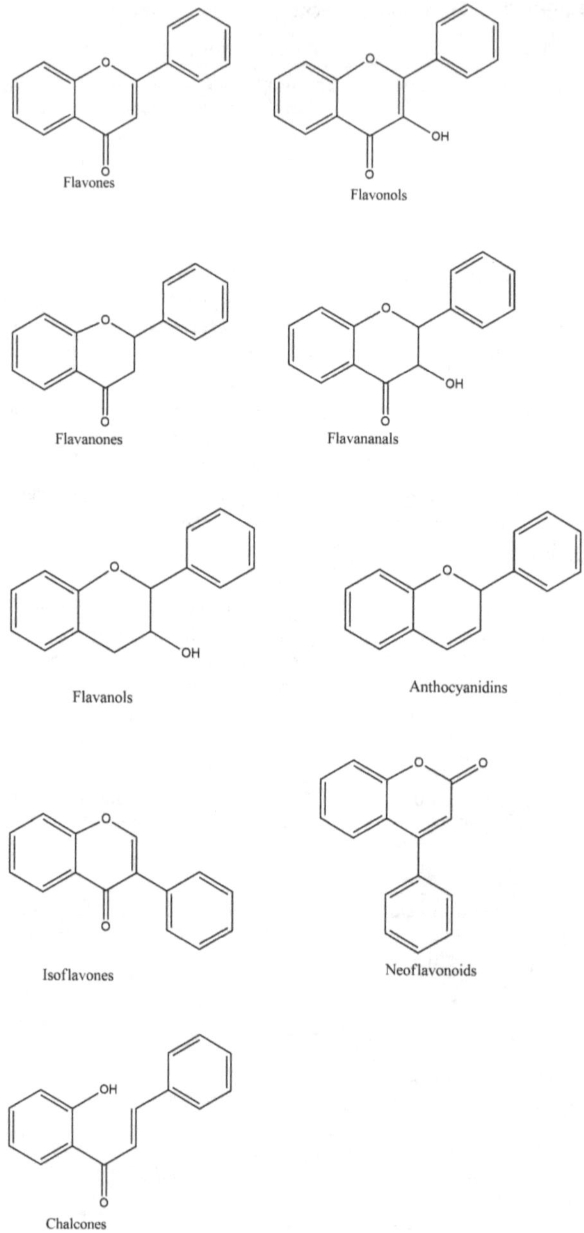

method is that it is simple and feasible, the detection time is fast, the reaction phenomenon is obvious, and no equipment is required. It is often used for the initial identification of flavonoids. Wu Yonglan [74] used $NaNO_2$-$Al(NO_3)_3$ with rutin as the standard sample to preliminarily identify the flavonoid compounds in Momordica charantia leaves. At the same time, ultraviolet spectroscopy and TLC can also be used. Guo Yanhua et al. [75] used ultraviolet spectroscopy to identify the type of flavonoids and determine their oxidation patterns.

3.3.2 Determine the Composition and Structure

Sun Yan [76] used LC-MS to analyze the composition of flavonoids in rape bee flower. Jiang Wenyue [77] used UPLC-LTQ/MS to analyze the composition of fenugreek flavonoids. Ji Lili [58] used reversed-phase high-performance liquid chromatography to analyze the flavonoids of Moringa oleifera leaves. Gong Wei [78] used NMR and IR to analyze the structure of specific hawthorn leaf flavonoid molecules. Lu Qiang [70] used ESI-MS and NMR to analyze the structure of flavonoids in litchi pulp.

3.3.3 Quantitative Calculation

Zheng Rong [79] used high-efficiency capillary electrophoresis diode array detection method (HPCE-DAD) to determine the specific content of various flavonoids in clove leaves. Zhang Yu et al. [80] used HPLC to identify the quality and content of flavonoids in Momordica charantia.

4 Alkaloids

Alkaloids are secondary metabolites originally defined as pharmacologically active compounds, primarily composed of nitrogen [81]. Alkaloids are a large group of secondary metabolites in plants. Currently, there are about 5000–7000 alkaloids isolated from plants, with a wide variety and complex structure. Studies have found that alkaloids can control blood [82].

4.1 Crude Extraction

There are many extraction methods for alkaloids, and most of the alkaloids are extracted by solvent extraction. According to the specific operation [83], it can be divided into dipping method, percolation method, decoction method, and heat reflux method. Xuan Guangshan et al. [57] extracted mulberry leaf alkaloids with 25% ethanol reflux. Yuan Pulong et al. [84] extracted three times with 85% ethanol under reflux, and then extracted the lotus leaf alkaloids with a lipophilic organic solvent (dichloromethane).

The new extraction technology mainly includes microwave-assisted extraction technology, ultrasonic extraction method, and supercritical fluid extraction method.

The principles of methods for alkaloids are the same as that for polysaccharides and flavonoids. Xiao Guqing et al. [85] conducted microwave-assisted extraction, ultrasonic-assisted extraction, and microwave -ultrasonic combined extraction on the alkaloids in Coptis chinensis. The experiment proved that the extraction effects of the three methods were very good, and the combined extraction was the best. Tong Ruofei et al. [86] used supercritical CO_2 to effectively extract Coptis chinensis alkaloids.

4.2 *Isolation and Purification*

4.2.1 Chromatography

Chromatography includes silica gel column chromatography, alumina column chromatography, high-performance liquid chromatography, and high-speed countercurrent chromatography separation methods. These chromatography methods have many applications. Chen Hongying [82] used high-speed countercurrent chromatography, multiple silica gel column chromatography, Sephadex LH-20 chromatography to separate Coptidis Rhizoma alkaloids. Yuan Pulong et al. [87] separated lotus leaf alkaloids by repeated silica gel column chromatography, Sephadex LH-20 chromatography, and ODS chromatography.

4.2.2 Resin Adsorption Method

Resin adsorption method mainly includes ion exchange resin and macroporous resin. The cation exchange resin is generally used in the purification of alkaloids and the hydrogen ions in the ion exchange resin can be exchanged with alkaloid salt cations [88]. Thus, it is separated from non-basic compounds and eluted with alkaline water to obtain alkaloids. Su Nan [89] successfully refined mulberry leaf alkaloids using cation exchange resin. Macroporous resins selectively adsorb organic substances from aqueous solutions (or other solutions) through physical adsorption, and their applications are becoming more and more widespread. Pan Shizhe et al. [90] used D101 macroporous resin to purify lotus leaf alkaloids.

4.2.3 Membrane Separation Method

Membrane separation technologies [88] include ultrafiltration (UF), microfiltration (MF), nanofiltration (NF), and reverse osmosis, of which the most used is ultrafiltration technology. Liu Guoyan et al. [91] explored the separation effects of UF1, UF2, and UF3 ultrafiltration membranes on mulberry leaf alkaloids and the concentration effects of NF1 and NF2 nanofiltration membranes on mulberry leaf alkaloids. The isolation effect was good.

4.2.4 Molecular Imprinting

Molecular imprinting technology [88] is a highly selective bionic technology that uses molecular imprinting polymer with molecular recognition as the stationary phase to separate, screen, and purify target molecules. It has the advantages of strong resistance to harsh environment, good stability, and long service life. Wang Ting [92] explored the separation of rosin-based polymer imprinted chromatography column for berberine.

4.3 Structural Elucidation

Alkaloids are secondary metabolites originally defined as pharmacologically active compounds, primarily composed of nitrogen [93]. Compounds like amino acids, peptides, proteins, nucleotides, nucleic acids, amines, and antibiotics are usually not called alkaloids. Compared with most other classes of natural compounds, alkaloids are characterized by a great structural diversity and there is no uniform classification of alkaloids. Alkaloids' classifications are based on similarity of the carbon skeleton (e.g., indole-, isoquinoline-, and pyridine-like) and/or biochemical precursor (ornithine, lysine, tyrosine, tryptophan, etc.) (Table 6.8).

4.3.1 Determine the Type of Compounds

The main method of primary identification of alkaloids is TLC detection. When testing alkaloids, alkaline silica gel plates can be used to make the results more accurate. Colorimetry can also be used to do a judgement. The nitrogen atoms in alkaloid bind to heavy metal atoms in Dragendorff reagents ($KBiI_4$) to form ionic bonds in complex compounds (BiI_3) (Alk·HI) which precipitate [94]. In the presence of nitric acid, tiourea and bismuth form a yellow bismuth complex. Alkaloids can also be directly reacted with some acid dyes to determine the type.

4.3.2 Determine the Composition and Structure

The comprehensive use of ESI-MS, TOF-MS, ^{1}H-NMR, ^{13}C-NMR, 1H-1HCOSY, HMBC, HMQC and other spectroscopy methods [95] makes it posssible to accurately analysis the bioidentify alkaloids. Powder ray diffraction is also an important method to study the composition and structure of substances. Zheng Xiaowei et al. [96] used TLC and X-ray diffraction to identify the Chinese patent medicine Huanglian Shangqing Pill.

Table 6.8 The main classifications of alkaloids

Alkaloids that originate from	Example
Ornithine	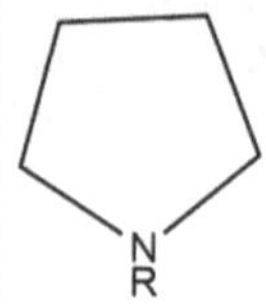
Lysine	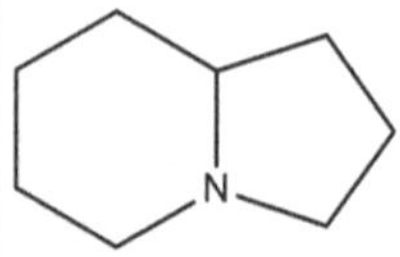
Tyrosine or Phenylalanine	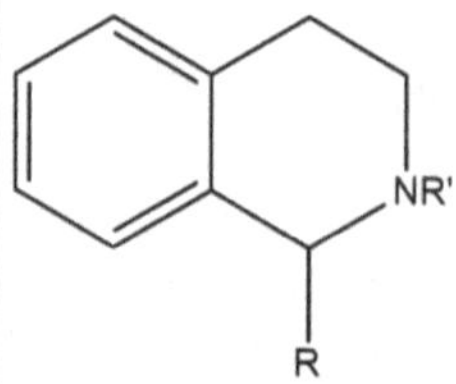
Tryptophan	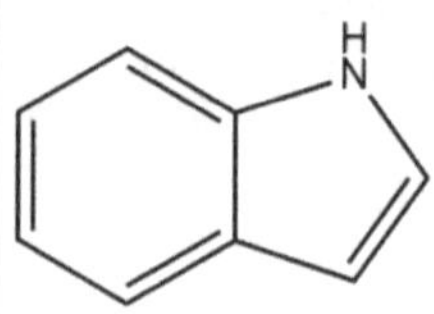
Terpenoids	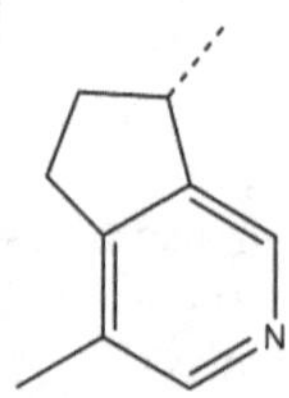 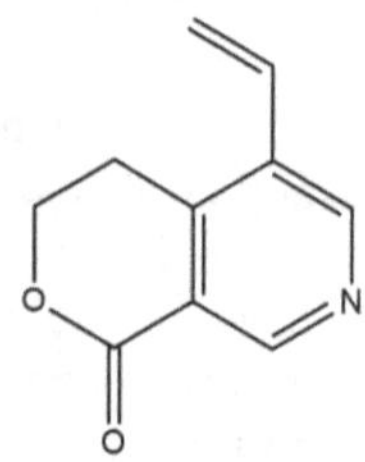

4.3.3 Quantitative Calculation

Chromatography In recent years, researchers have explored a variety of chromatographic techniques for alkaloid identification. Kim et al. [97] found that after dilute extraction of the total alkaloids obtained from mulberry leaves, the content can be determined well by pre-column derivatization reversed-phase high-performance liquid chromatography-fluorescence detection. Kimura et al. [98] established a hydrophilic interaction chromatography evaporative light scattering detection method (HILIC-ELSD) for the determination of mulberry leaf alkaloids. He Chenjie [99] successfully analyzed and identified the chemical constituents of Tianqi Jiangtang capsules based on UPLC-LTQ Orbitrap HRMS (UPLC-LTQ Orbitrap

HRMS) technology. Sheng Huoxin [100] established the fingerprint of Coptidis Rhizoma with HPLC-UV method, and qualitatively analyzed the main chromatographic peaks by HPLC-MS combined technology.

Chemical Method Liu Fan et al. [101] used ethanol and hydrochloric acid as the extraction solvent, purified the total alkaloids in mulberry leaves through a macroporous resin column, and determined the content of the total alkaloids by the Rayleigh salt colorimetric method, indicating that this method is stable and reliable. The Rayleigh salt colorimetric detection method is mainly used for alkalis with high water solubility, such as primary, secondary, and tertiary amine alkaloids and quaternary ammonium alkaloids.

5 Saponins

Saponins are phytochemicals that produce a foam when dissolved in water [102]. Their name derives from the same root as the word soap (Latin sapo=soap). Saponins are glycosides (the sugar part comprises the hydrophilic end). Two classes [102] are recognized based on the structure of their aglycone or sapogenin: steroidal saponins contain the characteristic four-ringed steroid nucleus, and triterpenoid saponins have a five-ringed structure.

5.1 Crude Extraction

Traditional extraction and separation methods, such as decoction method, dipping method, percolation method, reflux method, soxhlet extraction method, etc., have played an important role in the development of traditional Chinese medicine industry. Among them, the reflux method is the most widely used. Rui Wen et al. [103] investigated the effects of ethanol volume fraction, ethanol dosage, and extraction time on the process through a uniform design experiment to optimize reflux extraction of saponins. Zhu Hongyan et al. [104]optimized the ethanol reflux extraction method of jujube saponins by central combination design-response surface method. However, these methods have problems of long extraction time, large loss of effective components, and low extraction efficiency.

Modern extraction methods mainly include microwave extraction [105, 106], ultrasonic extraction [107], supercritical CO_2 fluid extraction [108], enzyme extraction [109], bionic extraction, etc. The biomimetic extraction method mainly simulates gastrointestinal environmental conditions in vitro and separates saponins based on the principle of drug metabolism in the body. Chen Xin et al. [110] proved that the extraction efficiency of the bionic extraction method is significantly higher than that of the traditional water extraction method.

5.2 *Isolation and Purification*

The purification methods of saponins mainly include solvent extraction method [111], high-speed countercurrent chromatography [112], macroporous resin method [113], organic solvent precipitation method [114], bubble membrane separation purification method [115], membrane separation purification method [116]. Among them, the D101 adsorptive macroporous resin method is the purification method in the 2015 edition of the Pharmacopoeia of the People's Republic of China [117]. The bubble membrane separation and purification method is to separate the foam from the liquid by blowing gas into the liquid containing the surface-active material, the material with the strongest surface activity is foamed first, and is adsorbed at the interface of the dispersed phase and the continuous phase to form a clear foam layer to achieve the purpose of separation and purification. This method is suitable for the separation and purification process of all surface-active substances. Zhang Rui et al. [115] used bubble membrane separation and purification method to extract ginsenosides with high yield. At the same time, a variety of technologies can also be used in combination, such as ultrasonic enhanced supercritical fluid extraction method, ultrasonic-silica gel column chromatography method, macroporous adsorption resin-silica gel column chromatography method, etc. [118].

5.3 *Structural Elucidation*

Saponins [119] consist of an aglycone with carbohydrate moieties (Fig. 6.3). The aglycone can be a triterpene or a steroid and can have a number of different substituents (-H, -COOH, $-CH_3$). In both the steroid and triterpenoid saponins, the carbohydrate side-chain is usually attached to the 3 carbon of the sapogenin. The number and type of carbohydrate moieties result in a considerable structural diversity of the saponins. Most carbohydrates [120] in saponins are hexoses (i.e., glucose, galactose), 6-deoxyhexoses (rhamnose), pentoses (arabinose, xylose), uronic acid (glucoronic acid), or carbohydrates with amino functionality (glucosamine). Saponins possess surface-active or detergent properties because the carbohydrate portion of the molecule is water-soluble, whereas the sapogenin is fat-soluble.

5.3.1 Determine the Type of Compounds

In general, you can first determine the type of compound by TLC and color reaction. Saponins generally use acetic anhydride-concentrated sulfuric acid method to distinguish triterpene saponins and steroidal saponins through color reaction [122].

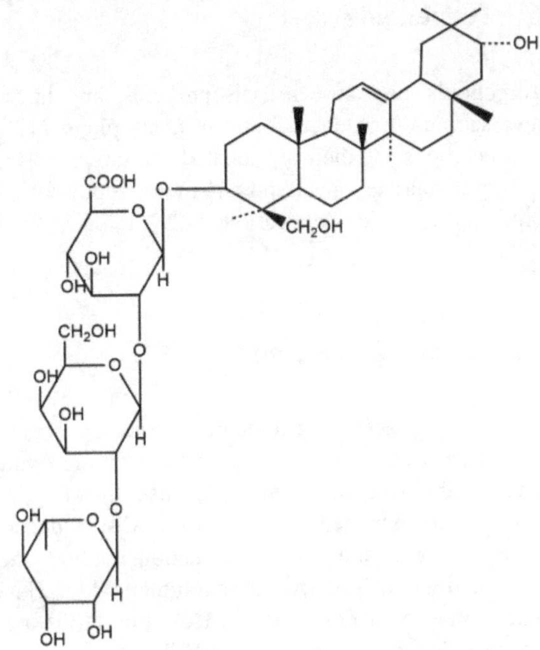

Fig. 6.3 Structure of a typical saponin (from soya beans) [121]

5.3.2 Determine the Composition and Structure

HPLC fingerprints provide a reference for the identification and quality evaluation, quality control, and the formulation of quality standards for saponins. Liu Jinfu et al. [123] established the HPLC fingerprint of momordica charantia. For analyzing the composition of saponins, various methods have been developed. Yu Bin [107] used LC-MS multi-stage mass spectrometry to show the composition of extracted bitter gourd saponins. Yu Feifei et al. [124] determined the content of five saponins in Huangqi and Jinqi Jiangtang tablets by HPLC-ELSD method. Du Zhengcai et al. [125] identified the types of saponins contained in Gesang Jiangtang capsules based on UPLC. Bahrami [126] used MALDI-MS/MS and ESI-MS/MS to analyze the obtained purified saponin to reveal the structure of isomeric saponins to contain multiple aglycones and/or sugar residues. For structure analysis, IR, MS, 1H-NMR, 13C-NMR can be used to accurately determine the structure of saponins [127].

5.3.3 Quantitative Calculation

For the quantitative calculation of saponins, commonly used methods include ultraviolet spectrophotometer, high-performance liquid chromatography, and liquid-mass spectrometry [122].

6 Terpenoids

Terpenoids, also known as isoprenoids, are the most numerous and structurally diverse natural products found in many plants [128], which are a general term for hydrocarbons and their oxygenated derivatives formed by the polymerization of two or more isoprene units. And terpenoids represent a highly diverse group of natural products with wide applications [129].

6.1 Crude Extraction

The main extraction methods of terpenes are solvent extraction, ultrasonic extraction, microwave extraction, supercritical fluid extraction, and headspace analysis. The solvent extraction method often uses the organic solvent reflux method. Xu Wen et al. [130] extracted terpenes from Alisma decoction with 85% ethanol reflux. Compared with the solvent extraction method, the ultrasonic extraction method [131] and the microwave extraction method [132] have higher extraction efficiency and shorter extraction time. Lu Hui et al. [133] used supercritical fluids to extract triterpenoids in Hedyotis diffusa Willd. Headspace analysis methods [134] include static headspace analysis and dynamic headspace analysis. Static headspace technology refers to placing the whole plant or some organs in the odor source tank, and volatile terpenoids are adsorbed through the adsorbent. The cost of this method is high. The extracted samples must be analyzed immediately and cannot be stored for a long time. Extracted substances can only be used for chemical analysis, not suitable for biological determination. The headspace dynamic analysis method is currently the most commonly used method for collecting plant volatile swimming compounds.

6.2 Isolation and Purification

Common purification methods for terpenes include macroporous adsorption resin method, silica gel column chromatography separation method, polyamide column chromatography separation method, alumina column chromatography separation method, and thin layer chromatography separation method. Xu Wen [130] purified and refined triterpenoid compounds in Alisma by macroporous adsorption resin. Li Ke [135] used thin layer chromatography and silica gel column chromatography to separate and purify the triterpenoids.

6.3 Structural Elucidation

The terpenoids, sometimes referred to isoprenoids, are derived from five-carbon isoprene units (Fig. 6.4) assembled and modified in thousands of ways. The molecular composition of open-chain terpenes conforms to the general formula $(C_5H_8)_n$ ($n \geq 2$). Meanwhile, most are multicyclic structures that differ from one another not only in functional groups but also in their basic carbon skeletons.

6.3.1 Determine the Type of Compounds

According to the basic physical and chemical constants of the compound, the type of compound is preliminarily judged. Among them, the color reaction is the most commonly used. Liebermann–Burchard reaction, Rosen-Heimer reaction, Salkowski reaction, Kahlenberg reaction, and Tschugaeff reaction can all have obvious color reaction with terpenoids [136]. In addition, the characteristic stains of TLC have become a more selective and predictive method [137]. Qu Jianhui [138] analyzed the types of triterpene compounds in Anli by LC-MS, with high sensitivity.

6.3.2 Determine the Molecular Weight and Structure

IR, MS, 1H-NMR, 13C-NMR can be used to accurately determine the primary structure of terpenes. Among them, high resolution mass spectrometry (NR-MS) has high accuracy. The higher-level structure can be measured by CD or ORD spectrum, NOEDS spectrum or 2D-NMR, NOESY and ROESY spectrum, X-ray crystal diffraction analysis, etc.

6.3.3 Quantitative Calculation

Quantitative calculations mainly include UV-visible spectrophotometry and various chromatographic methods. Liu Hui [139] determined the content of terpene lactone by HPLC-ELSD method. Cui Daming [140] determined the content of terpene lactones in the liposomes of *Ginkgo biloba* extract by RP-HPLC. In addition, Xu Wen [130] established ultraviolet-visible spectrophotometry and high-performance liquid chromatography to quantitatively determine the content of triterpenes in the extract of Alisma orientalis.

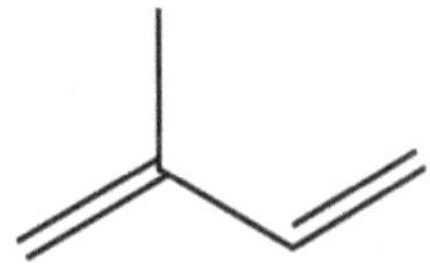

Fig. 6.4 The structure of isoprene units

7 Conclusion

The crude extraction of natural compounds is no longer just a simple solvent extraction, often requires auxiliary technical means. In the purification and identification of natural compounds, chromatography and spectroscopy play an important role. Natural hypoglycemic compounds have many types and complex structures, and most of the products are crude products. Therefore, understanding and mastering the separation and purification technology of hypoglycemic compounds is of great practical significance for making full use of the advantages of Chinese natural resources and effectively developing active ingredients contained in natural resources. A full grasp of the structure analysis methods of hypoglycemic compounds is benificial to the further understanding of the physical and chemical properties of the compounds and studying more hypoglycemic drugs.

References

1. Katiyar D, Singh V, Gilani SJ, Goel R, Grover P, Vats A (2015) Hypoglycemic herbs and their polyherbal formulations: a comprehensive review. Med Chem Res 24(1):1–21. https://doi.org/10.1007/s00044-014-1080-3
2. Wang LQ (2013) Establishing the discovery and evaluation system for development of α-glucosidase inhibitors from natural products. PhD, NanKai University
3. Gale EAM (2001) The discovery of Type 1 diabetes. Diabetes 50(2):217–226
4. Lozano R, Naghavi M, Foreman K, Lim S, Shibuya K (2012) Global and regional mortality from 235 causes of death for 20 age groups in 1990 and 2010: a systematic analysis for the Global Burden of Disease Study 2010. Lancet 380(9859):2095–2128. https://doi.org/10.1016/s0140-6736(12)61728-0
5. Ezuruike UF, Prieto JM (2014) The use of plants in the traditional management of diabetes in Nigeria: pharmacological and toxicological considerations. J Ethnopharmacol 155(2):857–924. https://doi.org/10.1016/j.jep.2014.05.055
6. Wu GJ, Liu D, Wan YJ, Huang XJ, Nie SP (2019) Comparison of hypoglycemic effects of polysaccharides from four legume species. Food Hydrocoll 90:299–304. https://doi.org/10.1016/j.foodhyd.2018.12.035
7. Rotenstein LS, Kozak BM, Shivers JP, Yarchoan M, Close J, Close KL (2012) The ideal diabetes therapy: what will it look like? How close are we? Clin Diabetes 30(2):44–53
8. WHO (2013) WHO traditional medicine strategy: 2014–2023
9. Luo HS (2013) Natural chemical composition for treating diabetes and its hypoglycemic mechanism. Sci Technol Inf 000(022):108
10. Lee SS, Lin HC, Chen CK (2008) Acylated flavonol monorhamnosides, a-glucosidase inhibitors, from *Machilus philippinensis*. Phytochemistry 69(12):2347–2353
11. Xie JH, Jin ML, Morris GA, Zha XQ, Chen HQ, Yi Y et al (2016) Advances on bioactive polysaccharides from medicinal plants. Crit Rev Food Sci Nutr 56:S60–S84
12. Yu Y, Shen MY, Song QQ, Xie JH (2018) Biological activities and pharmaceutical applications of polysaccharide from natural resources: a review. Carbohydr Polym 183:91–101. https://doi.org/10.1016/j.carbpol.2017.12.009
13. Li SP, Zhang GH, Zeng Q, Huang ZG, Wang YT, Dong TTX et al (2006) Hypoglycemic activity of polysaccharide, with antioxidation, isolated from cultured Cordyceps mycelia. Phytomedicine 13:428–433

14. Chen Y (2010) Hypoglycemic activity of ginseng aqueous extract. Northeast Normal University
15. Wu JJ, Shi SS, Wang HJ, Wang SC (2016) Mechanisms underlying the effect of polysaccharides in the treatment of type 2 diabetes: a review. Carbohydr Polym 144:474–494
16. Hou XT (2014) Research on chemical constituents and pharmacodynamics of sugarcane leaves. Guangxi Medical University
17. Wang F, Bao YF, Zhang Y, Weng ZB, Zhou ZJ, Shen XC (2018) Isolation and purification of polysaccharides from *Moringa oleifera* leaves and in vitro hypoglycemic activity. J Food Saf Qual (7):1592–1598
18. Wang HY, Liu Q, Dong YP (2018) Advances in studies on corn whisker polysaccharide and its hypoglycemic function. J Beijing Vocat Coll Agric 32(1)
19. Chen C, You LJ, Abbasi AM, Fu X, Liu RH (2015) Optimization for ultrasound extraction of polysaccharides from mulberry fruits with antioxidant and hyperglycemic activity in vitro. Carbohydr Polym:130
20. Zhong LX, Jiang ZY, Wang JN, Li XF, Xu LS (2019) Optimization of the extraction process of hawthorn polysaccharide and its hypoglycemic and hypolipidemic activity. Sci Technol Food Ind 40(13):119–124+147
21. Li QL (2019) Preparation of *Moringa oleifera* leaf extract, evaluation of its hypoglycemic effect and structure characterization of the main hypoglycemic ingredient. South China University of Technology
22. Ren ML (2012) Study on isolation, preparation and biological activity of bamboo leaves polysaccharides. Beijing Forestry University
23. Zhang ZM, Zhang YJ, Jin H, Li J, Zhu LY, Yao HY (2011) Response surface method to optimize the extraction process of pumpkin polysaccharide and its hypoglycemic effect. China Food Addit (01):45–49
24. Ma CH (2013) Study on efficient separation technology of active ingredients from *Schisandra chinensis*(Turcz.) *Baill.* and multistage utilization of medicinal resources. Northeast Forestry University
25. Zeng SM, Zhou X, Wu FH (2013) Optimization of extraction process of Fuqijiangtang granule. Acad J Guangdong Coll Pharm 029(1):43–46
26. Zhao Y, Xing JG, Wang XC, Xue GP, Dang XL (2010) Optimization of extraction process for compund kaliziran gel with orthogonal design. China J Chin Materia Medica (09):44–47. https://doi.org/10.4268/cjcmm20100909
27. Sun HF, Sun HF, Guo LQ, Guo XY, Cao L, Li JW (2011) Research on uniform design to optimize the extraction process of *Inonotus obliquus* polysaccharide. World Sci Technol; Modernization Tradit Chin Med (01):133–137
28. Zou S, Xu Y, Zhang Q (2015) Review on extraction and purification technology of polysaccharides from natural plants. Nat Prod Res Dev (8):181–189
29. Chen YY (2010) Purification, structure and anti-tumor mechanism of polysaccharides from *Inonotus obliqus*. Jiangnan University
30. Wang JL, Du WJ, Wang Q (2010) Comparative study of deproteinization process of exo-polysaccharide from Inonotus obliquus. J Jilin Agric Univ (6):35–40
31. Yu XR (2012) Study on extraction, separation, purification and the hypoglycemic activity of *Morus Nigra Linn* in Xinjiang. Xinjiang Medical University
32. Yang Q, Li XH, Wang LQ, Lu F, Zhang CH, Zheng YY, Xie H (2013) Physicochemical properties and antioxidant activity of Ginkgo Biloba l. Nut Mod Food Sci Technol (10):77–82
33. Gong YC (2010) Study on the hypoglycemic and hypolipidemic activities of *Mulberry* leaves and product development. Zhejiang Gongshang University
34. Gong GZ, Zhang XJ (2010) Decolorization of polysaccharide extracted from Eucommia ulmoides leaves. Guizhou Agric Sci 038(003):42–45
35. Alain M (2014) Extraction and fractionation of polysaccharides and B-phycoerythrin from the microalga *Porphyridium cruentum* by membrane technology. Algal Res 5:258–263

36. Li Q, Li WZ, Gao Q, Zou Y (2017) Hypoglycemic effect of Chinese yam (*Dioscorea opposita rhizoma*) polysaccharide in different structure and molecular weight. J Food Sci 82 (10–12):2487–2494
37. Zou S, Xu Y, Zhang Q (2015) Review on extraction and purification technology of polysaccharides from natural plants. Nat Prod Res Dev 27(08):181–189
38. Zhao H (2014) Purification, characterization and immunomodulatory effects of *Plantago depressa* polysaccharides. Carbohydr Polym 112:63–72
39. Xie MY, Yin JY, Nie SP (2017) Research progress on structural characterization of polysaccharides from natural resources. Chin J Food Sci 017(3):1–19
40. Yin J, Lin H, Li J (2012) Structural characterization of a highly branched polysaccharide from the seeds of *Plantago asiatica L.* Carbohydr Polym 87(4):2416–2424
41. Yang Q (2013) A study on speciality about extraction, separation and function of Ginkgo polysaccharides. PhD, ShenYang Agricultural University
42. Wu YL, Wang YL, Yang L (2011) Application of SEC - RI - MALLS technique on the study of lentinan molecular weight and distribution analysis of molecular weight. J Pharm Anal 031 (12):2256–2259
43. Liu YH, Tang X, Li YL, Zou Q, Wu XP, Chen T (2020) A review about the research of extraction, purification and structure analysis of active polysaccharide. Chin Natl Folk Med 29 (03):67–73
44. Hakomorl SI (1964) A rapid permethylation of glycolipid, and polysaccharide catalyzed by methylsulfinyl carbanion in dimethyl sulfoxide. J Biochem 55(205–208)
45. Harris P, Henry R, Blakeney AB (1984) An improved procedure for the methylation analysis of oligosaccharides and polysaccharides. Carbohyd Res 127(1):59–73
46. Needs PW, Selvendran RR (1993) Avoiding oxidative degradation during sodium hydroxide/methyl iodide- mediated carbohydrate methylation in dimethyl sulfoxide. Carbohyd Res 245 (1):1–10
47. Nie QX, Hu JL, Gao H, Fan LL, Chen HH, Nie SP (2019) Polysaccharide from Plantago asiatica L. attenuates hyperglycemia, hyperlipidemia and affects colon microbiota in type 2 diabetic rats. Food Hydrocoll 86:34–42. https://doi.org/10.1016/j.foodhyd.2017.12.026
48. Wu LQ, Jiang ZF, Zhang J (2012) Chromatographic analysis and atomic force microscope observation of polysaccharide extracted from Atractylodes macrocephala Koidz. Nat Prod Res Dev 024(5):631–634,609
49. Wang XM (2013) The effect of ultrasound on the structure, solution behavior and biological activity of *Ophiopogon japonicus* polysaccharide. Shaanxi Normal University
50. Chen XQ, Zhou Y, Ye Y, Cheng H, Yin JF (2008) Extraction and analysis of alkali-soluble tea polysaccharides. Chin J Appl Chem (12):1496–1498
51. Babu PVA, Liu DM (2009) In complementary and alternative therapies and the aging population
52. Sarian MN, Ahmed QU, Mat So'ad SZ, Alhassan AM, Murugesu S, Perumal V, Syed Mohamad SNA, Khatib A, Latip J (2017) Antioxidant and antidiabetic effects of flavonoids: a structure-activity relationship based study. Biomed Res Int 2017:8386065. https://doi.org/10.1155/2017/8386065
53. Khoza BS, Dubery IA, Byth-illing HA (2016) Optimization of pressurized hot water extraction of flavonoids from *Momordica foetida*, using UHPLC-qTOF-MS and multivariate chemometric approaches. Food Anal Methods 9(6):1480–1489
54. Piao CH, Liu LP, Chu Q, Zhou YN, Chai H, Wang YH (2014) Optimization of extraction technology of flavones in buckwheat shell with hot-water method and their antioxidant activity. J Jilin Agric Univ (6):92–95+107
55. Xiao GS, Wan LX, Xu YJ (2013) Study on the hypoglycemic effect of citrus peel flavonoids. Mod Food Sci Technol (4):10–13+91
56. Bao T, Wang Y, Sun C, Niu X, Yang H, Chen L, Su X, Chen W (2016) Optimization of flavonoids extraction from Tartary buckwheat rice and analysis of its hypoglycemic activity. Trans Chin Soc Agric Eng

57. Xuan GS, Pan SJ, Nan J (2011) Hypoglycemic effect of bioactive components in Mulberry leaves. Food Sci 000(007):323–326
58. Ji LL, Wang KY, Luo XB, Zhao L, Huang XL (2015) Optimization of extraction of total flavones from *Moriga oleifera* leaves by response surface methodology and its in vitro hypoglycemic activity. Nat Prod Res Dev (6):45–51
59. Li QZ (2015) Structure analysis of flavonoids of lotus seed heart and its hypoglycemic activity. Fujian Agriculture and Forestry University
60. Li Y, Zhou JZ, Wang G (2008) Study on extraction of flavonoids from Ginkgo biloba by ultrasonic and microwave. Food Sci Technol 6:153–155
61. Zhang SB, Ye J, Chen C (2012) Optimizing conditions for enzymatic extraction of flavonoids from lotus leaves. Food Sci (22):156–160
62. Wang ZY, Qu GX (2013) Optimization of flavonoids extraction from Opium graveolens L by supercritical CO2. Food Ferment Ind (04):242–246
63. Liu JX, Wang SX, Fan QS (2008) Extraction of total flavonoids from Ginkgo biloba leaves by alkaline solution acid precipitation. Anhui Agric Sci 36
64. Yin YG, Yan LN, Tao Y (2009) Extraction of flavonoids from pine needle by high intensity pulsed electric fields(PEF)and its in vitro anti-oxidative effect. Nat Prod Res Dev (6):124–127 +148
65. Gao F, Xu LP (2012) Membrane technology in the separation and purification of lotus leaf flavonoids. Liaoning Chem Ind 41(9):875–877
66. Sun T, Deng AX, Li ZH (2012) Study on the chemical constituents of flavonoids in Huangqi leaves. Chin J Pharm 47(20):1617–1620
67. Zhang P (2019) Study on extraction and purification of flavonoids, the active component of mulberry leaves in traditional Chinese medicine. Huazhong University of Science and Technology
68. Weng XD (2016) Study on the effect of mulberry root flavonoids on lowering blood sugar and its mechanism. Guangdong Pharmaceutical University
69. Ma L (2015) Study on extraction, separation and purification of flavonoids in Ginkgo leaves. Changan University
70. Lu Q (2015) Separation, purification, identification and hypoglycemic activity of phenolic substances in litchi pulp. Zhejiang University
71. Higuchi M (2014) In wheat and rice in disease prevention and health
72. Santos EL, Maia BHLNS, Ferriani AP, Teixeira SD (2017) Flavonoids: classification, biosynthesis and chemical ecology. Flavonoids - from biosynthesis to human health. https://doi.org/10.5772/67861
73. Saptarini NM, Herawati IE (2019) The colorimetric method for determination of total alkaloids and flavonoids content in Indonesian black nightshade (*Solanum nigrum* L.). J Adv Pharm Educ Res
74. Wu YL, Lei RH, Wen YZ (2009) The influence of general flavone extraction yield in Balsampear leaf under extraction condition. J Xiangnan Univ (5):70–72
75. Guo YH, Li J, Huang G, Wu WX, Bai JX, Zhang CH, Xiong JY (2020) Water extraction process and physicochemical properties of flavonoids in hawthorn. J Jianghan Univ (Nat Sci Ed) 48(01):29–36
76. Sun Y, Guo QX, Tong QY (2015) Study on hypoglycemic activities of flavonoids of Brassica napus L. bee pollen in vitro. Sci Technol Food Ind (12):117–121
77. Y JW (2015) Study on hypoglycemic components of Fenugreek and its hypoglycemic mechanism. Jilin University
78. Gong W (2015) Extraction and separation of effective components from Hawthorn leaves and its application. Chinese Academy of Agricultural Sciences
79. Zheng R (2014) A preliminary study about the hypoglycemic active ingredients and mechanisms of the Syringa veutina Kom leaves. Liaoning Medical College
80. Zhang Y, Guan J, Cui J, Zhao Y (2010) Quantitative analysis of aglycone of momordicoside L from Momordica charantia in different areas by HPLC. Chin J Tradit Chin Med 35(5):620–622

81. Ghashghaie J, Tcherkez G (2013) Metabolomics coming of age with its technological diversity. Adv Bot Res
82. Chen HY (2012) Isolation and antihyperglycemic effect of chemical constituents from *Rhizoma coptidis*. Southwest University
83. Zhang DH, Huang RS, Zuo L (2010) Progress on the method of extraction of alkaloids. Chin Wild Plant Resour 029(5):15–20
84. Zhang DH, Huang RS, Zuo L (2014) Alkaloids from lotus leaf and related bioactivities. Chin Patent Med 036(011):2330–2333
85. Xiao GQ, Long LP, Wang JL, Li H (2010) Comparison of microwave, ultrasonic wave and their combined extraction technology of the total alkaloid from Coptis Chinensis Franch. Chin J Spectrosc Lab (3):62–67
86. Tong RF, Zhang QS (2010) Optimization of process of alkaloids from Rhizoma Coptidis by supercritical CO2 extraction. Tianjin Pharm 22(05):71–73
87. Yuan PL, Chen L, Liu XY, Yu BW, Huang Q, Chen KX, Li YM, Jia Q (2014) Alkaloids from lotus leaf and related bioactivities. Chin Tradit Patent Med 036(011):2330–2333
88. Zhang DH, Huang RS, Zuo L, Geng ZH (2010) Progress on the method of extraction of alkaloids. Chin Wild Plant Resour 029(5):15–20
89. Su N (2014) Isolation and purification of Mulberry hypoglycemic active substance DNJ and the study about its pharmacological activity. Guangzhou University of Chinese Medicine
90. Pan SZ, Dan F, Fan JM, Jiang YH (2019) Optimization of extraction of alkaloid salts from lotus leaves by response surface methodology and purification. Fine Chem 36(8)
91. Liu GY, Xu X, Ye Q (2013) The effect of ultrafiltration and nanofiltration membrane technology on the separation and concentration of 1-deoxynojirimycin in Mulberry leaves was observed. J Culinary Sci Yangzhou Univ 30(04):55–60
92. Wang T (2013) Study on separation of alkaloids by rosin-based polymer imprinted column. Guangxi University for Nationalities
93. Richard T, Temsamani H, Cantos-Villar E, Monti J-P (2013) Chapter Two - application of LC–MS and LC–NMR techniques for secondary metabolite identification. In: Rolin D (ed) Advances in botanical research, vol 67. Academic Press, pp 67–98. https://doi.org/10.1016/B978-0-12-397922-3.00002-2
94. Christian GD, Dasgupta PK, Schug KA (2014) Analytical chemistry, 7th edn. Wiley, USA
95. Pan SJ (2011) Extraction and pharmacological investigation of hypoglycemic ingredients from Mulberryleaf and Balsam pear. Qingdao University of Science and Technology
96. Zheng XW, Xi NB, Wu YS, Lv Y (2005) The joint application of chromatography and x-ray diffraction technology in the identification of Chinese patent medicine. In: Proceedings of the 8th national symposium on traditional Chinese medicine and natural medicine and the 5th National Symposium on medicinal plants and phytopharmaceutics
97. Kim JW, Kim SU, Lee HS (2003) Determination of 1-DNJ in Morus alba L. leaves by derivation with 9-fluorenylmethyl chloroformate followed by reversed-phase high-performance liquid chromatography. J Chromatogr A 1002:93–99
98. Kimura T, Nakagawa K, Saito Y, Yamagishi K (2004) Determination of 1-deoxynojirimycin in mulberry leaves using hydrophilie interaction chromatography with evaporative light seattering detection. Agric Food Chem 52:1415–1418
99. He CJ, Peng F, Liu T (2018) Analysis of the chemical composition of Tianqi Jiangtang capsules based on UPLC-LTQ Orbitrap HRMS. Chin Herb Med 620(09):40–46
100. Sheng HX, Zhang JL, Sun SQ, Xu JT (2006) Quality analysis and evaluation of Rhizoma Coptidis under different cultivation conditions. Acta Pharm Sin (10):99–103
101. Liu F, Li PP, Liao ST, Xiao GS (2016) Determination of total alkaloids and 1-deoxynojirimycin in 98 Mulberry variety resources. China Sci Technol J 38(02)
102. Principles of herbal pharmacology (2013) In: Principles and practice of phytotherapy, 2nd edn
103. Rui W, Wu Y, Ji X, Meng Q, Feng YF (2013) Optimization of extraction process and component analysis of hypoglycemic effective parts from Anemarrhenae Rhizoma. Chin J Exp Tradit Med Formulae 19(22):4–8

104. Zhu HY, Lin HC, Yang H (2014) Study on the extraction process of total saponins of Choerospondias axillaris by central combination design-response surface method. J Food Saf Qual Inspection (11)
105. Li YL (2015) Dynamic microwave-assisted extraction of ginsenoside and polysaccharide from Ginseng. Wuhan Institute of Technology
106. Li YL, Zhang YF, Liu Z, Liu D, Chi NA (2015) Optimization of microwave assisted extraction of Ginsenosides and Polysac-charide by response surface methodology. China Med Her (16):48–53
107. Yu B, Ge BG, Song Y, Li DD, Ma XY, Wu MY (2013) Ultrasonic wave-surfactant synergistic extraction of momordica charantia L. J Agric Eng (15):302–308
108. Fan HX, Liu TT, Liu HC, Cui HY, Wang DW (2013) Supercritical extraction of ginsenoside and HPLC analysis. Food Sci 34(20):121–126
109. Zhang Y (2011) Study on the extraction of ginsenosides by Agaricus bisporus. Heilongjiang Sci Technol Inf (36):2–2
110. Chen X, Hu ZQ, Zhang HC, Sun Y (2012) Primary research on biomimetic extraction of ginsenoside saponins. China Pharm (19):30–32
111. Wang YS, Jin YP, Zheng PH, Liu JX, Xiao SY, Wang YP, Yuan ZJ (2014) Distribution relationship of ginsenosides in n-butanol phase and water phase. Special Wild Econ Anim Plant Res (4):26–31
112. Zhao YN, Wang ZL, Dai JG (2014) Preparation and quality assessment of high-purity ginseng total saponins by ion exchange resin combined with macroporous adsorption resin separation. Chin J Nat Med 2(5):382–392
113. Yao HY, Wang YH, Shen YJ, Li XY, Liu MP, Li G, Meng XC (2013) Study on purification process of ginsenosides with macroreticular resin YAO. Glob Tradit Chin Med (2):10–14
114. Ji HF, Li J, Zhang LW (2009) Purifying technology of the total saponins in bitter melon with solvent deposition method. Food Sci Technol 34(4):204–207
115. Zhang R (2012) Application of foam floatation and solid phase extraction in seperation and concentration of ginsenosides and hormones. Jilin University
116. Hui G (2010) The study on the interaction between ginsenoside molecules and biomembranes. Jilin University
117. Commission SP (2015) Pharmacopoeia of the People's Republic of China. China Medical Science and Technology Press, Beijing, pp 391–392
118. Dong W, Hao YG, He ZM, Zhao Y, Zhang LX (2014) Research progress in extraction and separation technology of Ginsenosides. Sci Technol Food Ind 035(17):366–369
119. Savage GP (2003) SAPONINS. In: Caballero B (ed) Encyclopedia of food sciences and nutrition, 2nd edn. Academic Press, Oxford, pp 5095–5098. https://doi.org/10.1016/B0-12-227055-X/01050-6
120. Boysen RI, Hearn MTW (2010) Modern methods in natural products chemistry. In: Comprehensive Natural Products II
121. Macrae R, Robinson RK, Sadler MJ (1993) Saponins, encyclopaedia of food science, food technology and nutrition. Academic Press
122. Zhao WT (2015) The extraction, purification, antioxidant and immunoenhancementin Chenopodium Quinoa WIlld Bran total saponins. Shanxi Agricultural University
123. Liu JF, Zhang PP, Zhang WJ, Zhang YC (2010) Study on fingerprints of saponins in Momordica Charantia L. by HPLC. Food Res Dev 31(6):76–80
124. Ding FF, Deng YR, Zhang XJ, Sun Y, Guo LP (2015) Simultaneous determination of five astragalosides in astragali radix and jinqi jiangtang tablet by HPLC-ELSD. J Chin Med Mater 038(1):156–159
125. Du ZC, Xia ZS, Wei M, Zhang MZ, Wei DT, Chen ZM, Wei LY, Hou XT, Deng JG (2019) Research on quality control of gesang jiangtang capsules based on UPLC and colorimetry. Chin Herb Med 50(21):5266–5272

126. Bahrami Y, Zhang W, Chataway T, Franco C (2014) Structure elucidation of five novel isomeric saponins from the viscera of the sea cucumber Holothuria lessoni. Mar Drugs 12 (8):4439–4473
127. Hu XA, Li X, Hu MM, Xu SM, Qiao RX, Ma PF (2019) Preparation and identification of gypenoside XLIX reference substance. Central South Pharmacy
128. Ludwiczuk A, Skalicka-Woźniak K, Georgiev MI (2017) Chapter 11 - Terpenoids. In: Badal S, Delgoda R (eds) Pharmacognosy. Academic Press, Boston, pp 233–266. https://doi.org/10.1016/B978-0-12-802104-0.00011-1
129. Bian GK, Ma T, Liu TG (2018) Chapter Five - in vivo platforms for terpenoid overproduction and the generation of chemical diversity. In: Methods in enzymology. https://doi.org/10.1016/bs.mie.2018.04.025
130. Xu W, Xu JT, Zhao WL, Qiu JF, Li XY, Luo FR, Huang MQ, Wu SS (2014) Study on the determination of triterpenoids in *Alisma orientale* extract. In: Academic Conference of Chinese Medicine Branch of Fujian Chinese Medicine Society
131. Liang WL, Xie DW, Ding G, Xu DH, Sun YC, Lian YP, Li YJ, Xiao W (2015) Quantitative and qualitative evaluation on tablets of Ginkgo biloba leaves using fingerprint and LC-MS analysis. China J Chin Materia Medica 40(9):1738
132. Liang J, Sun MY, Zhang T, Pei HH, Zhang MM, Fan GZ (2011) Optimization of extracting triterpene from phellinus igniarius by response surface methodology. Chin Agric Sci Bull 27 (10):235–238
133. Lu H, Jia XB, Shi F, Chen Y, Tan XB (2011) Supercritical CO_2 extraction of triterpenoids from Hedyotis Diffusa Willd. Chin Tradit Patent Med 33(8):1429–1432
134. Huang YZ, Luan JB (2014) Advances in techniques of extracting and analyzing plant terpenoids. Acta Agriculturae Shanghai (3):135–141
135. Li K (2013) Isolation and purification of triterpene chemical constituents in *Poria cocos*, structure identification and fingerprint analysis of *Poria cocos*. Hubei University of Traditional Chinese Medicine
136. Meng DL, Ji S, Zhang YC, Li N, Li X (2009) Isolation and identification of terpenoids and saccharides from Achyranthese bidentata Bl. J Shenyang Pharm Univ (5):348–352
137. Li JM (2015) Isolation, structures, and antitumor activity of Triterpenoids from *Pleurotus eryngii*. Tianjin University
138. Qu JH (2014) Separation and composition analysis of triterpenoids from *Pyrus ussuriensis* Maxim cv. Anli. Hebei Normal University of Science and Technology
139. Liu H, Huang YW, Guo Y, Xiong Y (2015) Determination of flavonol glycosides and terpene lactones in Ginkgo biloba extract from different pharmaceutical factories by HPLC. China J Tradit Chin Med Pharm (2)
140. Cui DM, Jin Y, Zhou YL, Li YH, Guo X, Teng LR, Zhang WY (2013) Quantitative determination of flavonol glycosides and terpene lactones from ginkgo leaf extract noisome. J Jilin Agric Univ 35(4)

Chapter 7
Structural Characterization and Health Effects of Polysaccharides from *Momordica charantia* on Diabetes Mellitus

Xuan Liu, Mingyue Shen, Rong Huang, and Jianhua Xie

Abstract *Momordica charantia* L. *(M. charantia),* known as an edible and medicinal crop, which is widely distributed in Asian countries. Recent research has demonstrated that extracts of *M. charantia* can effectively reduce glucose tolerance and the glucose content in serum, while polysaccharide was one of the main bioactive substances of *M. charantia, which* has various important biological activities, such as hypoglycemia, immunomodulatory, antioxidant, and antitumor activities, and more attention has been focused on studying its hypoglycemic activity. Moreover, it's necessary to isolate and identify the physical and chemical features of *M. charantia* polysaccharides as they play an important role in figuring out its hypoglycemic mechanism. This article is aimed at summarizing previous and current studies of the physical and chemical characterization, hypoglycemic activity, as well as hypoglycemic mechanism of *M. charantia* polysaccharides. The review will provide a useful reference material for further investigation and application of *M. charantia* polysaccharides in functional foods and medicine fields.

Keywords *Momordica charantia* · Polysaccharide · Hypoglycemic · Mechanism

1 Introduction

Hyperglycemia develops in conditions in which there is a low net insulin action. A severe degree of hyperglycemia may develop in these conditions if there is an obvious reduction in the glomerular filtration rate or a large intake of glucose. Diabetes mellitus (DM) is a chronic metabolic disease resulting from insufficient insulin production or tissue and organ insulin resistance, which will lead to hyperglycemia, hyperlipidemia, and hyperinsulinemia [1, 2]. In general, DM is classified into three types: type 1 (T1DM), type 2 (T2DM), and gestational DM (GDM). Only

X. Liu · M. Shen · R. Huang · J. Xie (✉)
State Key Laboratory of Food Science and Technology, Nanchang University, Nanchang, China
e-mail: jhxie@ncu.edu.cn

H. Chen, M. Zhang (eds.), *Structure and Health Effects of Natural Products on Diabetes Mellitus*, https://doi.org/10.1007/978-981-15-8791-7_7

about 10% of diabetes are T1DM and GDM, the remaining 90% are T2DM [3]. T1DM, which also called insulin-dependent DM, characterized by chronic insulin deficiency and hyperglycemia that develops due to autoimmune or other destruction of the pancreatic beta cells [4]. And data showed the incidence of T1DM was increasing. T2DM, characterized by insulin resistance (IR) and hyperglycemia, would cause a series of chronic complications, such as lipidic abnormality, cardiovascular disease, hepatic and renal failure [4–6]. Therefore, it's important to control blood glucose at a normal level for type 2 diabetes patients. It has reported that polysaccharides can reduce the risk of T2DM, such as astragalus polysaccharides [7], Ganoderma lucidum polysaccharides [8], azuki bean polysaccharide [9]. This review mainly focused on T2DM. The life quality of patients will be seriously influenced by a series of complications associated with diabetes, such as diabetic foot [10], atherosclerotic arterial disease [11], and diabetic nephropathy [12]. There are many conventional clinical medications used to treat diabetes, including sulfonylureas, thiazolidinediones, biguanides, alpha-glucosidase inhibitor [13], and insulin secretagogues, but most of them have negative effects [14] and diabetes is a chronic disease. Therefore, a more efficient, less negative effect and inexpensive treatment is urgently needed.

Recently, scientists have discovered a variety of active molecules from natural products. Natural polysaccharides have become a new research hotspot for their high bioactivity and low toxicity, including animal polysaccharides, plant polysaccharides, fungal polysaccharides, such as sea cucumber polysaccharide [15], pumpkin polysaccharide [16], hericium erinaceus polysaccharide [17]. It's reported that polysaccharides have great potential in the treatment of DM [18].

M. charantia is one of the representative plants of the *Cucurbitaceae* family and is widely planted in South Asia, Southeast Asia, China, and other places [19–21]. *M. charantia* has a long history of edible and medicinal use in China and has long been proven to be effective in treating diabetes [22]. Many studies have shown that *M. charantia* has a variety of biologically active ingredients, such as saponins, alkaloids, quinine-like proteins, polysaccharides, flavonoids, amino acids, fatty acids, and trace elements, etc. [22–24]. Among them, *M. charantia polysaccharides (MCP) have been* received growing attention for their diverse bioactivity, such as lowering blood sugar [11, 24, 25] and blood lipids [12], anti-oxidation [26, 27], anti-inflammatory [28], improving human immune regulation [26], and antitumor [22], the most significant is blood glucose-lowering activity.

The bioactivities *and* chemical structures of *polysaccharides are closely connected* [24, 29]. Therefore, it's necessary to figure out the hypoglycemic mechanism of *M. charantia* polysaccharides and the structural features of *M. charantia polysaccharide*, such as molecular weights, monosaccharide compositions, types of glycosyl linkage, types and polymerization degree of the branch.

2 Procedure of Extraction and Purification

Many studies are committed to find the best way to extract polysaccharides for further study their physicochemical characterizations and biological activities. Traditional hot water extraction, acid extraction, alkaline extraction, and microwave-assisted extraction are widely used in polysaccharide extraction [29]. The extraction efficiency is strongly related to extraction methods. For example, It is reported that the single factor test, orthogonal test design, and Box-Behnken design (BBD) were used to optimize the enzymolysis-ultrasonic assisted extraction (EUAE) process to study the simultaneous application of cellulose, pectinase, and trypsin in the extraction of MCP [30]. And the optimized extraction conditions were as follows: pH, 3.87; temperature, 52.02 °C; extraction time, 36.87 min, and the yield of MCP was 29.75 ± 0.48% under this condition.

Water extraction and alcohol preparation method were used in other study [31], after a series of deproteinization and depigmentation, crude polysaccharides were further purified through an ion-exchange chromatography (DEAE-32 column), which was used to separate neutral polysaccharide by gradient salt elution. The purified *M. charantia* polysaccharide was named PMC I. Moreover, the cellulase and neutral protease were used in the extraction to make the extraction process conditions milder, the yield of PMC I was significantly increased. Zhou et al. [32] also found that a crude polysaccharide of *M. charantia* was prepared by an enzymatic method with the addition of citrate buffer and cellulase, and then purified by DEAE-52 and dextran gel G-100 column chromatography to obtain a single component *M. charantia* polysaccharide (MCP I) of *M. charantia* polysaccharides with antioxidant activity. Chen et al. [33] concluded that the active polysaccharide MCP IIa was obtained by enzymolysis combined with water-soluble alcohol precipitation to extract water-soluble MCP which was purified by DEAE-52 cellulose anion-exchange and Sephadex G-100 gel filtration chromatography, and the protein was removed by the Sevage method. The same polysaccharide MCP IIa extraction and purification method is also used in Zhang's research [34]. Moreover, Ru et al. [24] selenized MCP IIa in order to obtain a selenylated *M. charantia* polysaccharide (Se-MCP II a), and further isolated by Sephadex G-100 column to obtain a single fraction named Se-MCP II a-1. And Sepharose-6B was used to purify the crude water-soluble pectic polysaccharide (PS) that was isolated from the green fruits of *M. charantia* [26]. Deng et al. [35] obtained two main components from MCP, named MCP1 and MCP2, which were purified by DEAE-52 cellulose anion-exchange chromatography.

3 Structural Features

Many studies showed that the biological activity of polysaccharides is closely related to their complex structural features, including relative molecular mass and purity, monosaccharide composition and type and ratio of uronic acid, type of glycosyl linkage, type of loop, linking order, linking site [36]. Many different polysaccharides isolated from *M. charantia* are shown in Table 7.1, including monosaccharide composition, molecular weight, structural characteristics and biological activity. Panda et al. studied PS in order to further clarify its structural characteristics and biological activity. Structural analysis indicated that the repeating unit of the pectin polysaccharides from *M. charantia* contained a backbone of four (1 → 4)–linked D-methyl galacturonate residues, out of which one residue was branched at O-2 position with terminal β-D-galactopyranosyl residue [26]. An acidic heteropolysaccharide *M. charantia* bioactive polysaccharide (MCBP) that isolated from *M. charantia* was constructed by Man, GalA, Rha, Glu, Gal, Xyl, Ara, with a molar ratio of 0.010: 0.145: 0.021: 0.383: 0.306: 0.049: 0.086. Its backbone consists of glucose and galactose, but its branches consist of rhamnose and arabinose [27]. The MCPIIa, isolated from *M. charantia* fruit, was studied by high-performance gel permeation chromatography. It found that MCPIIa was a homogeneous polysaccharide with a molecular weight of 13.0 kDa, and the composition of Rham, GalA, Gal, Xyl, Ara in a relative molar ratio of 12:3.05:19.89:5.95:56. Infrared (IR) spectrum, nuclear magnetic resonance (NMR), and Congo-red experiment found its existence linked at C1, C2, C3, and C5. A stable β-triple helix conformation was shown in aqueous solution, and a diamond crystal particle under a scanning electron microscope [33].

Subsequently, further research indicated that MCP II a linked to a large number of arabinofuranose, glucuronic acid, and xylopyranosyl residues through β-glycoside bonds [34]. The purified fractions (MCP1 and MCP2) of MCP comprised acid polysaccharides, MCP1 mainly contained glucose and galactose, and MCP2 contained glucose, mannose, and galactose. The Fourier transform-infrared (FTIR) spectroscopy indicated that MCP1 and MCP2 were mainly composed of D-glucose residues in the form of pyranose, and both α and β types are present [35]. A homogeneous polysaccharide MCP-A1 component with a molecular weight of 93.6 kDa was isolated and purified from M. charantia fruit. IR analysis showed the existence of a pyranose ring structure, β glycoside bonds, and mannose residues [41].

4 Hypoglycemic Activity

Hyperglycemia is a major feature of diabetes [25], many studies have shown that chronic hyperglycemia in diabetes was associated with long-term damage to various organs and dysfunction, and could easily lead to serious complications [42, 43]. In

Table 7.1 The summary of structure features and biological activities of Momordica charantia L. polysaccharides

No	Compound name	Monosaccharide composition	Molecular weight (Da)	Structural feature	Pharmacological properties	Reference
1	MCP4	Rham: Xyl: Gal: Ara with the molar ratio of 1:0.75:1.83:0.85	9.65×10^4	Acidic polysaccharides containing β-D-pyranoside bonds	–	[37]
2	MCP1	Man: Rha: GlcUA: GalUA: Glc: Gal: Xyl Ara with molar ratios of 1.03: 2.93: 1.00: 14.95: 2.16: 30.70: 2.85: 4.50	1.16×10^6	–	–	[38]
3	MCP2	Rha: GalUA: Gal: Xyl: Ara with respective molar ratios of 1.63:21.88:4.66:1.00:1.29	7.45×10^5	–	–	[38]
4	MCP I a	L-Rha:D-Xyl: D-Fru: D-gal in a molar ratio of 4.4:2.3:1:5.7		Intra- and inter-molecular hydrogen bonds exist, C-H bond stretching vibrations exist in molecules, -OH deformation vibrations, furanoside absorption peaks	Antioxidant	[39]
5	MCP1	Rib: Rha: Ara: Xyl: Man: Glu: Gal with molar ratio of 1.00:6.33:9.07:3.78:4.71:27.28:19.58	8.55×10^4	Sulfated heteropolysaccharides based on α-/β- D-glucopyranose backbone	Immunomodulation	[35]
6	MCP2	Rib: Rha: Ara: Xyl: Man: Glu: Gal with molar ratio of 1.86:1.00:8.92:9.62:34.18:44.20:23.61	4.41×10^5	Acid heteropolysaccharides based on α-/β- D-glucopyranose backbone	Immunomodulation	[35]
7	MCP	Ara: Xyl: Gal: Rha in a ratio of 1.00: 1.12: 4.07: 1.79		–	Antidiabetic, hypoglycemic	[25]
8	PS	D-gal and D-methyl GalU in a molar ratio of 1:4	2.00×10^5	Backbone is [→4)-α-D-GalpA6Me-(1] 3 → 4)-α-D-GalpA6Me-(1 → and the branched at O-2 position with terminal β-D-galactopyranosyl residue	Immunomodulation, antioxidant	[26]

(continued)

Table 7.1 (continued)

No	Compound name	Monosaccharide composition	Molecular weight (Da)	Structural feature	Pharmacological properties	Reference
9	MCBP	Man: GalA: Rha: Glc: Gal: Xyl: Ara with a molar ratio of 0.01: 0.15: 0.02: 0.38: 0.31: 0.05: 0.09	9.20×10^4	Acidic and branched heteropolysaccharide	Antioxidant	[27]
10	MCP-2	Rha: GalA: Gal: Xyl: Ara in a ratio of 1.19: 15.97 3.40: 0.73: 0.94	7.45×10^5	–	Immunomodulation, antitumor	[40]
11	MCP II a	L-Rha: D-GalA: D-gal: D-Xyl: L-Ara in a molar ratio of 12: 3.05: 19.89: 5.95: 56	1.30×10^4	β-Glycosidic linkages to a large number of arabinofuranose, glucuronic acid, and xylose pyran residues, containing β-D-glucopyranose, and the C1, C2 and C3, C5 positions were its binding sites	Antidiabetic, hypoglycemic	[33, 34]
12	MCP II a C	L-Rham: D-Ara: D-Xyl: D-Glc: D-gal: D-man: D-Fuc with a molar ratio of 3.40: 12.80: 4.50:15.20: 59.10: 3.80: 0.98	8.30×10^4	The chromium ions are linked to the polysaccharide's hydroxyl groups, and the flexibility of the three-dimensional structure of the polysaccharide increased	Antidiabetic, hypoglycemic	[6]
13	Se-MCP II a-1	D-man: L-Rha: D-GlucA: D-GalA: D-Glc: D-gal: D-Xyl: D-Ara: D-Fuc in the molar ratios of 1.94:5.81: 1.55:1.53:3.29:6.81: 0.09:76.02: 0.9	4.00×10^6 换成 4.00×10^4	–	Antioxidant, hypoglycemic	[24]

recent years, many scholars have carried out a lot of research on the hypoglycemic activity of polysaccharides. *Some researchers have demonstrated that some polysaccharides had shown hypoglycemic activity in drug-induced diabetic or normal mice, including sulfated rhamnose polysaccharides chromium(III) complex* [44], *hericium erinaceus polysaccharide* [18], *sargassum fusiforme polysaccharides* [45], etc.

M. charantia polysaccharide has various biological activities, its hypoglycemic activity has become a hotspot and has been widely studied in recent years [11, 29, 46]. A water-soluble polysaccharide that isolated from *M. charantia* possessed a remarkable hypoglycemic effect according to the results that it could significantly reduce fasting blood glucose, improve glucose tolerance and prevent body weight loss in alloxan-induction diabetic rats [25]. Another study showed that a water-soluble bitter gourd polysaccharide extracted from bitter gourd to prepared bitter gourd polysaccharide iron complex and proved that the complex has better blood glucose-lowering ability, better protects and repairs β cells, and promotes β-particles and mitochondria regeneration, improving the ability of β cells to release insulin, and significantly promoting blood glucose reduction [47]. Dong and Zhang [46] analyzed the hypoglycemic effect of bitter gourd polysaccharide and its hypoglycemic mechanism. It is speculated that the hypoglycemic mechanism of bitter melon polysaccharide may be performed by attenuating the damage of islet β cells by streptozotocin (STZ) as they found that the bitter gourd polysaccharide could significantly reduce blood glucose and fructosamine values in normal mice and STZ-induced diabetic mice, and increased glucose tolerance and liver glycogen content in diabetic model mice. Xu et al. found that both alkaline soluble polysaccharides (MCB) and water-soluble polysaccharides (MCW) that isolated from *M. charantia* can reduce fasting blood glucose in streptozotocin STZ-induced diabetic mice, and the MCB component has the best effect on reducing blood glucose. The results show that it cannot only significantly reduce fasting blood glucose in diabetic mice, but also help restore damaged pancreatic islet tissue, promote insulin secretion, and increase serum insulin levels in diabetic mice, thereby achieving the purpose of reducing blood glucose [48].

5 Mechanism of the Hypoglycemic Activity

5.1 *Protecting Islet β Cells and Promoting Insulin Secretion*

Islet β cell regulates the blood glucose levels by secreting insulin. Long-term exposure to hyperglycemia will impair the function of pancreatic islet β cell, resulting in absolute or relatively insufficient insulin secretion, and then triggering the onset and development of T2DM [1, 17, 49, 50]. Insulin is the only hormone in the body that reduces blood sugar [51] and plays an important role in maintaining blood sugar homeostasis. Therefore, protecting islet β cell and promoting insulin secretion is necessary to regulate blood sugar.

It was reported that the protective effect of polysaccharides on islet β cells may be related to its anti-diabetic mechanism [52]. *M. charantia* polysaccharides can protect the islet β cells, promote the recovery of diseased islet tissue and insulin secretion, thereby reducing blood sugar [22]. Wang et al. [11] suggested oral administration of *M. charantia* polysaccharides (PMC) could alleviate the STZ-induced organ tissues (kidney and pancreas), indicating the hypoglycemic mechanism of PMC might by repairing the pancreatic β cells in STZ-induced diabetic mouse model. Moreover, when treated with MCP II a, serum insulin levels and blood glucose levels were significantly increased and decreased respectively [34]. And the same result was found in Ru's research [24]. The histological analysis method was used by Zhang et al. [6] and they found that when treated with *M. charantia* polysaccharide-chromium (III) complex (MCP II a C) for 4 weeks, severe pancreatic lesions returned to a more normal appearance, and the number of islet β cells increased. These results indicated that MCP II a C can alleviate pancreatic islet lesions in diabetic mice, promote insulin secretion, and significantly reduce fasting blood glucose levels in diabetic mice in a dose-dependent manner.

5.2 *Increasing Sensitivity to Insulin*

Insulin resistance is a prominent feature of type 2 diabetes. When insulin resistance occurs, normal circulating concentrations of hormones cannot regulate the body's glucose homeostasis [12, 53, 54].

Peroxisome proliferator-activated receptor (PPAR), are a class of steroid hormone receptors, which are the main target molecules for improving insulin resistance [55, 56]. Results showed that *M. charantia* polysaccharides have a strong activation effect on PPAR [57]. It's reported that *M. charantia* polysaccharide could improve insulin resistance by activating PI3K-AKT pathway according to the mRNA expression levels of relative genes. Moreover, it could inhibit JNK pathway thus decreasing the oxidative stress level [58]. Fan et al. [59] found that *M. charantia* polysaccharide could stimulate glucose transporter 4 (GLUT4) to migrate to the cell membrane and mediate glucose uptake, and the activity of adenosine monophosphate-activated protein kinase (AMPK). The related mechanism is illustrated in Fig. 7.1 to understand the connection of insulin resistance with hyperglycemia under the action of *M. charantia* polysaccharides.

5.3 *Inhibiting the Activities of α-Amylase and α-Glucosidase*

α-Amylase and α-glycosidase are two key carbohydrate hydrolase enzymes in the human body. Their main function is to convert digestible carbohydrates into monosaccharides, which facilitates the blood absorption of sugars in the small intestine [60, 61]. Therefore, inhibiting the activity of α-amylase and α-glucosidase can

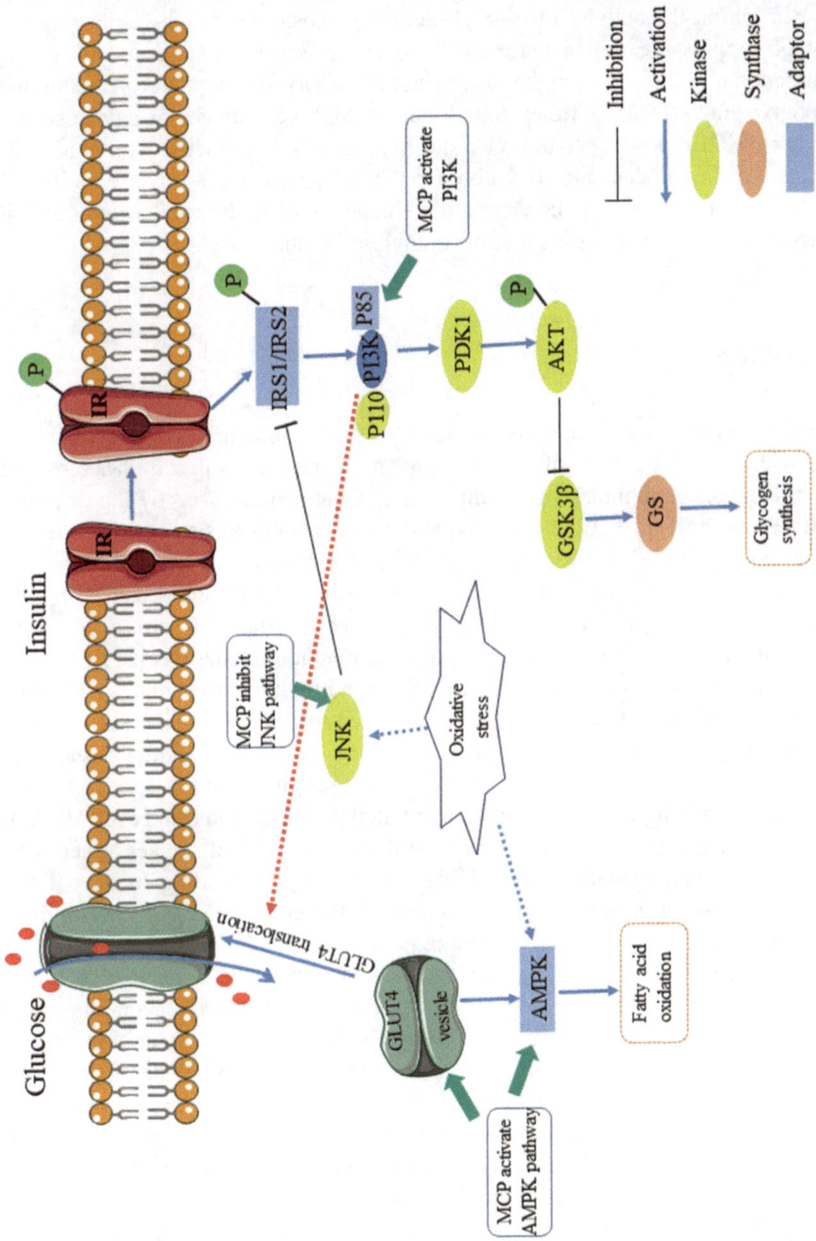

Fig. 7.1 The effects of *M. charantia* polysaccharides on the alleviation of insulin resistance

significantly improve the symptoms of abnormally elevated blood glucose in T2DM, and delay the conversion of glucose, fructose, and other monosaccharides into blood glucose [62–64]. Some studies have shown that some naturally-derived polysaccharides can inhibit the activity of α-amylase or α-glucosidase, thereby achieving the effect of suppressing the increase in blood sugar [65, 66]. Tang and Gan [27] confirmed that MCBP has a higher α-amylase inhibitory ability, therefore, exhibited a hypoglycemic effect. And they found that the high percentages of galactose and glucose in MCBP are associated with anti-hyperglycemic capacity. Yan et al. [67] demonstrated that BPS-F and BPS-I both exhibited significant α-amylase inhibitory activity within a certain concentration range, while the ability to inhibit α-glucosidase was weaker than that of α-amylase in vitro.

5.4 Others

It is reported that there is an association of low-grade inflammation and oxidative stress with T2DM because of the impairment of pancreatic β cell structure and insulin receptors. Inflammation is defined as a phenomena caused by the response to different pathological stimuli and tissue damage. Insulin Resistance and Atherosclerosis Study (IRAS) showed that IL-6 and TNF-α levels were linearly associated with insulin sensitivity [68]. It is well known that oxidative stress can be highly caused by excessive production of free radical species. And long-term exposure to superfluous free radical species will damage the function of islet β cells.

Oxidative stress is one of the risk factors caused by hyperglycemia [3, 34]. The endogenous antioxidant defenses can be destroyed by excessive generation of free radicals, along with an increased level of inflammation in surrounding tissues, the impairment of islet β cells, and a decrease in glucose utilization [17].

In a high oxidizing factor environment, internal proteins, lipids, and even DNA of cells will be damaged by exposure to superfluous free radical species. Therefore, hyperglycemia and oxidative stress form a vicious cycle, leading to islet β cell damage and insulin resistance, and ultimately to the occurrence and development of diabetes and its complications [50, 69]. Gong et al. found that bitter melon polysaccharides exhibited antioxidant activity in vitro. The detection of tumor cells confirmed that bitter melon polysaccharides have a direct scavenging effect on $O2^-$, NO, and $ONOO^-$, which can effectively inhibit the generation of free radicals and protect the antioxidant defense system [70]. Diabetic neuropathy is a diabetic complication caused by oxidative stress [71, 72]. The oral treatment of *M. charantia* polysaccharides can control the body weight, hyperglycemia, and hyperlipidemia of STZ-induced rats by up regulating the activity of antioxidant enzymes including glutathione peroxides (GSH), superoxide dismutase (SOD), and catalases CAT), reducing the malondialdehyde (MDA) content, thus restoring their total antioxidant capacity and mitigating the development of diabetic nephropathy through the suppression of oxidative stress to achieve the purpose of improving blood sugar levels [12]. This result is consistent with the results of Wang's group

[11]. Ru et al. [24] also demonstrated that Se-MCP II a-1 could increase the antioxidant enzyme activity and inhibit lipid peroxidation in diabetic mice induced by STZ. Moreover, fermented *M. charantia* polysaccharides significantly ameliorated oxidative stress in diabetic rats compared with non-fermented *M. charantia* polysaccharides [73]. It also showed that the anti-obesity properties of fermented M. charantia polysaccharide were revealed via relieving insulin resistance and fat accumulation of obese rats, which was associated with the regulation of lipid metabolism [74].

Related studies have shown that there's a strong correlation between type 2 diabetes and chronic inflammatory [75, 76]. This inflammation reduces the function of islet β cell and reduces the level and function of insulin secretion, which also brings insulin resistance [77]. Therefore, inhibiting islet β cell dysfunction and alleviating chronic inflammation of the islet β cell will alleviate the symptoms of type 2 diabetes and reduce the symptoms of hyperglycemia [78, 79]. Raish [28] found that *M. charantia* polysaccharides pretreatment via inhibiting the NF-κB signaling pathway to downregulation of proinflammatory cytokines and inflammatory genes and then ameliorates oxidative stress, hyperlipidemia, inflammation, and apoptosis in rats with isoproterenol-induced myocardial infarction. Subsequently, Raish et al. [80] further found the same mechanism of action ethanol-induced gastric ulcers in rats.

6 Summary and Future Perspectives

Diabetes is one of the diseases that threaten human health. Long-term high blood sugar can cause a series of complications. However, there hasn't been found a more effective way to cure diabetes, and commonly used western medicines often have serious side effects. Polysaccharides from the medicinal plants have attracted a lot of attention due to their significant bioactivities and low toxicity [81–84]. Therefore, polysaccharides with high efficiency and small side effects have been widely studied. It is well known that *M. charantia* has a long edible and medicinal history in China as a medicinal and edible plant [22, 85]. It has been found that *M. charantia* polysaccharides exhibited various biological activities, especially the *hypoglycemic activity*, and various *M. charantia* blood sugar lowering products have been favored by patients. *M. charantia* polysaccharide can be effectively extracted and purified by using the appropriate method. At the same time, by analyzing the structure of *M. charantia* polysaccharides, although the structure information is not fully explained, the monosaccharide composition, molecular weight, and some related primary structures of *M. charantia* polysaccharides have been basically grasped.

It is generally known that the relationship between the structural characteristics of polysaccharides and biological activity is inseparable. And considering the good hypoglycemic activity of *M. charantia* polysaccharides, further research on the exact structure-bioactivity relationships of *M. charantia* polysaccharides are required. Also, the mechanism may be diversified due to structural differences, which may be manifested in repairing islet β cells or inhibiting their apoptosis, enhancing insulin sensitivity, inhibiting key enzyme activities, and regulating signal pathways to reduce blood glucose. However, the comprehensive structures of *M. charantia* polysaccharides are still unclear due to the great structural diversity such as chain conformations and complexity of the polysaccharide molecules. So understanding the chemical structure of MCP is undoubtedly a huge challenge for future research.

As we all know, opportunities and challenges coexist in the research field of polysaccharides. Although there are many researches on the mechanism of lowering blood sugar in *M. charantia,* the exact mechanism of lowering blood sugar needs to be further studied. The study of hypoglycemic mechanism is mainly based on the research of single component, while the research on the synergy and mechanism of related active ingredients is lacking. For the future perspectives, the multiple biologically active ingredients of *M. charantia* can be used to coordinate with each other to make full use of biological effects, thus finding ways to further improve its hypoglycemic activity. At the same time, some modification methods can also be used to modify the *M. charantia* polysaccharides, such as selenization and sulfate to improve its hypoglycemic activity.

References

1. Muoio DM, Newgard CB (2008) Molecular and metabolic mechanisms of insulin resistance and β-cell failure in type 2 diabetes. Nat Rev Mol Cell Biol 9(3):193. https://doi.org/10.1038/nrm2327
2. Desai S, Tatke P (2015) Charantin: an important lead compound from Momordica charantia for the treatment of diabetes. J Pharmacogn Phytochem 3(6):163–166
3. Wu JJ, Shi SS, Wang HJ, Wang SH (2016) Mechanisms underlying the effect of polysaccharides in the treatment of type 2 diabetes: a review. Carbohydr Polym 144:474–494. https://doi.org/10.1016/j.carbpol.2016.02.040
4. Ozougwu JC, Obimba KC, Belonwu CD, Unakalamba CB (2013) The pathogenesis and pathophysiology of type 1 and type 2 diabetes mellitus. J Physiol Pathophysiol 4(4):46–57
5. Tahrani AA, Bailey CJ, Del Prato S, Barnett AH (2011) Management of type 2 diabetes: new and future developments in treatment. Lancet 378(9786):182–197. https://doi.org/10.1016/S0140-6736(11)60207-9
6. Zhang C, Huang M, Hong R, Chen HM (2019) Preparation of a *Momordica charantia* L. polysaccharide chromium (III) complex and its anti-hyperglycemic activity in mice with streptozotocin-induced diabetes. Int J Biol Macromol 122:619–627. https://doi.org/10.1016/j.ijbiomac.2018.10.200
7. Wang N, Zhang DL, Mao XQ, Zou F, Jin H, Ouyang JP (2009) *Astragalus* polysaccharides decreased the expression of PTP1B through relieving ER stress induced activation of ATF6 in a rat model of type 2 diabetes. Mol Cell Endocrinol 307(1–2):89–98. https://doi.org/10.1016/j.mce.2009.03.001
8. Chen MY, Xiao D, Liu W, Song YF, Zhou B et al (2019) Intake of *Ganoderma lucidum* polysaccharides reverses the disturbed gut microbiota and metabolism in type 2 diabetic rats. Int J Biol Macromol. https://doi.org/10.1016/j.ijbiomac.2019.11.047
9. Wu GJ, Bai ZY, Wan YJ, Shi HF, Huang X et al (2020) Antidiabetic effects of polysaccharide from azuki bean (*Vigna angularis*) in type 2 diabetic rats via insulin/PI3K/AKT signaling pathway. Food Hydrocoll 101:105456. https://doi.org/10.1016/j.foodhyd.2019.105456

10. Thomas PB, Robertson DH, Chawla NV (2018) Predicting onset of complications from diabetes: a graph based approach. Appl Netw Sci 3(1):48. https://doi.org/10.1007/s41109-018-0106-z
11. Wang Q, Wu XY, Shi FL, Liu Y (2019) Comparison of antidiabetic effects of saponins and polysaccharides from *Momordica charantia* L. in STZ-induced type 2 diabetic mice. Biomed Pharmacother 109:744–750. https://doi.org/10.1016/j.biopha.2018.09.098
12. Raish M, Ahmad A, Jan BL, Alkharfy KM, Ansari MA et al (2016) *Momordica charantia* polysaccharides mitigate the progression of STZ induced diabetic nephropathy in rats. Int J Biol Macromol 91:394–399. https://doi.org/10.1016/j.ijbiomac.2016.05.090
13. Ghadge AA, Kuvalekar AA (2017) Controversy of oral hypoglycemic agents in type 2 diabetes mellitus: novel move towards combination therapies. Diabetes Metab Syndr Clin Res Rev 11: S5–S13. https://doi.org/10.1016/j.dsx.2016.08.009
14. Wang PC, Zhao S, Yang BY, Wang QH, Kuang HX (2016) Anti-diabetic polysaccharides from natural sources: a review. Carbohydr polym 148:86–97. https://doi.org/10.1016/j.carbpol.2016.02.060
15. Zhao FQ, Liu QB, Cao J, Xu YS, Pei Z et al (2020) A sea cucumber (*Holothuria leucospilota*) polysaccharide improves the gut microbiome to alleviate the symptoms of type 2 diabetes mellitus in Goto-Kakizaki rats. Food Chem Toxicol 135:110886. https://doi.org/10.1016/j.fct.2019.110886
16. Liu GM, Liang L, Yu GY, Li QH (2018) Pumpkin polysaccharide modifies the gut microbiota during alleviation of type 2 diabetes in rats. Int J Biol Macromol 115:711–717. https://doi.org/10.1016/j.ijbiomac.2018.04.127
17. Cai WD, Ding ZC, Wang YY, Yang Y, Zhang HN et al (2020) Hypoglycemic benefit and potential mechanism of a polysaccharide from *Hericium erinaceus* in streptozotoxin-induced diabetic rats. Process Biochem 88:180–188. https://doi.org/10.1016/j.procbio.2019.09.035
18. Huang M, Wang FQ, Zhou XH, Yang HX, Wang YZ (2015) Hypoglycemic and hypolipidemic properties of polysaccharides from *Enterobacter cloacae* Z0206 in KKAy mice. Carbohydr polym 117:91–98. https://doi.org/10.1016/j.carbpol.2014.09.008
19. Dandawate PR, Subramaniam D, Padhye SB, Anant S (2016) Bitter melon: a panacea for inflammation and cancer. Chin J Nat Med 14(2):81–100. https://doi.org/10.1016/S1875-5364(16)60002-X
20. Trakoon-osot W, Sotanaphun U, Phanachet P, Porasuphatana S, Udomsubpayakul U, Komindr S (2013) Pilot study: hypoglycemic and antiglycation activities of bitter melon (*Momordica charantia* L.) in type 2 diabetic patients. J Pharm Res 6(8):859–864. https://doi.org/10.1016/j.jopr.2013.08.007
21. Krishnaiah D, Sarbatly R, Nithyanandam R (2011) A review of the antioxidant potential of medicinal plant species. Food Bioprod Process 89(3):217–233. https://doi.org/10.1016/j.fbp.2010.04.008
22. Jia S, Shen MY, Zhang F, Xie JH (2017) Recent advances in *Momordica charantia*: functional components and biological activities. Int J Mol Sci 18(12):2555. https://doi.org/10.3390/ijms18122555
23. Horax R, Hettiarachchy N, Chen PY (2014) Characteristics and functionality enhancement by glycosylation of bitter melon (*Momordica charantia*) seed protein. J Food Sci 79(11):C2215–C2221. https://doi.org/10.1111/1750-3841.12680
24. Ru Y, Liu KX, Kong XY, Li XY, Shi X, Chen HM (2020) Synthesis of selenylated polysaccharides from *Momordica charantia* L. and its hypoglycemic activity in streptozotocin-induced diabetic mice. Int J Biol Macromol 152:295–304. https://doi.org/10.1016/j.ijbiomac.2020.02.288
25. Xu X, Shan B, Liao CH, Xie JH, Wen PW, Shi JY (2015) Anti-diabetic properties of *Momordica charantia* L. polysaccharide in alloxan-induced diabetic mice. Int J Biol Macromol 81:538–543. https://doi.org/10.1016/j.ijbiomac.2015.08.049
26. Panda BC, Mondal S, Devi KSP, Maiti TK, Khatua S et al (2015) Pectic polysaccharide from the green fruits of *Momordica charantia* (Karela): structural characterization and study of

immunoenhancing and antioxidant properties. Carbohydr Res 401:24–31. https://doi.org/10.1016/j.carres.2014.10.015
27. Tan HF, Gan CY (2016) Polysaccharide with antioxidant, α-amylase inhibitory and ACE inhibitory activities from *Momordica charantia*. Int J Biol Macromol 85:487–496. https://doi.org/10.1016/j.ijbiomac.2016.01.023
28. Raish M (2017) *Momordica charantia* polysaccharides ameliorate oxidative stress, hyperlipidemia, inflammation, and apoptosis during myocardial infarction by inhibiting the NF-κB signaling pathway. Int J Biol Macromol 97:544–551. https://doi.org/10.1016/j.ijbiomac.2017.01.074
29. Zhang F, Lin L, Xie JH (2016) A mini-review of chemical and biological properties of polysaccharides from *Momordica charantia*. Int J Biol Macromol 92:246–253. https://doi.org/10.1016/j.ijbiomac.2016.06.101
30. Fan T, Hu JG, Fu LD, Zhang LJ (2015) Optimization of enzymolysis-ultrasonic assisted extraction of polysaccharides from *Momordica charabtia* L. by response surface methodology. Carbohydr Polym 115:701–706. https://doi.org/10.1016/j.carbpol.2014.09.009
31. Wu JD (2005) Studies on the enzyme hydrolysis of soluble polysaccharide from *Momordica charantia*. J Anhui Agric Sci 32(1):73–74
32. Zhou R, Tu NY, Cong M, Kan GS, Chen HM (2013) Purification and hepatoprotective effect of *Momordica charantia* polysaccharides. Food Sci Technol 38(2):173–176
33. Chen HM, Guo LV, Zhang SZ, Liu JM, Kan GS et al (2015) Purification and structural analysis of a polysaccharide from bitter gourd (*Momordica charantia*). Food Sci 36(10):90–94. https://doi.org/10.7506/spkx1002-6630-201510018
34. Zhang C, Chen HM, Bai WQ (2018) Characterization of *Momordica charantia* L. polysaccharide and its protective effect on pancreatic cells injury in STZ-induced diabetic mice. Int J Biol Macromol 115:45–52. https://doi.org/10.1016/j.ijbiomac.2018.04.039
35. Deng YY, Yi Y, Zhang LF, Zhang RF, Zhang Y et al (2014) Immunomodulatory activity and partial characterisation of polysaccharides from *Momordica charantia*. Molecules 19(9):13432–13447. https://doi.org/10.3390/molecules190913432
36. Huang R, Xie JH, Yu Y, Shen MY (2019) Recent progress in the research of yam mucilage polysaccharides: isolation, structure and bioactivities. Int J Biol Macromol. https://doi.org/10.1016/j.ijbiomac.2019.11.095
37. Dong Y, Xu B, Lu Q, Zha Q (2005) Studies on the isolation, purification and composition of *Momordica charantia* L. polysaccharide. Food Sci 26(11):115–119
38. Wu XG, Wang ZG, Cai Y, Yang Y, Wang M (2011) Extraction, isolation, purification and composition of *Momordica charantia* polysaccharide. Chin J Bioproc Eng 9(1):19–23. https://doi.org/10.1007/s11518-011-5154-1
39. Chen HM, Yang JY, Kan GS, Ren DM (2011) Isolation, purification and composition analysis of an antioxidant polysaccharide fraction from Momordica charantia. Food Sci 32(1):S367–S368
40. Zhang Y, Wang W, Cai Y, Wang M, Yang Y et al (2017) Purification, structural characterization and immunomodulatory and antitumor activities of a polysaccharide isolated from *Momordica charantia* L. Nangjing Univ Tradit Chin Med 33. https://doi.org/10.14148/j.issn.1672-0482.2017.0033
41. Liu GL, Ning ZX, Guo HH (2010) Isolation and characterization of polysaccharides from *Momordica charantia* L. Fruits Food Sci 31(3):30–34. https://doi.org/10.1111/j.1365-2583.2010.01002.x
42. Balaji R, Duraisamy R, Kumar MP (2019) Complications of diabetes mellitus: a review. Drug Invent Today 12:1
43. Giri B, Dey S, Das T, Sarkar M, Banerjee J et al (2018) Chronic hyperglycemia mediated physiological alteration and metabolic distortion leads to organ dysfunction, infection, cancer progression and other pathophysiological consequences: an update on glucose toxicity. Biomed Pharmacother 107:306–328. https://doi.org/10.1016/j.biopha.2018.07.157

44. Ye H, Shen ZP, Cui JF, Zhu YJ, Li Y et al (2019) Hypoglycemic activity and mechanism of the sulfated rhamnose polysaccharides chromium (III) complex in type 2 diabetic mice. Bioorg Chem 88:102942. https://doi.org/10.1016/j.bioorg.2019.102942
45. Jia RB, Li ZR, Wu J, Ou ZR, Zhu Q et al (2020) Physicochemical properties of polysaccharide fractions from *Sargassum fusiforme* and their hypoglycemic and hypolipidemic activities in type 2 diabetic rats. Int J Biol Macromol 147:428–438. https://doi.org/10.1016/j.ijbiomac.2019.12.243
46. Dong Y, Zhang HH (2008) Studies on components with antihyperglycemic effect of Momordica charantia L. polysaccharides. Acta Nutr Sinica (1):60–62. https://doi.org/10.1016/S1005-8885(08)60096-5
47. Du GF, Chen HM, Kan GS, Gao S, Guo LW (2017) Study on preparation of Momordica charantia polysaccharide-iron complex and its hypoglycemic activities in diabetic mice. Sci Technol Food Ind 09:311–314. https://doi.org/10.13386/j.issn1002-0306.2017.09.060
48. Xu B, Dong Y, Zhang HH, Cui HL, Lu Q (2006) The antihyperglycemic effects of polysaccharide from *Momordica charantia* in STZ-induced diabetic mice. Acta Nutr Sinica 28 (5):401–403
49. Bai WQ, Zhang C, Chen HM (2018) Transcriptomic analysis of *Momordica charantia* polysaccharide on streptozotocin-induced diabetic rats. Gene 675:208–216. https://doi.org/10.1016/j.gene.2018.06.106
50. Elangovan A, Subramanian A, Durairaj S, Ramachandran J, Lakshmanan DK et al (2019) Antidiabetic and hypolipidemic efficacy of skin and seed extracts of *Momordica cymbalaria* on alloxan induced diabetic model in rats. J Ethnopharmacol 241:111989. https://doi.org/10.1016/j.jep.2019.111989
51. Pessin JE, Saltiel AR (2000) Signaling pathways in insulin action: molecular targets of insulin resistance. J Clin Invest 106(2):165–169. https://doi.org/10.1172/JCI10582
52. Liu YT, Sun J, Rao SQ, Su YJ, Yang YJ (2013) Antihyperglycemic, antihyperlipidemic and antioxidant activities of polysaccharides from catathelasma ventricosum in streptozotocin-induced diabetic mice. Food Chem Toxicol 57:39–45. https://doi.org/10.1016/j.fct.2013.03.001
53. Sridhar MG, Vinayagamoorthi R, Suyambunathan VA, Bobby Z, Selvaraj N (2008) Bitter gourd (*Momordica charantia*) improves insulin sensitivity by increasing skeletal muscle insulin-stimulated IRS-1 tyrosine phosphorylation in high-fat-fed rats. Br J Nutr 99 (4):806–812. https://doi.org/10.1017/s000711450783176x
54. Matzinger M, Fischhuber K, Heiss EH (2018) Activation of Nrf2 signaling by natural products-can it alleviate diabetes? Biotechnol Adv 36(6):1738–1767. https://doi.org/10.1016/j.biotechadv.2017.12.015
55. Guo L, Tabrizchi R (2006) Peroxisome proliferator-activated receptor gamma as a drug target in the pathogenesis of insulin resistance. Pharmacol Ther 111(1):145–173. https://doi.org/10.1016/j.pharmthera.2005.10.009
56. Vidal-Puig AJ, Considine RV, Jimenez-Liñan M, Werman A, Pories WJ et al (1997) Peroxisome proliferator-activated receptor gene expression in human tissues. Effects of obesity, weight loss, and regulation by insulin and glucocorticoids. J Clin Invest 99(10):2416–2422. https://doi.org/10.1172/JCI119424
57. He XY, Liu ZH (2007) Study on HTS anti-diabetic antagonists from Momordica charantia L. Food Sci 28(2):313–316. https://doi.org/10.1007/s11767-005-0212-9
58. Zhang Y, Dong Y, Zhu Y, Bai J (2016) Regulation pathway of polysaccharides from Momordica charantia L. to the insulin resistant of hepg2 cell. J Chin Inst Food Sci Technol 016(012):10–16. https://doi.org/10.16429/j.1009-7848.2016.12.002
59. Fan M, Kim EK, Choi YJ, Tang Y, Moon SH (2019) The role of Momordica charantia in resisting obesity. Int J Environ Res Public Health 16(18). https://doi.org/10.3390/ijerph16183251
60. Wang C, Li WW, Chen ZQ, Gao XD, Yuan G et al (2018) Effects of simulated gastrointestinal digestion in vitro on the chemical properties, antioxidant activity, α-amylase and α-glucosidase

inhibitory activity of polysaccharides from Inonotus obliquus. Food Res Int 103:280–288. https://doi.org/10.1016/j.foodres.2017.10.058

61. Perez JL, Jayaprakasha GK, Patil BS (2019) Metabolite profiling and in vitro biological activities of two commercial bitter melon (*Momordica charantia* Linn.) cultivars. Food Chem 288:178–186. https://doi.org/10.1016/j.foodchem.2019.02.120
62. Ye F, Shen Z, Xie M (2002) Alpha-glucosidase inhibition from a Chinese medical herb (*Ramulus mori*) in normal and diabetic rats and mice. Phytomedicine 9(2):161–166. https://doi.org/10.1078/0944-7113-00065
63. Magaji UF, Sacan O, Yanardag R (2020) Alpha amylase, alpha glucosidase and glycation inhibitory activity of Moringa oleifera extracts. S Afr J Bot 128:225–230. https://doi.org/10.1016/j.sajb.2019.11.024
64. Zhang XL, Li GL, Wu D, Yu YX, Hu N et al (2020) Emerging strategies for the activity assay and inhibitor screening of alpha-glucosidase. Food Funct 11(1):66–82. https://doi.org/10.1039/C9FO01590F
65. Yue JY, Xu J, Cao JQ, Zhang XS, Zhao YQ (2017) Cucurbitane triterpenoids from Momordica charantia L. and their inhibitory activity against α-glucosidase, α-amylase and protein tyrosine phosphatase 1B (PTP1B). J Funct Foods 37:624–631. https://doi.org/10.1016/j.jff.2017.07.041
66. Trinh BT, Staerk D, Jäger AK (2016) Screening for potential α-glucosidase and α-amylase inhibitory constituents from selected Vietnamese plants used to treat type 2 diabetes. J Ethnopharmacol 186:189–195. https://doi.org/10.1016/j.jep.2016.03.060
67. Yan JK, Wu LX, Qiao ZR, Cai WD, Ma H (2019) Effect of different drying methods on the product quality and bioactive polysaccharides of bitter gourd (*Momordica charantia* L.) slices. Food Chem 271:588–596. https://doi.org/10.1016/j.foodchem.2018.08.012
68. Festa A, D'Agostino R, Howard G, Mykkanen L, Tracy RP, Haffner SM (2000) Chronic subclinical inflammation as part of the insulin resistance syndrome: the insulin resistance atherosclerosis study (IRAS). Circulation 102(1):42–47. https://doi.org/10.1161/01.CIR.102.1.42
69. Robertson RP, Harmon JS (2006) Diabetes, glucose toxicity, and oxidative stress: a case of double jeopardy for the pancreatic islet β cell. Free Radic Biol Med 41(2):177–184. https://doi.org/10.1016/j.freeradbiomed.2005.04.030
70. Gong JJ, Sun FM, Li YH, Zhou XL, Duan FG et al (2015) *Momordica charantia* polysaccharides could protect against cerebral ischemia/reperfusion injury through inhibiting oxidative stress mediated c-Jun n-terminal kinase 3 signaling pathway. Neuropharmacology 91:123–134. https://doi.org/10.1016/j.neuropharm.2014.11.020
71. Figueroa-Romero C, Sadidi M, Feldman EL (2008) Mechanisms of disease: the oxidative stress theory of diabetic neuropathy. Rev Endocr Metab Disord 9(4):301–314. https://doi.org/10.1007/s11154-008-9104-2
72. Kasznicki J, Kosmalski M, Sliwinska A, Mrowicka M, Drzewoski J et al (2012) Evaluation of oxidative stress markers in pathogenesis of diabetic neuropathy. Mol Biol Rep 39(9):8669–8678. https://doi.org/10.1007/s11033-012-1722-9
73. Gao H, Wen JJ, Hu JL, Nie QX, Chen HH et al (2018) Polysaccharide from fermented *Momordica charantia* L. with Lactobacillus plantarum NCU116 ameliorates type 2 diabetes in rats. Carbohydr Polym 201:624–633. https://doi.org/10.1016/j.carbpol.2018.08.075
74. Wen JJ, Gao H, Hu JL, Nie QX, Chen HH et al (2019) Polysaccharides from fermented *Momordica charantia* ameliorate obesity in high-fat induced obese rats. Food Funct 10(1):448–457. https://doi.org/10.1039/C8FO01609G
75. Donath MY, Shoelson SE (2011) Type 2 diabetes as an inflammatory disease. Nat Rev Immunol 11(2):98–107. https://doi.org/10.1038/nri2925
76. Dong JL, Cai FL, Shen RL, Liu YQ (2011) Hypoglycemic effects and inhibitory effect on intestinal disaccharidases of oat beta-glucan in streptozotocin-induced diabetic mice. Food Chem 129(3):1066–1071. https://doi.org/10.1016/j.foodchem.2011.05.076

77. Bai J, Zhu Y, Dong Y (2016) Response of gut microbiota and inflammatory status to bitter melon (*Momordica charantia* L.) in high fat diet induced obese rats. J Ethnopharmacol 194:717–726. https://doi.org/10.1016/j.jep.2016.10.043
78. Imai Y, Dobrian AD, Morris MA, Nadler JL (2013) Islet inflammation: a unifying target for diabetes treatment? Trends Endocrinol Metab 24(7):351–360. https://doi.org/10.1016/j.tem.2013.01.007
79. Yudkin JS (2007) Inflammation, obesity, and the metabolic syndrome. Horm Metab Res 39 (10):707–709. https://doi.org/10.1055/s-2007-985898
80. Raish M, Ahmad A, Ansari MA, Alkharfy KM, Ali N et al (2018) *Momordica charantia* polysaccharides ameliorate oxidative stress, inflammation, and apoptosis in ethanol-induced gastritis in mucosa through NF-kB signaling pathway inhibition. Int J Biol Macromol 111:193–199. https://doi.org/10.1016/j.ijbiomac.2018.01.008
81. Xie JH, Jin ML, Morris GA, Zha XQ, Chen HQ et al (2016) Advances on bioactive polysaccharides from medicinal plants. Crit Rev Food Sci Nutr 56(Suppl 1):S60–S84. https://doi.org/10.1080/10408398.2015.1069255
82. Huang L, Shen M, Morris GA, Xie J (2019) Sulfated polysaccharides: immunomodulation and signaling mechanisms. Trends Food Sci Technol 92:1–11. https://doi.org/10.1016/j.tifs.2019.08.008
83. Yu Y, Shen M, Song Q, Xie J (2018) Biological activities and pharmaceutical applications of polysaccharide from natural resources: a review. Carbohydr Polym 183:91–101. https://doi.org/10.1016/j.carbpol.2017.12.009
84. Xie JH, Tang W, Jin ML, Li JE, Xie MY (2016) Recent advances in bioactive polysaccharides from *Lycium barbarum* L., *Zizyphus jujuba* Mill, *Plantago* spp., and *Morus* spp.: structures and functionalities. Food Hydrocoll 60:148–160. https://doi.org/10.1016/j.foodhyd.2016.03.030
85. Peter EL, Mtewa AG, Nagendrappa PB, Kaligirwa A, Sesaazi CD (2020) Systematic review and meta-analysis protocol for efficacy and safety of *Momordica charantia* L. on animal models of type 2 diabetes mellitus. Syst Rev 9(1):1–9. https://doi.org/10.1186/s13643-019-1265-4

Chapter 8
Effects of Polysaccharides on Reducing Blood Glucose Based on Gut Microbiota Alteration

Min Zhang and Liyuan Yun

Abstract Type 2 diabetes mellitus (T2DM) is an endocrine and metabolic disease with insulin resistance and insulin deficiency, and the structural changes of gut microbiota play a very important role in the occurrence and development of T2DM. Gut microbiota is an important part of intestinal micro-ecosystem. Long-term high-sugar and high-fat diet can change the intestinal microenvironment, especially the structure of gut microbiota. Gut bacteria affect the body's absorption of sugars and energy, as well as regulating the production of lipopolysaccharide and short-chain fatty acid. The changes of gut microbiota can induce low-grade chronic inflammation, affect bile acid metabolism, and lead to destruction and apoptosis of islet cells and insulin resistance. It is noteworthy that natural polysaccharides have been widely used in regulating gut microbiota with many advantages, such as good stability, nontoxicity, and safety. Thus, polysaccharide-regulated composition changes of gut microbiota have become a new target for prevention and treatment of type 2 diabetes.

Keywords Type 2 diabetes mellitus · Reducing blood glucose · Polysaccharides · Gut microbiota

1 Introduction

Over 90% of diabetes mellitus cases are type 2 diabetes mellitus (T2DM) [1]. T2DM, characterized by a metabolic disorder resulting from a lack of insulin secretion or insulin action or both, has been a major global health focus and often associated with obesity, systemic complications, and mortality [2]. Over the past few decades, the prevalence of T2DM has increased in many countries. It is estimated

M. Zhang (✉) · L. Yun
Tianjin Agricultural University, Tianjin, China

State Key Laboratory of Food Nutrition and Safety (Tianjin University of Science and Technology), Tianjin, China

H. Chen, M. Zhang (eds.), *Structure and Health Effects of Natural Products on Diabetes Mellitus*, https://doi.org/10.1007/978-981-15-8791-7_8

that by 2035, the number of adult patients worldwide will rise to 592 million [3, 4]. Many factors, besides genetic and environmental factors such as lifestyle, can influence its emergence and development [5, 6]. Therefore, the epidemic of the disease has become a major global public health problem.

Gut microbiota is closely related to the association and development of diseases [7, 8]. At present, the study of gut microbiota is one of the most popular research directions in biomedical field and has been listed as one of the ten breakthroughs in human technology [1, 9]. Gut microbiota influences many aspects of host physiology, including diet, vitamin production, and the pathogenesis of disease [10]. A lot of studies have shown that intestinal microflora plays a key role in host health [11, 12], metabolic phenotype [13], nutrient intake and absorption [14, 15], and immune system regulation [16]. The dysbiosis of gut microbiota is associated with several disorders including obesity, irritable bowel syndrome [17], abnormal immune responses [18], and diabetes type 1 and type 2 [19, 20].

Polysaccharide is a kind of macromolecular substance that exists naturally in plants, microorganisms, algae, and animals [21]. In the past decade, more and more researchers have paid attention to the isolation of polysaccharides from natural resources such as plants, mushrooms, bacteria, and seaweeds, mainly due to their pharmacological activities [22, 23], such as antitumor, anti-inflammatory, immunostimulatory, lowering blood glucose and lipid [24, 25], and regulating intestinal microbiota [26]. A chronically irregular diet could lead to glucose metabolism disorder, which was manifested by gradual decline of glucose tolerance over time, leading to elevated fasting blood glucose or impaired glucose tolerance [27]. The clinical manifestations of glycometabolism-related diseases are characterized by hyperglycemia or hypoglycemia, which can eventually lead to serious diseases such as diabetes, obesity, and malnutrition [28]. People prefer dietary intervention instead of medicine to treat glycometabolism-related disorders, as dietary intervention possessed less side effect and more safety. Therefore, in this review, we collected the recent literature regarding the role of polysaccharides in the treatment of type 2 diabetes by regulating gut microbiota and summarized the most reliable findings.

2 Effects Mechanism of Polysaccharides Regulating Blood Glucose

Polysaccharides are polymeric carbohydrate molecules that are linked by long chains of monosaccharide units by glycosidic bonds and produce monosaccharides or oligosaccharides when hydrolyzed. Their structure varies from linear to highly branch. In recent years, more and more studies have shown that polysaccharide has obvious antidiabetic effect with almost no side effects or adverse drug reactions [29, 30]. *Astragalus polysaccharide* (APS), which is extracted from the root of *Astragalus membranaceus*, is a water-soluble heteroglycan. Ye et al. had showed that APS can lower body weight, blood glucose, and triglyceride levels by downregulating the ubiquitination levels of IRS-1 and its nuclear expression, which are the important regulators of insulin signal transduction [31]. Bin et al. also showed that APS can promote glucose uptake and improve insulin resistance in 3T3-L1 adipocytes through miR-721-PPAR-γ-PI3K/AKT-GLUT4 signaling pathway [32]. In this research, the mechanisms underlying polysaccharide are classified in Table 8.1.

3 Gut Microbiota and Blood Glucose Regulation

Microecological studies have found that there is a certain relationship between the number of gut microbiota and the occurrence and development of diabetes mellitus. When the blood glucose level was relatively high (>20 mmol/L, mice), the number of beneficial bacteria (*bifidobacteria*, *lactobacillus*, etc.) was decreased in the intestine. While after drug treatment, the blood glucose level decreased (<14 mmol/L, mice) and the number of beneficial bacteria increased. These results indicated that the number of gut microbiota were negatively correlated with the blood glucose level [84]. Several metabolites produced by gut microbiota, including SCFAs, amino acids, and bile acids, which involved in the regulation of host metabolism and gut integrity [85]. The potential mechanisms of microbiota effects on metabolism in the T2DM patient were collected as follows.

3.1 Regulation of Inflammation

T2DM is a chronic inflammatory disease accompanied with the increase of cytokines that upregulate insulin targets including liver, adipose tissue, and muscle, resulting in insulin resistance and leading to the disease occurrence [86]. The imbalance of cytokine network and enhancement of inflammatory activation signal are associated with low-degree inflammation. Intestinal microbial molecules, especially lipopolysaccharides (LPS), which stimulate the inflammatory network response and play a role in its inflammatory response. For example, *Bacteroides fragilis*, *Akkermansia muciniphila*, and *Lactobacillus plantarum* L. can induce IL-10 to improve the glucose metabolism, as the overexpression of this cytokine in muscle may protect against aging-related insulin resistance [87]. In a recent report, the study indicated that *Lactobacillus reuteri* GMNL-263 could significantly inhibited the expression of glucose, insulin, leptin, C-peptide, GLP-1, inflammatory IL-6, and TNF-γ in adipose tissue, and the expression of PPAR-γ and GLUT4 genes in high-fructose fed rats [88]. Short-chain fatty acids (SCFAs) are organic fatty acids with less than 6 carbon atoms in the carbon chain. They are not only an important energy source for intestinal epithelial cells but also affect the permeability and oxidative stress of intestinal mucosal barrier and intestinal epithelial cells. SCFAs play an

Table 8.1 Hypoglycemic activity and mechanism of polysaccharides

Effects	Effects mechanism	Types of polysaccharides	Ref.
Targeting β-cell dysfunction	Reducing blood glucose and lipids	*Acacia tortilis* polysaccharide	[33]
		Lachnum calyculiforme polysaccharide	[34]
		Mactra veneriformis polysaccharide	[35]
		Schisandra chinensis Baill. Polysaccharide	[36]
		Saccharina japonica polysaccharide	[37]
		Trametes gibbosa polysaccharide	[38]
		Inula britannica flower polysaccharide	[39]
		Corn silk polysaccharide	[40]
		Pumpkin polysaccharide	[41]
		Phellinus linteus polysaccharide	[42]
		Talinum triangulare polysaccharide	[43]
		Astragalus polysaccharides	[44]
		Chinese Yam (*Dioscorea opposita* rhizoma) polysaccharide	[45]
		Mulberry leaf polysaccharide	[46]
		Ganoderma atrum polysaccharide	[47, 48]
		Ganoderma lucidum fruiting bodies polysaccharide	[49, 50]
		Pumpkin fruit polysaccharide	[51]
		Cucumaria frondosa polysaccharide	[52]
		Lycium barbarum L. polysaccharide	[53]
	Targeting β-cell dysfunction with an antioxidative effect	*Salvia miltiorrhiza* polysaccharide Bunge polysaccharide	[54]

(continued)

Table 8.1 (continued)

Effects	Effects mechanism	Types of polysaccharides	Ref.
		Grifola frondosa polysaccharide	[55, 56]
		Catathelasma ventricosum polysaccharide	[57]
		Ophiopogon japonicus polysaccharide	[58]
		Bacillus licheniformis Levan polysaccharide	[59]
		Panax ginseng C.A. Meyer (ginseng) polysaccharide	[60]
		Pleurotus florida polysaccharide	[61]
	Targeting β-cell dysfunction with an anti-inflammatory effect	*Angelica sinensis* polysaccharide	[62]
		Pseudostellaria heterophylla polysaccharide	[63]
		Pleurotus sajor-caju polysaccharide	[64]
	Targeting β-cell dysfunction with anti-inflammatory and antioxidative effects	*Ramulus mori* polysaccharide	[65]
		Misgurnus anguillicaudatus polysaccharide	[66]
		Rehmannia glutinosa polysaccharide (Gaertn.) DC. polysaccharide	[67]
		Hedysarum polysaccharide *Polybotrys* polysaccharide	[52]
Inhibiting α-amylase and α-glucosidase		*Opilia amentacea* Roxb. fruit polysaccharide	[68]
		Blackcurrant fruits polysaccharide	[69]
		Inonotus obliquus polysaccharide	[70]
		Camellia sinensis polysaccharide	[71]
		Diaphragma juglandis fructus polysaccharide	[72]

(continued)

Table 8.1 (continued)

Effects	Effects mechanism	Types of polysaccharides	Ref.
Targeting insulin enhancement	PI3 K/Akt pathway activators	*Ophiopogon japonicus* polysaccharide	[73]
		Administering mulberry leaf polysaccharide	[74]
		Acaudina molpadioides polysaccharide	[75]
		Liriope spicata var. *prolifera* polysaccharides	[76]
		Grifola frondosa polysaccharide	[77]
	MAPK pathway modulators	*Acaudina molpadioides* polysaccharide	[78]
		Ramulus mori polysaccharide	[79]
Targeting glucose metabolism in noninsulin-dependent pathways	Inhibiting glucose uptake	*Lycium barbarum* L. polysaccharide	[80]
	5′-AMP-activated protein kinase activators	*Pleurotus sajor-caju* polysaccharide	[64]
		Astragalus membranaceus polysaccharide	[76]
		Enterobacter cloacae Z0206 polysaccharide	[81]
	Pancreatic duodenal homeobox-1 modulators	Mulberry leaf polysaccharide	[82]
	Increasing liver glycogen content, inhibiting its decomposition	*Ganoderma lucidum* polysaccharides	[83]

anti-inflammatory role by balancing the expression of inflammatory mediators, which involved in regulating the functions of leukocytes and adipocytes, reducing the expression and production of cytokines and chemokines, and sequentially reducing the release of TNF-α induced by LPS [26]. In addition, SCFAs can inhibit leukocyte chemotaxis and reduce inflammatory infiltration of adipose tissue by downregulation of adhesion molecules of tissue endothelial cells.

3.2 *Maintain Intestinal Mucosal Permeability*

Intestinal mucosa plays an important role in nutrient absorption and barrier function regulation, which can prevent gut microbiota from shifting. The integrity of intestinal mucosa is achieved by tight intercellular connections, mucous secretion, antigen peptides, and immunoprotein secretion. Zonulin protein is a regulator of gut. The concentration of Zonulin is related to the ability of osmotic pressure of the stomach and intestine [89]. In a previous study, 23 men were supplemented with 6 probiotic strains, and the results showed that probiotics could enhance intestinal mucosal integrity [90]. Glucagon-like peptide-2 (GLP-2) plays an important role in the intestinal mucosal integrity, the proliferation of intestinal epithelial cells, and the prevention of apoptosis. Prebiotics enhance the expression of GLP-2, while prebiotic antagonists eliminate the effect of GLP-2. Changing the composition of gut microbiota, the expression of GLP-2 can be increased, which is useful to maintain the stability of intestinal mucosa and improve the intestinal barrier function of diabetic patients. Therefore, gut microflora plays a key role in main intestinal mucosal permeability.

3.3 *Control the Metabolism of Bile Acids, Short-Chain Fatty Acids, and Glucose*

Short-chain fatty acids (SCFAs) with a variety of functional and beneficial effects on host physiology, including maintaining intestinal barrier, participating in host signal transduction, and promoting energy metabolism [91]. SCFA is the main metabolite produced by the fermentation of indigestible fatty bacteria in the colon and is composed primarily of acetate, propionate, and butyrate. Meanwhile, butyric acid produced by *Clostridiales* (*Roseburia* and *Faecalibacterium prausnitzii*) can significantly regulate glucose and lipid metabolism [28]. Wang et al. [26] had reported that *Ophiopogon japonicus* polysaccharide could regulate metabolic disorder of SCFAs and reduce diabetic symptom. Gut microflora may also affect T2DM by affecting glucose homeostasis and insulin resistance in major metabolic organs such as liver, muscle, and fat. T2DM is influenced by gut hormones, which influence the digestion of sugars and control this process. Bile acids play an important role in glucose metabolism as signaling molecules and cell receptor ligands [92]. Gut microflora can activate bile acid receptors to reduce the expression of bile acid synthase, which in turn helps inhibit bacterial colonization and growth.

4 Effects of Polysaccharides on Intestinal Microbiota

Polysaccharides are composed of more than 10 monosaccharide molecules, which are linked by glycosidic bonds and can be extracted from a variety of natural products. It was demonstrated that polysaccharide plays a key role in the management of glycometabolism-related diseases. For example, polysaccharide from tuberous roots of Ophiopogon japonicus ameliorated HFD-induced insulin resistance and glucose tolerance and regulated the gut microbiota dysbiosis in HFD-fed mice decreasing Proteobacteria and type-2 diabetes-enriched taxa [26]. Zhang et al. reported that intragastric administration of pumpkin polysaccharide in alloxan-induced diabetic rabbits for 21 days could reduce the levels of blood glucose, TC, TG, and glycosylated hemoglobin [93]. Liu et al. also indicated that after treating male Wistar rats with pumpkin polysaccharide for 4 weeks, the type 2 diabetes were alleviated by improving the insulin tolerance and decreasing the levels of serum glucose (GLU), total cholesterol (TC), and low-density lipoprotein (LDL-C), while increasing the levels of high-density lipoprotein (HDL-C). Simultaneously, pumpkin polysaccharide also enriches the SCFAs-producing gut microbiota [94]. A number of researches also showed that several indigestible polysaccharides have multiple benefits for diabetes and obesity, such as *Ganoderma lucidum*, *Astragalus membranaceus*, *Polygonatum kingianum*, *Schisandra chinensis*, and *Ophiopogon japonicas* [95]. These polysaccharides could normalize blood glucose and insulin levels in obese/diabetic animal models by modulating gut microbiota composition, lowering colonic pH, increasing production of SCFAs, and stimulating colonic immune responses. In addition to proving that polysaccharides can improve glycometabolism-related diseases such as diabetes, obesity, and malnutrition, its mechanism has been further deepened.

4.1 Relationship Between Polysaccharides and Gut Microbiota

The carbon source for the growth of gut microbiota generally comes from the intake of indigestible carbohydrates such as nonstarch polysaccharides in the diet, which are resistant to stomach acids and mammalian enzymes but ferment partially or completely in large intestine [96]. The effect of nonstarch polysaccharides on human health is self-evident, because these ingredients can improve composition or metabolism of gut microbiota, especially enriching healthy microbiota and the production of SCFAs, maintaining gut health and preventing the development of colonic diseases. For example, supplementation with laminarin in rats fed, the cecum weight, and cecal concentrations of SCFAs were significantly increased, and laminarin as a prebiotic agent promoted the growth of *Bifidobacterium* spp. and *Lactobacillus* spp. in the gut [97, 98]. Dietary fucoidan not only increased the abundance of *Lactobacillus* spp. in experimental animals [99] but also increased intestinal concentrations of SCFAs. Meanwhile, fucoidan has been used in Russia in combination with commercial probiotics, such as *Bifidobacterium longum* B379M and *Bifidobacterium bifidum* 791B [100]. *Artemisia sphaerocephala* Krasch polysaccharide alleviated the body weight gain in HFD-fed mice and dyslipidemia, significantly ameliorated metabolic endotoxemia, and reversed the gut dysbiosis induced by HFD [101]. The improvement effects of different polysaccharides on gut microbiota are listed in Table 8.2.

Table 8.2 Effects of natural polysaccharides on gut microbiota

Natural polysaccharides	Changes in gut microbiota	Changes in SCFAs	Changes in blood glucose	Monosaccharide composition	Mw kDa	Ref.
Ganoderma lucidum polysaccharides	Aerococcus↓Ruminococcus↓Corynebactrium Proteus↓ Blautia↑Dehalobacterium↑Parabacteroides, Bacteroides↑	Acetic acid↑propionic acid↑ Butyric acid↑	FBG↓	Mannose, glucose, galactose, rhamnose, arabinose in the molar ratio of 3.16:16.17:3.74:1.65:1	13.7	[102]
Lycium barbarum L. leaf polysaccharides	Blautia↓ Marvinbryantia↓Parasutterella↓Prevotellaceae_NK3B31_group↓Blautia, Ruminococcus_1↓Coprococcus_2↓	–	FBG↓	–	–	[5]
Holothuria leucospilota polysaccharide	Clostridium↑Turicibacter↑Allobaculum↑Ruminococcus↑	Acetic↑butyric acid↑pentanoic acid↑	FBG↓	Rhamnose, fucose, glucuronic acids, galactose, glucose, and xylose in the ratio were 39.08%, 35.72%, 10.72%, 8.43%, 4.23%, and 1.83%	52.80, 12.01	[103, 104]
Grifola frondosa polysaccharides	Porphyromonas gingivalis↑Akkermansia muciniphila↑Lactobacillus acidophilus↑Tannerella forsythia↑Bacteroides↑Roseburia intestinalis↑	–	FBG↓	L-arabinose, D-mannose, and D-glucose with molar ratios of 3.79:1.00:49.70	1.26×10^4	[105]
Ganoderma lucidum polysaccharides	Enterococcus↑	–	FBG↓	–	–	[106]
Sargassum confusum seaweed oligosaccharides	Lactobacillus↑Clostridium XIVa↑Allobaculum↓Bacteroides↓Clostridium IV↓	–	FBG↓	Galactose, fucose, arabinose, xylose, and glucose	–	[107]
Grifola frondosa polysaccharides	Alistipes↑ Streptococcus↓Enterococcus↓Staphylococcus↓ Aerococcus↓	Acetic↑butyric acid↑pentanoic acid↑	FBG↓	Mannose, rhamnose, glucuronic acid, galacturonic acid, glucose, galactose, and fucose with molar ratio of 25.49:5.18:9.49:7.30:27.59:15.02:9.92	18.18	[108]
Oat β-glucan	Bifidobacterium↑ Lactobacillus↑	–		–	–	[109]

4.2 Polysaccharides Improve Glucometabolism by Gut Microbiota

It had been confirmed that gut microbiota is closely associated with diseases related to glucose metabolism, such as obesity, diabetes, and malnutrition. Supplementation with some polysaccharides can improve gut microbiota disorder and regulate glucose metabolism-related diseases through different mechanisms [28]. Nielsen et al. found that diabetes and impaired glucose tolerance are linked to obesity, and the abundance of phylum Firmicutes and class Clostridia in the human gut was significantly reduced in diabetes [110]. Chen et al. indicated that *Ganoderma lucidum* polysaccharides (GLP) treatment could reduce the abundance of harmful bacteria, such as *Aerococcus*, *Ruminococcus*, *Corynebactrium*, and *Proteus*, and increase the level of *Blautia*, *Dehalobacterium*, *Parabacteroides*, and *Bacteroides*. The research also showed that GLP could restore the disordered gut microbiota to a normal level in T2DM rats and modify host metabolites to achieve its antidiabetic effects [102]. *Holothuria leucospilota* polysaccharide treatment increased short-chain fatty acid-producing bacteria and reduced opportunistic bacterial pathogen in the feces of Goto-Kakizaki (GK) rats [104]. Therefore, it may be an effective method to modify specific bacteria related to glycometabolism disorder by polysaccharides.

5 Conclusions and Future Prospects

T2DM is a global health problem with high morbidity and mortality and no complete cure for type 2 diabetes. Therefore, more and more attention has been paid to the screening of antidiabetic components including polysaccharides from natural sources. In recent years, it has been increasingly recognized that T2DM is associated with gut microbiome composition and function. Fortunately, natural polysaccharides can improve diabetes by modifying gut microflora. Different polysaccharides from different sources are beneficial to different gut microflora. This review indicated that the intake of polysaccharides will increase the number of bacteria in the gut tract, and the relationship between polysaccharides, gut microflora, and blood glucose regulation provides a new perspective for understanding the mechanism of polysaccharide on T2DM disease. Furthermore, a need to further elucidate the structures of polysaccharides to determine the relationship between three-dimensional structures and functions of different polysaccharides.

References

1. Wang N, Zhu F, Chen L, Chen K (2018) Proteomics, metabolomics and metagenomics for type 2 diabetes and its complications. Life Sci 212:194–202

2. Franks PW, Mccarthy MI (2016) Exposing the exposures responsible for type 2 diabetes and obesity. Science 354(6308):69–73
3. Guariguata L, Whiting DR, Hambleton I, Beagley J, Linnenkamp U et al (2014) Global estimates of diabetes prevalence for 2013 and projections for 2035. Diabetes Res Clin Pr 103(2):137–149
4. Gilor C, Niessen SJM, Furrow E, Dibartola SP (2016) What's in a name? Classification of diabetes mellitus in veterinary medicine and why it matters. J Vet Intern Med 30(4):927–940
5. Zhao X-Q, Guo S, Lu Y-Y, Hua Y, Zhang F et al (2020) Lycium barbarum L. leaves ameliorate type 2 diabetes in rats by modulating metabolic profiles and gut microbiota composition. Biomed Pharmacother 121:109559
6. Boyle JP, Thompson TJ, Gregg EW, Barker LE, Williamson DF (2010) Projection of the year 2050 burden of diabetes in the US adult population: dynamic modeling of incidence, mortality, and prediabetes prevalence. Popul Health Metr 8(1):29
7. Gholizadeh P, Mahallei M, Pormohammad A, Varshochi M, Ganbarov K et al (2019) Microbial balance in the intestinal microbiota and its association with diabetes, obesity and allergic disease. Microb Pathog 127:48–55
8. Thomas RM, Jobin C (2020) Microbiota in pancreatic health and disease: the next frontier in microbiome research. Nat Rev Gastro Hepat 17(1):53–64
9. Gu J-F, Su S-L, Guo J-M, Zhu Y, Zhao M et al (2017) The aerial parts of Salvia miltiorrhiza Bge. strengthen intestinal barrier and modulate gut microbiota imbalance in streptozocin-induced diabetic mice. J Funct Foods 36:362–374
10. Xu X, Xu P, Ma C, Tang J, Zhang X (2013) Gut microbiota, host health, and polysaccharides. Biotechnol Adv 31(2):318–337
11. Abreu TM (2010) Toll-like receptor signalling in the intestinal epithelium: how bacterial recognition shapes intestinal function. Nat Rev Immunol 10(2):131–144
12. Frank DN, Amand ALS, Feldman RA, Boedeker EC, Pace NR (2007) Molecular-phylogenetic characterization of microbial community imbalances in human inflammatory bowel diseases. Proc Natl Acad Sci U S A 104(34):13780–13785
13. Li M, Wang B, Zhang M, Rantalainen M, Zhao L (2008) Symbiotic gut microbes modulate human metabolic phenotypes. Proc Natl Acad Sci U S A 105(6):2117–2122
14. Allen RH, Stabler SP (2008) Identification and quantitation of cobalamin and cobalamin analogues in human feces. Am J Clin Nutr 87(5):1324
15. Goodman AL, Mcnulty NP, Zhao Y, Leip D, Mitra RD et al (2009) Identifying genetic determinants needed to establish a human gut symbiont in its habitat. Cell Host Microbe 6 (3):279–289
16. Lee YK, Menezes JS et al (2011) Proinflammatory T-cell responses to gut microbiota promote experimental autoimmune encephalomyelitis. Proc Natl Acad Sci 108(Suppl 1):4615–4622
17. Kerckhoffs APM, Ben-Amor K, Samsom M, Van Der Rest ME, De Vogel J et al (2011) Molecular analysis of faecal and duodenal samples reveals significantly higher prevalence and numbers of Pseudomonas aeruginosa in irritable bowel syndrome. J Med Microbiol 60 (2):236–245
18. Lee YK, Mazmanian SK (2010) Has the microbiota played a critical role in the evolution of the adaptive immune system? Science 330(6012):1768–1773
19. Blaser MJ, Falkow S (2009) What are the consequences of the disappearing human microbiota. Nat Rev Microbiol 7(12):887
20. Gurung M, Li Z, You H, Rodrigues R, Jump DB et al (2020) Role of gut microbiota in type 2 diabetes pathophysiology. EBioMedicine 51:102590
21. Bahramzadeh S, Tabarsa M, You S, Li C, Bita S (2019) Purification, structural analysis and mechanism of murine macrophage cell activation by sulfated polysaccharides from Cystoseira indica. Carbohydr Polym 205:261–270
22. Stephanie B, Eric D, Sophie FM, Christian B, Yu G (2010) Carrageenan from Solieria chordalis (Gigartinales): structural analysis and immunological activities of the low molecular weight fractions. Carbohydr Polym 81(2):448–460

23. Yu Y, Shen M, Song Q, Xie J (2017) Biological activities and pharmaceutical applications of polysaccharide from natural resources: a review. Carbohydr Polym 183:91–101
24. Zou W, Li J, Vinogradov E, Cox A (2019) Removal of cell wall polysaccharide in pneumococcal capsular polysaccharides by selective degradation via deamination. Carbohydr Polym 218:199–207
25. Nie C, Zhu P, Ma S, Wang M, Hu Y (2018) Purification, characterization and immunomodulatory activity of polysaccharides from stem lettuce. Carbohydr Polym 188:236–242
26. Wang H-Y, Guo L-X, Hu W-H, Peng Z-T, Wang C et al (2019) Polysaccharide from tuberous roots of Ophiopogon japonicus regulates gut microbiota and its metabolites during alleviation of high-fat diet-induced type-2 diabetes in mice. J Funct Foods 63:103593
27. Micha R, Penalvo JL, Cudhea F, Imamura F, Rehm CD et al (2017) Association between dietary factors and mortality from heart disease, stroke, and type 2 diabetes in the United States. JAMA 317(9):912–924
28. Fang Q, Hu J, Nie Q, Nie S (2019) Effects of polysaccharides on glycometabolism based on gut microbiota alteration. Trends Food Sci Technol 92:65–70
29. Zheng Y, Bai L, Zhou Y, Tong R, Zeng M et al (2019) Polysaccharides from Chinese herbal medicine for anti-diabetes recent advances. Int J Biol Macromol 121:1240–1253
30. Wu J, Shi S, Wang H, Wang S (2016) Mechanisms underlying the effect of polysaccharides in the treatment of type 2 diabetes: a review. Carbohydr Polym 144:474–494
31. Ye Y, Deng T, Wan XY, Ouyang JP, Liu M et al (2014) The role of quantitative changes in the expression of insulin receptor substrate-1 and nuclear ubiquitin in abnormal glycometabolism in the livers of KKay mice and the relative therapeutic mechanisms of Astragalus polysaccharide. Int J Mol Med 33(2):341–350
32. Ke B, Ke X, Wan X (2017) Astragalus polysaccharides attenuates TNF-α-induced insulin resistance via suppression of miR-721 and activation of PPAR-γ and PI3K/AKT in 3T3-L1 adipocytes. Am J Transl Res 9(5):2195
33. Bhateja PK, Singh R (2014) Antidiabetic activity of Acacia tortilis (Forsk.) Hayne ssp. raddiana polysaccharide on streptozotocin-nicotinamide induced diabetic rats. Biomed Res Int 2014:572013
34. Ming Y, Qiu T, Peng W, Chen W-X, Ye Y-W et al (2011) Purification, characterization and hypoglycemic activity of extracellular polysaccharides from Lachnum calyculiforme. Carbohydr Polym 86(1):285–290
35. Wang L, Hao WU, Chang N, Zhang K (2011) Anti-hyperglycemic effect of the polysaccharide fraction isolated from Mactra veneriformis. Front Chem Sci Eng 5(2):238–244
36. Duchartre Y, Kim YM, Kahn M (2016) The Wnt signaling pathway in cancer. Crit Rev Oncol Hemat 99:141–149
37. Wang J, Jin W, Zhang W, Hou Y, Zhang H et al (2013) Hypoglycemic property of acidic polysaccharide extracted from Saccharina japonica and its potential mechanism. Carbohydr Polym 95(1):143–147
38. Sawa M, Masuda M, Yamada T (2016) Targeting the Wnt signaling pathway in colorectal cancer. Expert Opin Ther Tar 20(4):419
39. Hong T, Zhao J, Dong M, Meng Y, Mu J et al (2012) Composition and bioactivity of polysaccharides from Inula britannica flower. Int J Biol Macromol 51(4):550–554
40. Zhao W, Yin Y, Yu Z, Liu J, Chen F (2012) Comparison of anti-diabetic effects of polysaccharides from corn silk on normal and hyperglycemia rats. Int J Biol Macromol 50 (4):1133–1137
41. Jin H, Zhang Y-J, Jiang J-X, Zhu L-Y, Chen P et al (2013) Studies on the extraction of pumpkin components and their biological effects on blood glucose of diabetic mice. J Food Drug Anal 21(2):184–189
42. Zhao, Liao ZS, Wu XQ et al (2014) Isolation, purification, and structural features of a polysaccharide from Phellinus linteus and its hypoglycemic effect in alloxan-induced diabetic mice. J Food Sci 79:H1002–H1010

43. Xu W, Zhou Q, Yin J-J et al (2015) Anti-diabetic effects of polysaccharides from Talinum triangulare in streptozotocin (STZ)-induced type 2 diabetic male mice. Int J Biol Macromol 72:575–579
44. Ruderman NB, Carling D, Prentki M, Cacicedo JM (2013) AMPK, insulin resistance, and the metabolic syndrome. J Clin Invest 123(7):2764–2772
45. Li Q, Li W, Gao Q, Zou Y (2017) Hypoglycemic effect of Chinese Yam (Dioscorea opposita rhizoma) polysaccharide in different structure and molecular weight. J Food Sci 82(2):1–8
46. Zhang Y, Ren C, Lu G, Cui W, Mu Z et al (2014) Purification, characterization and anti-diabetic activity of a polysaccharide from mulberry leaf. Regul Toxicol Pharm 70(3):687–695
47. Zhu K, Nie S, Gong D, Xie M (2014) Effect of polysaccharide from Ganoderma atrum on the serum metabolites of type 2 diabetic rats. Food Hydrocolloid 53:31–36
48. Zhu K, Nie S, Li C, Lin S, Xing M et al (2013) A newly identified polysaccharide from Ganoderma atrum attenuates hyperglycemia and hyperlipidemia. Int J Biol Macromol 57:142–150
49. Zheng J, Yang B, Yu Y, Chen Q, Huang T et al (2012) Ganoderma lucidum polysaccharides exert anti-hyperglycemic effect on streptozotocin-induced diabetic rats through affecting β-cells. Comb Chem High Throughput Screen 15(7):542–550
50. Ricketts CJ, Hill VK, Linehan WM (2014) Tumor-specific hypermethylation of epigenetic biomarkers, including SFRP1, predicts for poorer survival in patients from the TCGA Kidney Renal Clear Cell Carcinoma (KIRC) Project. PLos One 9(1):e85621
51. Zhu H-Y, Chen G-T, Meng G-L, Xu J-L (2015) Characterization of pumpkin polysaccharides and protective effects on streptozotocin-damaged islet cells. Chin J Nat Med 13(3):199–207
52. Hu F, Li X, Zhao L, Feng S, Wang C (2010) Antidiabetic properties of purified polysaccharide from Hedysarum polybotrys. Can J Physiol Pharm 88(1):64–72
53. Lin B, Hong H, Jiang X, Li C, Zhu S et al (2017) WNT inhibitory factor 1 promoter hypermethylation is an early event during gallbladder cancer tumorigenesis that predicts poor survival. Gene 62:42–49
54. Yun L, Wu T, Li Q, Zhang M (2018) Dietary supplementation with purified wheat germ glycoprotein improve immunostimulatory activity in cyclophosphamide induced Balb/c mice. Int J Biol Macromol 118(Pt A):1267–1275
55. Hong L, Qin W, Shuzhen G, Juncheng H, Hanju S et al (2013) The protective effect of MT-α-glucan against streptozotocin (STZ)-induced NIT-1 pancreatic β-cell damage. Carbohydr Polym 92(2):1211–1217
56. Lei H, Zhang M, Wang Q, Guo S, Wu W (2013) MT-α-glucan from the fruit body of the maitake medicinal mushroom Grifola frondosa (higher basidiomyetes) shows protective effects for hypoglycemic pancreatic β-cells. Int J Med Mushrooms 15(4):373–381
57. Liu Y, Sun J, Rao S, Su Y, Yang Y (2013) Antihyperglycemic, antihyperlipidemic and antioxidant activities of polysaccharides from Catathelasma ventricosum in streptozotocin-induced diabetic mice. Food Chem Toxicol 57:39–45
58. Chen X, Jin J, Tang J, Wang Z, Wang J et al (2011) Extraction, purification, characterization and hypoglycemic activity of a polysaccharide isolated from the root of Ophiopogon japonicus. Carbohydr Polym 83(2):749–754
59. Dahech I, Belghith KS, Hamden K, Feki A, Belghith H et al (2011) Antidiabetic activity of levan polysaccharide in alloxan-induced diabetic rats. Int J Biol Macromol 49(4):742–746
60. Sun C, Chen Y, Li X, Tai G, Fan Y et al (2014) Anti-hyperglycemic and anti-oxidative activities of ginseng polysaccharides in STZ-induced diabetic mice. Food Funct 5(5):845
61. Ganeshpurkar A, Kohli S, Rai G (2014) Antidiabetic potential of polysaccharides from the white oyster culinary-medicinal mushroom Pleurotus florida (higher basidiomycetes). Int J Biol Macromol 16(3):207
62. Hotamisligil GS (2000) Molecular mechanism of insulin resistance and the role of the adipocyte. Int J Obes Relat Metab Disord 24(Suppl 4(Suppl 3)):S23–S27

63. Hu J, Pang W, Chen J, Bai S, Wu X (2013) Hypoglycemic effect of polysaccharides with different molecular weight of Pseudostellaria heterophylla. BMC Complement Altern Med 13 (1):267
64. Kanagasabapathy G, Chua KH, Malek SNA, Vikineswary S, Kuppusamy UR (2014) AMP-activated protein kinase mediates insulin-like and lipo-mobilising effects of β-glucan-rich polysaccharides isolated from Pleurotus sajor-caju (Fr.), Singer mushroom, in 3T3-L1 cells. Food Chem 145:198–204
65. Guo C, Liang T, He Q, Wei P, Zheng N et al (2013) Renoprotective effect of Ramulus mori polysaccharides on renal injury in STZ-diabetic mice. Int J Biol Macromol 62:720–725
66. Zhou J, Yan J, Bai Z, Li K, Huang K (2015) Hypoglycemic activity and potential mechanism of a polysaccharide from the loach in streptozotocin-induced diabetic mice. Carbohydr Polym 121:199–206
67. Zhou J, Xu G, Yan J, Li K, Bai Z et al (2015) Rehmannia glutinosa (Gaertn.) DC. polysaccharide ameliorates hyperglycemia, hyperlipemia and vascular inflammation in streptozotocin-induced diabetic mice. J Ethnopharmacol 164:229–238
68. Kasipandi M, Manikandan A, Sreeja PS, Suman T, Saikumar S et al (2019) Effects of in vitro simulated gastrointestinal digestion on the antioxidant, α-glucosidase and α-amylase inhibitory activities of water-soluble polysaccharides from Opilia amentacea roxb fruit. LWT 111:774–781
69. Xu Y, Guo Y, Gao Y, Niu X, Wang L et al (2018) Seperation, characterization and inhibition on α-glucosidase, α-amylase and glycation of a polysaccharide from blackcurrant fruits. LWT 93:16–23
70. Wang C, Gao X, Santhanam RK, Chen Z, Chen Y et al (2018) Effects of polysaccharides from Inonotus obliquus and its chromium (III) complex on advanced glycation end-products formation, α-amylase, α-glucosidase activity and H2O2-induced oxidative damage in hepatic L02 cells. Food Chem Toxicol 116:335–345
71. Wang Y, Yang Z, Wei X (2010) Sugar compositions, α-glucosidase inhibitory and amylase inhibitory activities of polysaccharides from leaves and flowers of Camellia sinensis obtained by different extraction methods. Int J Biol Macromol 47(4):534–539
72. Meng Q, Chen F, Xiao T, Zhang L (2019) Inhibitory effects of polysaccharide from Diaphragma juglandis fructus on α-amylase and α-d-glucosidase activity, streptozotocin-induced hyperglycemia model, advanced glycation end-products formation, and H2O2-induced oxidative damage. Int J Biol Macromol 124:1080–1089
73. Wang L-Y, Wang Y, Xu D-S, Ruan K-F, Feng Y et al (2012) MDG-1, a polysaccharide from Ophiopogon japonicus exerts hypoglycemic effects through the PI3K/Akt pathway in a diabetic KKAy mouse model. J Ethnopharmacol 143(1):347–354
74. Ren C, Zhang Y, Cui W, Lu G, Wang Y et al (2015) A polysaccharide extract of mulberry leaf ameliorates hepatic glucose metabolism and insulin signaling in rats with type 2 diabetes induced by high fat-diet and streptozotocin. Int J Biol Macromol 72:951–959
75. Chen Z, Ma T, Huang C, Zhang L, Lv X et al (2013) JMiR-27a modulates the MDR1/P-glycoprotein expression by inhibiting FZD7/beta-catenin pathway in hepatocellular carcinoma cells[J]. Cell Signal 25(12):2693–2701
76. Liu Y, Wan L, Xiao Z, Wang J, Wang Y et al (2013) Antidiabetic activity of polysaccharides from tuberous root of Liriope spicata var. prolifera in KKAy mice. Evid-Based Compl Alt:1–11
77. Ma X, Zhou F, Chen Y, Zhang Y, Hou L et al (2014) A polysaccharide from Grifola frondosa relieves insulin resistance of HepG2 cell by Akt-GSK-3 pathway. Glycoconj J 31(5):355–363
78. Hu S, Xu H, Chen R, Wang J, Li Z et al (2014) Activation of PKB and ERK, but not PI3K, is involved in fucosylated chondroitin sulphate from Acaudina molpadioides induced glucose uptake. J Funct Foods 10:385–396
79. He C, Wang L, Zhang J, Xu H (2017) Hypoxia-inducible microRNA-224 promotes the cell growth, migration and invasion by directly targeting RASSF8 in gastric cancer. Mol Cancer 16 (1):35

80. Ono K, Kuwabara Y, Han J (2011) MicroRNAs and cardiovascular diseases. FEBS J 278 (10):1619–1633
81. Huang M, Wang F, Zhou X, Yang H, Wang Y (2015) Hypoglycemic and hypolipidemic properties of polysaccharides from Enterobacter cloacae Z0206 in KKAy mice. Carbohydr Polym 117:91–98
82. Zhou XX, Wang X (2013) Role of microRNAs in chronic lymphocytic leukemia (Review). Mol Med Rep 8(3):719
83. Xiao C, Wu Q-P, Cai W, Tan J-B, Yang X-B et al (2012) Hypoglycemic effects of Ganoderma lucidum polysaccharides in type 2 diabetic mice. Arch Pharm Res 35(10):1793–1801
84. Wang KC, Yang YW, Liu B, Sanyal A, Corces-Zimmerman R et al (2011) A long noncoding RNA maintains active chromatin to coordinate homeotic gene expression. Nature 472 (7341):120–124
85. Caesar R (2019) Pharmacologic and nonpharmacologic therapies for the gut microbiota in type 2 diabetes. Can J Diabetes 43(3):224–231
86. Sun Y (2011) Structure and biological activities of the polysaccharides from the leaves, roots and fruits of Panax ginseng C.A. Meyer: an overview. Carbohydr Polym 85(3):490–499
87. Plovier H, Everard A, Druart C, Depommier C, Van Hul M et al (2017) A purified membrane protein from Akkermansia muciniphila or the pasteurized bacterium improves metabolism in obese and diabetic mice. Nat Med 23(1):107–113
88. Hsieh F-C, Lee C-L, Chai C-Y, Chen W-T, Lu Y-C et al (2013) Oral administration of Lactobacillus reuteri GMNL-263 improves insulin resistance and ameliorates hepatic steatosis in high fructose-fed rats. Nutr Metab 10(1):35
89. Backhed F, Ding H, Wang T, Hooper LV, Koh GY et al (2004) The gut microbiota as an environmental factor that regulates fat storage. Proc Natl Acad Sci U S A 101 (44):15718–11523
90. Lamprecht M, Bogner S, Schippinger G, Steinbauer K, Fankhauser F et al (2012) Probiotic supplementation affects markers of intestinal barrier, oxidation, and inflammation in trained men; a randomized, double-blinded, placebo-controlled trial. J Int Soc Sport Nutr 9(1):45
91. Park J, Goergen CJ, Hogenesch H, Kim CH (2016) Chronically elevated levels of short-chain fatty acids induce T cell-mediated ureteritis and hydronephrosis. J Immunol 1502046
92. Prawitt J, Abdelkarim M, Stroeve JHM, Popescu I, Staels B (2011) Farnesoid X receptor deficiency improves glucose homeostasis in mouse models of obesity. Diabetes 60 (7):1861–1871
93. Zhang Y, Chen P, Zhang Y, Jin H, Zhu L et al (2013) Effects of polysaccharide from pumpkin on biochemical indicator and pancreatic tissue of the diabetic rabbits. Int J Biol Macromol 62:574–581
94. Liu G, Liang L, Yu G, Li Q (2018) Pumpkin polysaccharide modifies the gut microbiota during alleviation of type 2 diabetes in rats. Int J Biol Macromol 115:711–717
95. Zheng Y, Bai L, Zhou Y et al (2018) Polysaccharides from Chinese herbal medicine for anti-diabetes recent advances. Int J Biol Macromol 121:1240–1253
96. Kaoutari AE, Armougom F, Gordon JI, Raoult D, Henrissat B (2013) The abundance and variety of carbohydrate-active enzymes in the human gut microbiota. Nat Rev Microbiol 11 (7):1–8
97. Lam KL, Chi-Keung CP (2013) Non-digestible long chain beta-glucans as novel prebiotics. Bioact Carbohydr Diet Fibre 2(1):45–64
98. Kuda T, Enomoto T, Yano T (2009) Effects of two storage β-1,3-glucans, laminaran from Eicenia bicyclis and paramylon from Euglena gracili, on cecal environment and plasma lipid levels in rats. J Funct Foods 1(4):399–404
99. Lynch MB, Sweeney T, Callan JJ, O'sullivan JT, O'doherty JV (2010) The effect of dietary Laminaria derived laminarin and fucoidan on intestinal microflora and volatile fatty acid concentration in pigs. Livest Sci 2010(133):4

100. Zaporozhets TS, Besednova NN, Kuznetsova TA, Zvyagintseva TN, Makarenkova ID et al (2014) The prebiotic potential of polysaccharides and extracts of seaweeds. Russ J Mar Biol 40 (1):1–9
101. Li J, Pang B, Shao D, Jiang C, Hu X et al (2019) Artemisia sphaerocephala Krasch polysaccharide mediates lipid metabolism and metabolic endotoxaemia in associated with the modulation of gut microbiota in diet-induced obese mice. Int J Biol Macromol
102. Chen M, Xiao D, Liu W, Song Y, Zou B et al (2019) Intake of Ganoderma lucidum polysaccharides reverses the disturbed gut microbiota and metabolism in type 2 diabetic rats. Int J Biol Macromol
103. Yuan Y, Li C, Zheng Q, Wu J, Zhu K et al (2018) Effect of simulated gastrointestinal digestion in vitro on the antioxidant activity, molecular weight and microstructure of polysaccharides from a tropical sea cucumber (Holothuria leucospilota). Food Hydrocolloid 89:735–741
104. Zhao F, Liu Q, Cao J, Xu Y, Pei Z et al (2020) A sea cucumber (Holothuria leucospilota) polysaccharide improves the gut microbiome to alleviate the symptoms of type 2 diabetes mellitus in Goto-Kakizaki rats. Food Chem Toxicol 135:110886
105. Chen Y, Liu D, Wang D, Lai S, Zhong R et al (2019) Hypoglycemic activity and gut microbiota regulation of a novel polysaccharide from Grifola frondosa in type 2 diabetic mice. Food Chem Toxicol 126:295–302
106. Xu S, Dou Y, Ye B, Wu Q, Wang Y et al (2017) Ganoderma lucidum polysaccharides improve insulin sensitivity by regulating inflammatory cytokines and gut microbiota composition in mice. J Funct Foods 38:545–552
107. Yang C-F, Lai S-S, Chen Y-H, Liu D, Liu B et al (2019) Anti-diabetic effect of oligosaccharides from seaweed Sargassum confusum via JNK-IRS1/PI3K signalling pathways and regulation of gut microbiota. Food Chem Toxicol 131:110562
108. Guo W-L, Deng J-C, Pan Y-Y, Xu J-X, Hong J-L et al (2019) Hypoglycemic and hypolipidemic activities of Grifola frondosa polysaccharides and their relationships with the modulation of intestinal microflora in diabetic mice induced by high-fat diet and streptozotocin. Int J Biol Macromol
109. Zhang PP, Hu XZ, Zhen HM, Xu C, Fan MT (2012) Oat beta-glucan increased ATPases activity and energy charge in small intestine of rats. J Agric Food Chem 60(39):9822–9827
110. Larsen N, Vogensen FK, Van Den Berg FWJ, Nielsen DS, Andreasen AS et al (2010) Gut microbiota in human adults with type 2 diabetes differs from non-diabetic adults. PLos One 5 (2):e9085

Chapter 9
An Overview of Polysaccharides and the Influence Factors of Hypoglycemic Activity

Haixia Chen and Yajie Wang

Abstract Polysaccharides are the most abundant biopolymers in nature, which are widely found in animals and plants, microorganisms, bacteria, fungi, and seaweeds. They are natural products with many bioactivities, such as immune regulation, antioxidant, hypoglycemia, anti-tumor. Hypoglycemic polysaccharides have become the focus of attention in recent years. With the further research on the relationship between polysaccharides and diabetes, studies have indicated that some polysaccharides showed the advantages of good safety, exact curative effect, no significant side effects, improving the complications of diabetes. This review summarizes recent researches on the natural polysaccharides and their determinant factors. In view of the significant performance of polysaccharides in traditional medicine and food industry, it is important to summarize the determinant factors of polysaccharides hypoglycemic activity.

Keywords Polysaccharides · Hypoglycemic · Polysaccharides complex · Determinant factors

1 Introduction

Polysaccharides are a kind of important natural macromolecular substance, which are widely exist in higher plants, animals, and microbial cells [1]. It is a long chain polymer composed of more than 10 monosaccharide molecules connected together by glycosidic bond with the general formula $C_x(H_2O)_y$ [2]. In natural system, polysaccharides play an important role in biological functions such as energy storage, structural support, defense, and antigenicity. They constitute life activities and act as one of the four basic substances [3]. Polysaccharides have been studied for more than a century, and there are currently hundreds of active polysaccharides.

H. Chen (✉) · Y. Wang
Tianjin Key Laboratory for Modern Drug Delivery and High-Efficiency, School of Pharmaceutical Science and Technology, Tianjin University, Tianjin, China
e-mail: chenhx@tju.edu.cn

H. Chen, M. Zhang (eds.), *Structure and Health Effects of Natural Products on Diabetes Mellitus*, https://doi.org/10.1007/978-981-15-8791-7_9

According to their sources, they can be divided into five categories: fungal polysaccharides, plant polysaccharides, algae polysaccharides, animal polysaccharides, and bacterial polysaccharides [1].

1.1 Plant Polysaccharides

In recent decades, polysaccharides extracted from plants have been found to have important and special physiological activities. These polysaccharides participate in various activities of life cells and have a variety of biological functions, such as participating in immune regulation of organisms, reducing blood sugar, blood lipids, antivirus, anti-tumor, anti-fatigue, anti-aging, and so on [3]. Due to the limitations and adverse reactions of western medicine in the treatment of diabetes, such as hypoglycemia and complications caused by long-term use, it has become a trend to find natural active substances and develop functional and nutritious health foods. At present, polysaccharides have been successfully extracted from nearly 100 kinds of plants and widely used in the research and development of medicine and health food. Many plant polysaccharides have been reported to exhibit an effect of reducing blood sugar, such as tea polysaccharides, corn silk polysaccharides, Atractylodes polysaccharides, yam polysaccharide, and so on [4]. Ganoderma atrum polysaccharide has obvious hypoglycemic effect on hypoglycemic mice induced by streptozotocin against pancreatic damage [5]. Polysaccharide from azuki bean could significantly increase the expression of insulin receptor, insulin receptor substrate-1, phosphoinositide 3-kinase, protein kinase B, and glucose transporter-2, which indicated that the glucose metabolism is activating the PI3K/AKT signaling pathway [6].

1.2 Fungal Polysaccharides

Edible fungal polysaccharides possess many biological activities, such as anti-tumor, immune regulation, hypoglycemic, anti-inflammatory, and so on. At present, the research on β-glucan in fungal polysaccharides is relatively in-depth, while the research on the structure and activity of other types of fungal polysaccharides is less [5, 7]. Mushroom polysaccharides consist of D-glucose monomers, two glucose units combined together via the α-glycosidic bond or β-glycosidic bond [8]. The study of hypoglycemic activity shows that these galactosans and glucans have certain hypoglycemic activity, and their activities are different according to their structures [9]. It has been found that the mechanism of hypoglycemic activity of fungal polysaccharides was related to the improvement of intestinal flora. The results enrich the structural information of fungal polysaccharides and provide an important theoretical basis for the research and development of the structure–activity relationship of fungal polysaccharides [10]. The polysaccharides from Gomphidiaceae rutilus fruiting bodies possess the hypoglycemic activity by promoting insulin stimulated glucose uptake in high glucose and fatty acid treated hepatocytes in an autophagic manner [11]. The results enrich the structural information of fungal polysaccharides and provide an important theoretical basis for the research and development of the structure–activity relationship of fungal polysaccharides.

1.3 Algae Polysaccharides

The source of algal polysaccharides can be divided into four categories: brown algal polysaccharides, red algal polysaccharides, green algal polysaccharides, and blue algal polysaccharides, of which the former two have been studied more popular [12]. With the development of marine drugs, seaweed polysaccharides were considered as valuable natural health food [13]. The physiological activities of polysaccharides are closely related to its structure. The relationship between the hypoglycemic effect of algal polysaccharide and its structure has not been fully elucidated [14]. The structure–activity relationship of polysaccharides refers to the relationship between the primary structure, advanced structure, physical and chemical properties of polysaccharides and their biological activities. Red algal polysaccharide mainly consists of agar, carrageenan, and agar carrageenan intermediate polysaccharide, which are all galactosans combined with galactose [15]. Galactan is a kind of colloidal polysaccharides, which belongs to soluble dietary fiber and has high viscosity. Therefore, it can delay gastric emptying, small intestine digestion, and carbohydrate absorption, thus preventing the rapid rise of blood sugar [16]. Marine-derived polysaccharides with low molecular weight and low toxicity, thus, it could enhance insulin sensitivity. For instance, Alginate is a natural polysaccharide identified from various marine brown seaweeds [17]. Fucoidin is a linear polymer composed of glucuronic acid. Fucoidin is mainly composed of l-fucose-4-sulfate. Fucoidin is dextran composed of glucose. Dextran can be divided into α-glucan and β-glucan. In the study of the structure and diabetes activity of α-glucan, it was found that when the O-3 position of α-glucan with 1,6-linked α-glucose as the main chain contained branched chain, the polysaccharide showed hypoglycemic effect [18].

1.4 Animal Polysaccharides

Animal polysaccharides are widely found in animal organisms. Some active polysaccharides in animals can significantly reduce blood sugar, such as sea cucumber, *cordyceps sinensis*, propolis, loach. *Corbicula fluminea* is a fresh-water bivalve mollusk, which displayed anti-diabetic activities by activation of the PI3K/Akt signaling pathway in streptozotocin induced diabetic rats [19]. The polysaccharide of *Beauveria bassiana* has hypoglycemic effect on STZ diabetic mice [20]. Its hypoglycemic effect may be related to the scavenging oxygen free radicals and its anti-lipid peroxidation pathway, rather than through promoting insulin secretion by β cells of islets of Langerhans.

1.5 Bacterial Polysaccharides

PHEA is a polysaccharide purified from *Paecilomyces hepiali* fermented mycelium, which has anti-diabetic and anti-nephritic activities via Nrf2-meadited NF-κB signaling in db/db mice [11]. Z0206 bacterial polysaccharides could reduce the blood glucose by inhibiting the process of gluconeogenesis together with increasing the liver's glucose intake and utilization of blood glucose [21].

2 Determinant Factors of Hypoglycemic Polysaccharides

2.1 Preparation of Polysaccharides

The extraction process of polysaccharide is the basis of its follow-up research work, so the study of natural active polysaccharide is the first choice of the best extraction process [22]. The commonly used extraction methods of polysaccharide include water extraction, enzymolysis, ultrasonic extraction, and microwave extraction. When the extraction rate of a single extraction method is low, several methods are usually used to assist extraction to get a higher extraction rate. Enzymolysis-ultrasonic assisted extraction is an important method employed to extract polysaccharides from corn silk. The optimized enzymolysis condition was employed with the cellulose content of 7.5% for 150 min at 55 °C and ultrasonic pretreatment was employed with liquid–solid ratio of 31.8 for 34.2 min at 66.3 °C [23]. Digested polysaccharides under different simulated salivary, gastric, and intestinal conditions showed various physicochemical properties and bioactivities. Li et al. concluded that there was no change in the oral phase and parts of peptide bonds were broken down in gastric digestion. But the morphological, molecular weight, and molar ratio of monosaccharide were significantly changed, which might be reason for the improvement of α-amylase inhibitory effect [24]. Wang et al. also indicated the enzyme inhibitory activities of polysaccharides from *Inonotus obliquus* improved after intestinal digestion [25]. Chen et al. compared four extraction methods, including hot-water extraction (HWE), ultrasound-assisted extraction (UE) and microwave-assisted extraction (ME), and high-pressure extraction (HPE). The yield of polysaccharides was highest by HWE, while the hypoglycemic activity was elucidated that was significant high by UE as the lowest molecular weight [26]. The physicochemical properties of tea polysaccharides was changed and the yield and the antioxidant property was improved by extrusion treatment, which might be helpful for the tea polysaccharide related products in the food industry [27]. Following are the preparation methods that influence the hypoglycemic activity of polysaccharides (Table 9.1).

Table 9.1 Preparation of polysaccharides that influence the hypoglycemic activity

Methods	Resources of polysaccharides	Changes	Effects	Reference
Simulated digestion in vitro	Corn silk	Morphological, molecular weight, molar ratio of monosaccharide	α-amylase inhibition ↑	[24]
	Inonotus obliquus	Molecular weight (*Mw*), disrupting the aggregates and glycosidic bonds	α-amylase and α-glucosidase inhibition ↑	[25]
Enzymolysis-ultrasonic assisted extraction	Corn silk	Molecular ratio of monosaccharide, *Mv*, morphological and conformation	Yield ↑	[23]
Alkali-extraction	Arabinoxylans	Higher solubility and free ferulic acid level	α-amylase inhibition ↑	[28]
	Ginger stems and leaves	Higher uronic acid content, sulfate content and lower molecular weight, including higher free carboxyl and hydroxyl groups	α-glucosidase inhibition ↑	[29]
	Apocynum venetum leaves	Reverse the gut microbiota dysbiosis	Decreased the levels of fasting blood glucose, reverse the gut microbiota dysbiosis	[15]
Chitosan–whey protein nanoparticles	*Tamarindus indica* L.seeds	Prevention of hydrolysis	Maintain antitrypsin activity in the gastric	[30]
Ultrasonic assisted extraction	*Schizophyllum commune*	Lowest molecular weight	α-amylase and α-glucosidase inhibition ↑	[26]
Artificial selenium-enriched	Green tea polysaccharides	C-1 and C-6 position in the form of the Se-H bond	Glycosidase inhibitory activity ↑	[31]
Liquid fermentation	*Coprinus comatus*	Monosaccharides composition: Galactose and xylose	α-amylase activity ↑	[32]
Citric acid (pH 3.0) solvent extraction	*Mentha haplocalyx*	Higher content of uronic acid and its lower molecular weight, and thus its higher content of free carboxyl and hydroxyl groups	Inhibitory activities on α-glucosidase and α-amylase ↑	[33]

2.2 Molecular Weight

The biological characteristics of polysaccharides are related to their composition and structure, in which molecular weight plays an important role [29, 34]. At present, the activity of polysaccharide is closely related to its molecular weight [35]. A large number of studies have shown that antioxidant and anticoagulant activities are the characteristics of low molecular weight polysaccharides. Xue et al. obtained soluble dietary fiber homogeneous subfractions via stepwise ethanol precipitation from *Lentinula edodes* and studied the structural characterization, physiochemical and functional properties for first time. It has been indicated that the subfractions of four LESDF had different molecular weights [36]. The molecular weight of three polysaccharides separated by ultrafiltration was 42.73, 62.16, and 92.09 kDa, respectively. Among these, polysaccharides in 62.16 kDa molecular weight displayed appreciable hypoglycemic activity by regulating PI3K/Akt signaling pathway and promoted the translocation of GLUT4 then increased the glucose uptake in L6 cells [37]. The molecular weight (Mw) of polysaccharides from *Inonotus obliquus* was steadily decreased from 105.02 kDa to 88.97 kDa and 36.74 kDa after simulated digestion, which might be the reason of the improvement of hypoglycemic activity polysaccharides [25].

2.3 Structure–Activity Relationship

2.3.1 Monosaccharide Composition

Monosaccharide composition is the key factor affecting the activity of polysaccharides. The monosaccharide composition of bacterial, fungal, and algal polysaccharides is moderately complex between animals and plants [38]. Studies showed that polysaccharides possessed lower uronic acid content (13.18%), and higher neutral sugar content (76.73%) displayed more significant antioxidant activities and inhibitory effects on the proliferation of human prostate [23].

2.3.2 Molecular Modification of Polysaccharides

The molecular modification of polysaccharides is to modify its structure by physical and chemical means to obtain higher or new biological activity, such as sulfation, desulfuration, and acetylation of polysaccharides [30, 39]. Sulfated polysaccharide is a kind of polysaccharide with sulfate group, including natural and synthetic sulfated polysaccharide derivatives extracted from plants [40]. Chen et al. compared sulfated, acetylated, and carboxymethylated derivatives of corn silk polysaccharides (CSPS), among which the carboxymethylated CSPS showed the highest inhibitory effect on α-amylase. The reason might be the hyperbranched conformation and acid polysaccharides content [41]. Zhu et al. concluded that the C-6 hydroxyl groups in polysaccharides were replaced by selenium esters, and most of them replaced the C-1 and C-6 hydroxyls in the form of Se-H bonds. Selenium-enriched methods displayed significant effects on the structural characteristics of selenium, hypoglycemic activity, and the binding sites or modes of selenium in tea polysaccharides [31].

2.3.3 Polysaccharide Complex

Flavonoids are easily to form van der Waals forces and hydrogen bonds with polysaccharides, which can improve the biological activity of polysaccharides, such as α-amylase and α-glucosidase inhibition [42]. Studies indicated that the selenium polysaccharide of natural green tea and found that the selenium polysaccharide of natural green tea can activate glucokinase, enhance the activity of liver glucokinase, and play a role in lowering blood sugar [43]. It has been studied that the structure of pumpkin polysaccharide and found that the copper in polysaccharide was in a stable chelating state. Copper salt could promote the activity of enzyme and reduce blood sugar. Therefore, metal ions also play an important role in lowering blood glucose, which can modify the natural polysaccharide to enhance or make it have hypoglycemic activity [44]. *Inonotus obliquus* polysaccharides chromium (III) complex showed better regulation capacity of the postprandial hyperglycemia than original poly polysaccharides [45]. Polysaccharides and flavonoids derived from corn silk could interact by van der Waals forces and hydrogen bonding and improve the α-amylase and α-glucosidase inhibition activities [42]. A novel and water-soluble polysaccharide–protein complex (PSP) separated from *Corbicula fluminea* was demonstrated that it could inhibit α-glucosidase and α-amylase, reduce fasting blood glucose levels, and elevate glucose tolerance [19]. In the alkaline condition, the purified Astragalus polysaccharide was complexed with 3-valent chromium ion to obtain the chromium complex of Astragalus polysaccharide [46]. Li et al. formulated a new nanoparticle for PTX delivery with zein as the core together with green tea polysaccharides. When the concentration of TPS was up to 0.2% and the amount of PTX:zein = 1:10, the nanoparticle spherical surface and the encapsuled PTX were in an amorphous form rather than crystalline form [47].

3 Mechanism of Hypoglycemic Polysaccharides

3.1 β-Cell Dysfunction

Type 2 diabetes mellitus (T2DM) is characterized by relative or absolute deficiency of insulin secretion which leads to the imbalance of blood glucose regulation [48]. The β cell dysfunction and insulin resistance are inherently related with their interrelation for triggering the pathogenesis of diabetes mellitus. The β-cells can alleviate the blood glucose disorder caused by insulin resistance by increasing insulin secretion. β-cell dysfunction can be caused by oxidative stress, cytokine induced inflammation, and excessive intake of saturated fat and free fatty acids. β-cell dysfunction and T2DM that follow this period of β-cell compensation may result from an inadequate expansion of the β cell mass and inadequate production of insulin [49]. Although the exact mechanisms underlying T2DM are not clear, the role of lipo toxicity and glucotoxicity in the development of chronic diseases like obesity and T2DM has been recognized [50]. In histopathological observation, polysaccharide from azuki bean could inhibit β-cell apoptosis and protect the integrity of islet structure [51].

3.2 Inhibiting α-Amylase and α-Glucosidase

The two most important carbohydrate hydrolases in human body are pancreatic α-amylase and small intestine α-glucosidase [2]. α-amylase can hydrolyze alpha-1,4-sugar bonds to generate smaller oligosaccharides and disaccharides, while the role of alpha-glucosidase is to hydrolyze disaccharides into monosaccharides [52]. When digestible carbohydrates enter the body, they are digested into glucose and fructose, which are easily absorbed by the small intestine [53]. As the inhibition of α-amylase and α-glucosidase activities can significantly slow down the conversion of carbohydrate to blood glucose and reduce the improvement of postprandial blood glucose, which is an effective way to control diabetes [54]. The structure of TPS was analyzed, and there was no significant change in TPS structure after salivary digestion, while obvious changes in TPS structure after gastrointestinal digestion. After the digestion, the reducing sugar content was increased from 0.650 ± 0.011 to 1.594 ± 0.078 mM, and the central molecular weight was decreased from 540.57 to 375.35 kDa [24]. Polysaccharides have good α-glucosidase and α-amylase inhibitory activities, which may be due to the interaction between total free carboxyl and hydroxyl groups and amino acid residues in the enzyme [55].

3.3 Related Signaling Pathways

Insulin is an important factor in regulating blood glucose. When combined with its receptor, insulin can activate a series of complex physiological activities in mammalian cells, including promoting glucose influx, glycogen synthesis, glycolysis and inhibiting glucose production in liver [56, 57]. Insulin can bind to insulin receptor and the insulin receptors are phosphorylated into IRS1 and IRS2. Then insulin can bind and activate phosphatidylinositol 3 kinase (PI3K)/protein kinase B (Akt/PKB) pathway and RAS mitogen activated protein kinase (MAPK) pathway [58]. Tea polysaccharides (400 mg/kg) had the most efficient effect on the reduction of serum insulin level, thus, it could ameliorate the insulin resistance caused by T2DM [59]. Akt is a key multi effect kinase, which mediates most of the biological functions of insulin in glucose metabolism by inactivating glycogen synthase kinase (GSK), activating phosphofructokinase and promoting translocation of glucose transporter 4 (GLUT4). MAPK signal transduction pathways include extracellular signal regulated kinase (ERK1/2) signal that regulates cell growth and differentiation, and c-jun NH2 terminal kinase (JNK) and p38 MAPK signaling pathways that regulate the stress response of inflammation and apoptosis. The over-activation of

MAPKs leads to insufficiently insulin secretion and related to the apoptosis of pancreatic β-cells [60, 61]. The expressions of PI3Kp85/p-Akt/GLUT4 were upregulated by TPS, which indicated the involvement of the PI3K/Akt signal pathway in the hypoglycemic mechanism of TPS. Results suggested that TPS could ameliorate the T2MD and might be a promising candidate functional food or medicine for T2DM treatment [59]. *Inonotus obliquus* polysaccharides could increase the protein expressions of PI3K-p85, p-Akt (ser473), and GLUT4, which suggested that the anti-diabetic mechanism might be associated with activating PI3K and Akt phosphorylation and the translocation of GLUT4 in high fat diet and STZ-induced type 2 diabetic mice [62]. The involvement of the PI3K/Akt signal pathway was also be demonstrated as the hypoglycemic mechanism of tea polysaccharides and corn silk polysaccharides [37, 59]. Anti-diabetic nephropathy effects of *Termitornyces albuminosus* polysaccharides are related to the modulation of NF-κB signaling pathway, as they could reduce the expression levels of NF-κB db/db mice [63]. Wu et al. illustrated polysaccharide from azuki bean could regulate glucose metabolism by increasing the expression of insulin receptor (INSR), insulin receptor substrate-1 (IRS-1), phosphoinositide 3-kinase (PI3K), protein kinase B (AKT), and glucose transporter-2 (GLUT-2) [51].

3.4 Others

The recovery of endogenous antioxidant system was also associated with hypoglycemic activity. Wang et al. indicated that *Inonotus obliquus* polysaccharides decreased the oxidative damage induced by H_2O_2 in L02 cells [45]. Fatty acid, phospholipid, glycerolipid, amino acid and bile acid metabolism are determined obviously disordered in diabetic rats. *Dendrobium officinale* polysaccharide can improve the metabolism disorders of lipid, bile acids, and amino acids [64]. Diabetic patients will have intestinal Dysbiosis, and human intestinal microflora plays a fundamental role in diseases such as obesity and diabetes. Polysaccharide from *Cyclocarya paliurus* leaves increased *Ruminococcaceae* UCG-005, which is a key genus for preventing diabetes [65]. Type 2 *Apocynum venetum* leaves polysaccharides could increase the abundance of genera *Odoribacter, Anaeroplasma, Parasutterella,* and *Muribaculum,* while decrease genera *Enterococcus, Klebsiella,* and *Aerococcus*, thus protect the environment of intestinal flora in diabetic mice. These data provided new sights for a promising anti-diabetic nutraceutical [15].

4 Conclusion and Prospect

Diabetes mellitus has been considered as a common and potentially preventable public health problem. In 2030, the number of people with diabetes is estimated to increase from 117 million in 2000 to 366 million [66]. Because there are many causes and complicated pathogenesis of diabetes mellitus, it is not easy for single target chemicals to achieve satisfactory therapeutic effect [67]. Polysaccharide plays an active role in the treatment of diabetes and its complications. Because of its small adverse reactions, it has become a hot spot in the development of hypoglycemic drugs and health products, and has a good research prospect [68, 69]. The biological activity of plant polysaccharides is affected by many factors, such as relative molecular weight, molecular main chain composition, chemical modification, and so on [70, 71]. To explore the structure of polysaccharides is the premise of studying the multiple activities of polysaccharides, but the complex and irregular structure of polysaccharides brings certain difficulties to the research work. In addition, the separation and purification of polysaccharides are complicated, which limits the development and utilization of polysaccharides.

At present, the experiment of polysaccharide only stays at the research level of biological activity, the systematic research of multi-target effect is not perfect, the relationship between quantity and effect is not clear, and the basic research method is relatively backward, which hinders the research and development of polysaccharide drugs [72]. Therefore, it is necessary to further explore the research methods and hypoglycemic mechanism of polysaccharides, find more targets of interaction, and provide new ideas for the research of hypoglycemic effect of polysaccharides and the development of natural hypoglycemic drugs.

References

1. Wang PC, Zhao S, Yang BY, Wang QH, Kuang HX (2016) Anti-diabetic polysaccharides from natural sources: a review. Carbohydr Polym 148:86–97. https://doi.org/10.1016/j.carbpol.2016.02.060
2. Wu J, Shi S, Wang H, Wang S (2016) Mechanisms underlying the effect of polysaccharides in the treatment of type 2 diabetes: a review. Carbohydr Polym 144:474–494. https://doi.org/10.1016/j.carbpol.2016.02.040
3. He X, Wang X, Fang J, Chang Y, Ning N, Guo H, Huang L, Huang X, Zhao Z (2017) Polysaccharides in *Grifola frondosa* mushroom and their health promoting properties: a review. Int J Biol Macromol 101:910–921. https://doi.org/10.1016/j.ijbiomac.2017.03.177
4. He S, Tang M, Zhang Z, Liu H, Luo M, Sun H (2020) Hypoglycemic effects of phenolic compound-rich aqueous extract from water dropwort (*Oenanthe javanica* DC.) on streptozotocin-induced diabetic mice. New J Chem 44(14):5190–5200. https://doi.org/10.1039/c9nj05533a
5. Li L, Li RC, Song YH, Wu WY, Yin SH, Fu WW, Wu RT, Li W (2019) Effects of a Ganoderma atrum polysaccharide against pancreatic damage in streptozotocin-induced diabetic mice. Food Funct 10(11):7227–7238. https://doi.org/10.1039/c9fo01990a
6. Wu G, Bai Z, Wan Y, Shi H, Huang X, Nie S (2020) Antidiabetic effects of polysaccharide from azuki bean (*Vigna angularis*) in type 2 diabetic rats via insulin/PI3K/AKT signaling pathway. Food Hydrocoll 101:105456
7. Ru Y, Liu K, Kong X, Li X, Shi X, Chen H (2020) Synthesis of selenylated polysaccharides from Momordica charantia L. and its hypoglycemic activity in streptozotocin-induced diabetic mice. Int J Biol Macromol 152:295–304. https://doi.org/10.1016/j.ijbiomac.2020.02.288

8. Jiang X, Meng W, Li L, Meng Z, Wang D (2020) Adjuvant therapy with mushroom polysaccharides for diabetic complications. Front Pharmacol 11:168. https://doi.org/10.3389/fphar.2020.00168
9. Khan BM, Cheong KL, Liu Y (2019) Pumpkin polysaccharides: purification, characterization and hypoglycemic potential. Int J Biol Macromol 139:842–849. https://doi.org/10.1016/j.ijbiomac.2019.08.053
10. Wang L, Li C, Huang Q, Fu X (2020) Polysaccharide from *Rosa roxburghii* Tratt fruit attenuates hyperglycemia and hyperlipidemia and regulates colon microbiota in diabetic db/db mice. J Agric Food Chem 68(1):147–159. https://doi.org/10.1021/acs.jafc.9b06247
11. Hu W, Wang J, Guo W, Liu Y, Guo Z, Miao Y, Wang D (2020) Studies on characteristics and anti-diabetic and -nephritic effects of polysaccharides isolated from *Paecilomyces hepiali* fermentation mycelium in db/db mice. Carbohydr Polym 232:115766. https://doi.org/10.1016/j.carbpol.2019.115766
12. Yang S, Qu Y, Zhang H, Xue Z, Liu T, Yang L, Sun L, Zhou Y, Fan Y (2020) Hypoglycemic effects of polysaccharides from Gomphidiaceae rutilus fruiting bodies and their mechanisms. Food Funct 11(1):424–434. https://doi.org/10.1039/c9fo02283j
13. Wang C, Gao X, Chen Z, Chen Y, Chen H (2017) Preparation, characterization and application of polysaccharide-based metallic nanoparticles: a review. Polymers (Basel) 9(12):689. https://doi.org/10.3390/polym9120689
14. Yoo SH, Lee YE, Chung JO, Rha CS, Hong YD, Park MY, Shim SM (2020) Addition of flavonols and polysaccharides as excipient ingredients into epicatechin rich green tea extract inhibited free radical formation and glucose uptake. Food Funct 11(4):3105–3111. https://doi.org/10.1039/c9fo03020d
15. Yuan Y, Zhou J, Zheng Y, Xu Z, Li Y, Zhou S, Zhang C (2020) Beneficial effects of polysaccharide-rich extracts from *Apocynum venetum* leaves on hypoglycemic and gut microbiota in type 2 diabetic mice. Biomed Pharmacother 127:110182. https://doi.org/10.1016/j.biopha.2020.110182
16. Zhang H, Zhao J, Shang H, Guo Y, Chen S (2020) Extraction, purification, hypoglycemic and antioxidant activities of red clover (*Trifolium pratense* L.) polysaccharides. Int J Biol Macromol 148:750–760. https://doi.org/10.1016/j.ijbiomac.2020.01.194
17. Xing M, Cao Q, Wang Y, Xiao H, Zhao J, Zhang Q, Ji A, Song S (2020) Advances in research on the bioactivity of alginate oligosaccharides. Mar Drugs 18(3):144. https://doi.org/10.3390/md18030144
18. Zhang R, Zhang X, Tang Y, Mao J (2020) Composition, isolation, purification and biological activities of Sargassum fusiforme polysaccharides: a review. Carbohydr Polym 228:115381. https://doi.org/10.1016/j.carbpol.2019.115381
19. Wang YY, Zhu J, Ma H, Ding ZC, Li L, Yan JK (2019) Antidiabetic activity of a polysaccharide-protein complex from Asian clam (*Corbicula fluminea*) in streptozotoxin-induced diabetic rats and its underlying mechanism. Food Funct 10(9):5574–5586. https://doi.org/10.1039/c9fo01341e
20. Zhu J, Chen Z, Zhou H, Yu C, Han Z, Shao S, Hu X, Wei X, Wang Y (2020) Effects of extraction methods on physicochemical properties and hypoglycemic activities of polysaccharides from coarse green tea. Glycoconj J 37(2):241–250. https://doi.org/10.1007/s10719-019-09901-2
21. Akhtar HMS, Abdin M, Hamed YS, Wang W, Chen G, Chen D, Chen C, Li W, Mukhtar S, Zeng X (2019) Physicochemical, functional, structural, thermal characterization and α-amylase inhibition of polysaccharides from chickpea (*Cicer arietinum* L.) hulls. Lwt 113:108265. https://doi.org/10.1016/j.lwt.2019.108265
22. Cai W-D, Ding Z-C, Wang Y-Y, Yang Y, Zhang H-N, Yan J-K (2020) Hypoglycemic benefit and potential mechanism of a polysaccharide from *Hericium erinaceus* in streptozotoxin-induced diabetic rats. Process Biochem 88:180–188. https://doi.org/10.1016/j.procbio.2019.09.035

23. Chen S, Chen H, Tian J, Wang J, Wang Y, Xing L (2014) Enzymolysis-ultrasonic assisted extraction, chemical characteristics and bioactivities of polysaccharides from corn silk. Carbohydr Polym 101:332–341. https://doi.org/10.1016/j.carbpol.2013.09.046
24. Li W, Wang C, Yuan G, Pan Y, Chen H (2018) Physicochemical characterisation and α-amylase inhibitory activity of tea polysaccharides under simulated salivary, gastric and intestinal conditions. Int J Food Sci Technol 53(2):423–429. https://doi.org/10.1111/ijfs.13600
25. Wang C, Li W, Chen Z, Gao X, Yuan G, Pan Y, Chen H (2018) Effects of simulated gastrointestinal digestion in vitro on the chemical properties, antioxidant activity, alpha-amylase and alpha-glucosidase inhibitory activity of polysaccharides from Inonotus obliquus. Food Res Int 103:280–288. https://doi.org/10.1016/j.foodres.2017.10.058
26. Chen Z, Yin C, Fan X, Ma K, Yao F, Zhou R, Shi D, Cheng W, Gao H (2019) Characterization of physicochemical and biological properties of Schizophyllum commune polysaccharide extracted with different methods. Int J Biol Macromol 156:1425–1434. https://doi.org/10.1016/j.ijbiomac.2019.11.183
27. Zhang X, Chen H, Zhang N, Chen S, Tian J, Zhang Y, Wang Z (2013) Extrusion treatment for improved physicochemical and antioxidant properties of high-molecular weight polysaccharides isolated from coarse tea. Food Res Int 53(2):726–731. https://doi.org/10.1016/j.foodres.2012.08.011
28. Chen H, Chen Z, Fu Y, Liu J, Lin S, Zhang Q, Liu Y, Wu D, Lin D, Han G, Wang L, Qin W (2019) Structure, antioxidant, and Hypoglycemic activities of Arabinoxylans extracted by multiple methods from triticale. Antioxidants (Basel) 8(12):584. https://doi.org/10.3390/antiox8120584
29. Chen X, Chen G, Wang Z, Kan J (2020) A comparison of a polysaccharide extracted from ginger (*Zingiber officinale*) stems and leaves using different methods: preparation, structure characteristics, and biological activities. Int J Biol Macromol 151:635–649. https://doi.org/10.1016/j.ijbiomac.2020.02.222
30. Matias LLR, Costa ROA, Passos TS, Queiroz JLC, Serquiz AC, Maciel BLL, Santos PPA, Camillo CS, Goncalves C, Amado IR, Pastrana L, Morais AHA (2019) Tamarind trypsin inhibitor in chitosan-whey protein nanoparticles reduces fasting blood glucose levels without compromising Insulinemia: a preclinical study. Nutrients 11(11):2770. https://doi.org/10.3390/nu11112770
31. Zhu J, Yu C, Han Z, Chen Z, Wei X, Wang Y (2019) Comparative analysis of existence form for selenium and structural characteristics in artificial selenium-enriched and synthetic selenized green tea polysaccharides. Int J Biol Macromol. https://doi.org/10.1016/j.ijbiomac.2019.11.022
32. Cao H, Ma S, Guo H, Cui X, Wang S, Zhong X, Wu Y, Zheng W, Wang H, Yu J, Ma L, Chun-Chao H (2019) Comparative study on the monosaccharide compositions, antioxidant and hypoglycemic activities in vitro of intracellular and extracellular polysaccharides of liquid fermented *Coprinus comatus*. Int J Biol Macromol 139:543–549. https://doi.org/10.1016/j.ijbiomac.2019.08.017
33. Fang C, Chen G, Kan J (2020) Comparison on characterization and biological activities of Mentha haplocalyx polysaccharides at different solvent extractions. Int J Biol Macromol 154:916–928. https://doi.org/10.1016/j.ijbiomac.2020.03.169
34. Pan Y, Wang C, Chen Z, Li W, Yuan G, Chen H (2017) Physicochemical properties and antidiabetic effects of a polysaccharide from corn silk in high-fat diet and streptozotocin-induced diabetic mice. Carbohydr Polym 164:370–378. https://doi.org/10.1016/j.carbpol.2017.01.092
35. Chen YC, Huang SD, Tu JH, Yu JS, Nurlatifah AO, Chiu WC, Su YH, Chang HL, Putri DA, Cheng HL (2020) Exopolysaccharides of *Bacillus amyloliquefaciens* modulate glycemic level in mice and promote glucose uptake of cells through the activation of Akt. Int J Biol Macromol 146:202–211. https://doi.org/10.1016/j.ijbiomac.2019.12.217
36. Xue Z, Chen Y, Jia Y, Wang Y, Lu Y, Chen H, Zhang M (2019) Structure, thermal and rheological properties of different soluble dietary fiber fractions from mushroom *Lentinula*

edodes (Berk.) Pegler residues. Food Hydrocoll 95:10–18. https://doi.org/10.1016/j.foodhyd.2019.04.015

37. Guo Q, Chen Z, Santhanam RK, Xu L, Gao X, Ma Q, Xue Z, Chen H (2019) Hypoglycemic effects of polysaccharides from corn silk (*Maydis stigma*) and their beneficial roles via regulating the PI3K/Akt signaling pathway in L6 skeletal muscle myotubes. Int J Biol Macromol 121:981–988. https://doi.org/10.1016/j.ijbiomac.2018.10.100
38. Deng Y, Huang L, Zhang C, Xie P, Cheng J, Wang X, Liu L (2020) Novel polysaccharide from *Chaenomeles speciosa* seeds: structural characterization, alpha-amylase and alpha-glucosidase inhibitory activity evaluation. Int J Biol Macromol 153:755–766. https://doi.org/10.1016/j.ijbiomac.2020.03.057
39. Ding Q, Zhang B, Zheng W, Chen X, Zhang J, Yan R, Zhang T, Yu L, Dong Y, Ma B (2019) Liupao tea extract alleviates diabetes mellitus and modulates gut microbiota in rats induced by streptozotocin and high-fat, high-sugar diet. Biomed Pharmacother 118:109262. https://doi.org/10.1016/j.biopha.2019.109262
40. Dou Z, Chen C, Fu X (2019) The effect of ultrasound irradiation on the physicochemical properties and α-glucosidase inhibitory effect of blackberry fruit polysaccharide. Food Hydrocoll 96:568–576. https://doi.org/10.1016/j.foodhyd.2019.06.002
41. Chen S, Chen H, Tian J, Wang Y, Xing L, Wang J (2013) Chemical modification, antioxidant and alpha-amylase inhibitory activities of corn silk polysaccharides. Carbohydr Polym 98 (1):428–437. https://doi.org/10.1016/j.carbpol.2013.06.011
42. Guo Q, Ma Q, Xue Z, Gao X, Chen H (2018) Studies on the binding characteristics of three polysaccharides with different molecular weight and flavonoids from corn silk (Maydis stigma). Carbohydr Polym 198:581–588. https://doi.org/10.1016/j.carbpol.2018.06.120
43. Duan GL, Yu XB (2019) Isolation, purification, characterization, and antioxidant activity of low-molecular-weight polysaccharides from *Sparassis latifolia*. Int J Biol Macromol 137:1112–1120. https://doi.org/10.1016/j.ijbiomac.2019.06.177
44. Ganesan AR, Saravana Guru M, Balasubramanian B, Mohan K, Chao Liu W, Valan Arasu M, Abdullah Al-Dhabi N, Duraipandiyan V, Ignacimuthu S, Sudhakar MP, Seedevi P (2020) Biopolymer from edible marine invertebrates: a potential functional food. J King Saud Univ Sci 32(2):1772–1777. https://doi.org/10.1016/j.jksus.2020.01.015
45. Wang C, Gao X, Santhanam RK, Chen Z, Chen Y, Xu L, Wang C, Ferri N, Chen H (2018) Effects of polysaccharides from *Inonotus obliquus* and its chromium (III) complex on advanced glycation end-products formation, alpha-amylase, alpha-glucosidase activity and H2O2-induced oxidative damage in hepatic L02cells. Food Chem Toxicol 116(Pt B):335–345. https://doi.org/10.1016/j.fct.2018.04.047
46. Guo WL, Deng JC, Pan YY, Xu JX, Hong JL, Shi FF, Liu GL, Qian M, Bai WD, Zhang W, Liu B, Zhang YY, Luo PJ, Ni L, Rao PF, Lv XC (2020) Hypoglycemic and hypolipidemic activities of Grifola frondosa polysaccharides and their relationships with the modulation of intestinal microflora in diabetic mice induced by high-fat diet and streptozotocin. Int J Biol Macromol 153:1231–1240. https://doi.org/10.1016/j.ijbiomac.2019.10.253
47. Li S, Wang X, Li W, Yuan G, Pan Y, Chen H (2016) Preparation and characterization of a novel conformed bipolymer paclitaxel-nanoparticle using tea polysaccharides and zein. Carbohydr Polym 146:52–57. https://doi.org/10.1016/j.carbpol.2016.03.042
48. Han X, Song C, Feng X, Wang Y, Meng T, Li S, Bai Y, Du B, Sun Q (2020) Isolation and hypoglycemic effects of water extracts from mulberry leaves in Northeast China. Food Funct 11 (4):3112–3125. https://doi.org/10.1039/d0fo00012d
49. Weiwei L, Guoqi Y, Yuxiang P, Cong W, Haixia CJFP (2017) Network pharmacology studies on the bioactive compounds and action mechanisms of natural products for the treatment of diabetes mellitus: a review. Front Pharmacol 8:74
50. Huang JH, Zhou RR, He D, Chen L, Yang YY, Xie HL, Zhang SH, Zhao CX, Huang LQ (2020) Rapid identification of Lilium species and polysaccharide contents based on near infrared spectroscopy and weighted partial least square method. Int J Biol Macromol 154:182–187. https://doi.org/10.1016/j.ijbiomac.2020.03.109

51. Wu G, Bai Z, Wan Y, Shi H, Huang X, Nie S (2020) Antidiabetic effects of polysaccharide from azuki bean (*Vigna angularis*) in type 2 diabetic rats via insulin/PI3K/AKT signaling pathway. Food Hydrocoll 101:105456. https://doi.org/10.1016/j.foodhyd.2019.105456
52. Apostolidis E, Lee CM (2010) In vitro potential of *Ascophyllum nodosum* phenolic antioxidant-mediated α-glucosidase and α-amylase inhibition. J Food Sci 75(3):H97–H102
53. Ji X, Hou C, Yan Y, Shi M, Liu Y (2020) Comparison of structural characterization and antioxidant activity of polysaccharides from jujube (*Ziziphus jujuba* mill.) fruit. Int J Biol Macromol 149:1008–1018. https://doi.org/10.1016/j.ijbiomac.2020.02.018
54. Jia RB, Li ZR, Wu J, Ou ZR, Zhu Q, Sun B, Lin L, Zhao M (2020) Physicochemical properties of polysaccharide fractions from Sargassum fusiforme and their hypoglycemic and hypolipidemic activities in type 2 diabetic rats. Int J Biol Macromol 147:428–438. https://doi.org/10.1016/j.ijbiomac.2019.12.243
55. Fang C, Chen G, Kan J (2020) Comparison on characterization and biological activities of Mentha haplocalyx polysaccharides at different solvent extractions. Int J Biol Macromol 154:916–928
56. Ktari N, Bkhairia I, Nasri M, Ben Salah R (2020) Structure and biological activities of polysaccharide purified from Senegrain seed. Int J Biol Macromol 144:190–197. https://doi.org/10.1016/j.ijbiomac.2019.12.087
57. Li L, Xu JX, Cao YJ, Lin YC, Guo WL, Liu JY, Bai WD, Zhang YY, Ni L, Liu B, Rao PF, Lv XC (2019) Preparation of Ganoderma lucidum polysaccharidechromium (III) complex and its hypoglycemic and hypolipidemic activities in high-fat and high-fructose diet-induced pre-diabetic mice. Int J Biol Macromol 140:782–793. https://doi.org/10.1016/j.ijbiomac.2019.08.072
58. Li Y, Li S, He S, Yue Y, Ni Y (2020) Structural analysis and antidiabetic activity study of three acidic-type polysaccharides from Crepis crocea (Lam.) Babc. Nat Prod Res:1–6. https://doi.org/10.1080/14786419.2020.1756805
59. Li S, Chen H, Wang J, Wang X, Hu B, Lv F (2015) Involvement of the PI3K/Akt signal pathway in the hypoglycemic effects of tea polysaccharides on diabetic mice. Int J Biol Macromol 81:967–974. https://doi.org/10.1016/j.ijbiomac.2015.09.037
60. Ren Y, Liu S (2020) Effects of separation and purification on structural characteristics of polysaccharide from quinoa (*Chenopodium quinoa* willd). Biochem Biophys Res Commun 522(2):286–291. https://doi.org/10.1016/j.bbrc.2019.10.030
61. Trempolec N, Dave-Coll N, Nebreda AR (2013) SnapShot: p38 MAPK signaling. Cell 152 (3):656–656.e651. https://doi.org/10.1016/j.cell.2013.01.029
62. Wang J, Wang C, Li S, Li W, Yuan G, Pan Y, Chen H (2017) Anti-diabetic effects of Inonotus obliquus polysaccharides in streptozotocin-induced type 2 diabetic mice and potential mechanism via PI3K-Akt signal pathway. Biomed Pharmacother 95:1669–1677. https://doi.org/10.1016/j.biopha.2017.09.104
63. Yang C, Feng Q, Liao H, Yu X, Liu Y, Wang D (2019) Anti-diabetic nephropathy activities of polysaccharides obtained from *Termitornyces albuminosus* via regulation of NF-kappaB signaling in db/db mice. Int J Mol Sci 20(20):5205. https://doi.org/10.3390/ijms20205205
64. Chen H, Nie Q, Hu J, Huang X, Huang W, Nie S (2020) Metabolism amelioration of Dendrobium officinale polysaccharide on type II diabetic rats. Food Hydrocoll 102:105582. https://doi.org/10.1016/j.foodhyd.2019.105582
65. Li Q, Hu J, Nie Q, Chang X, Fang Q, Xie J, Li H, Nie S (2020) Hypoglycemic mechanism of polysaccharide from Cyclocarya paliurus leaves in type 2 diabetic rats by gut microbiota and host metabolism alteration. Sci China Life Sci. https://doi.org/10.1007/s11427-019-1647-6
66. Adams GG, Imran S, Wang S, Mohammad A, Kok S, Gray DA, Channell GA, Morris GA, Harding SE (2011) The hypoglycaemic effect of pumpkins as anti-diabetic and functional medicines. Food Res Int 44(4):862–867. https://doi.org/10.1016/j.foodres.2011.03.016
67. Shang H, Zhao J, Guo Y, Zhang H, Duan M, Wu H (2020) Extraction, purification, emulsifying property, hypoglycemic activity, and antioxidant activity of polysaccharides from comfrey. Ind Crop Prod 146. https://doi.org/10.1016/j.indcrop.2020.112183

68. Park J, Jang H-J (2017) Anti-diabetic effects of natural products an overview of therapeutic strategies. Mol Cell Toxicol 13(1):1–20. https://doi.org/10.1007/s13273-017-0001-1
69. Simpson R, Morris GA (2014) The anti-diabetic potential of polysaccharides extracted from members of the cucurbit family: a review. Bioact Carbohydr Diet Fibre 3(2):106–114. https://doi.org/10.1016/j.bcdf.2014.03.003
70. Xie J-H, Tang W, Jin M-L, Li J-E, Xie M-Y (2016) Recent advances in bioactive polysaccharides from *Lycium barbarum* L., Zizyphus jujuba mill, Plantago spp., and Morus spp.: structures and functionalities. Food Hydrocoll 60:148–160. https://doi.org/10.1016/j.foodhyd.2016.03.030
71. Zhang F, Lin L, Xie J (2016) A mini-review of chemical and biological properties of polysaccharides from *Momordica charantia*. Int J Biol Macromol 92:246–253. https://doi.org/10.1016/j.ijbiomac.2016.06.101
72. Ma L, Chen H, Zhang Y, Zhang N, Fu L (2012) Chemical modification and antioxidant activities of polysaccharide from mushroom Inonotus obliquus. Carbohydr Polym 89(2):371–378. https://doi.org/10.1016/j.carbpol.2012.03.016

Chapter 10
Plant Secondary Metabolites with α-Glucosidase Inhibitory Activity

Yanfang Su, Bing Gao, Tao Qin, Zhijing Gao, Wei Wang, and Jie Zhang

Abstract Plant-derived compounds with α-glucosidase inhibitory activity reported in the past four decades were summarized, mainly covering flavonoids, N-containing compounds, and terpenoids. Among these inhibitors, eighty-eight flavonoids account for the majority, and N-containing compounds including eleven aromatic amides, 1-deoxynojirimycin (DNJ), and 16 derivatives or structural analogues of DNJ, which are members of polyhydroxylated piperidine or pyrrolidine alkaloids, or polyhydroxy-nor-tropane alkaloids, are the second largest group and have the most specific structures, and twelve terpenoids without glycosylation have also been reported to show inhibitory activity. The plant sources and reported IC_{50} values of these inhibitors are presented in tables, and their structure characteristics and isolation methods are also briefly summarized. Medicinal plants are proved to be important and potential sources to discover α-glucosidase inhibitors. Whereas, low content and time-consuming extraction limit the application of plant-derived α-glucosidase inhibitors in the treatment of diabetes. Chemical modification and SAR study of natural α-glucosidase inhibitors are of importance, and standard plant extracts have potential use for the future prevention and treatment of diabetes.

Keywords α-glucosidase inhibitors · Diabetes · Flavonoids · N-containing compounds · Terpenoids

1 Introduction

Type 2 diabetes usually occurs in adults when enough insulin cannot be produced or insulin-resistance is developed in body. In the past three decades the prevalence of type 2 diabetes has risen dramatically in countries of all income levels, particularly in low- and middle-income countries [1]

Y. Su (✉) · B. Gao · T. Qin · Z. Gao · W. Wang · J. Zhang
Tianjin Key Laboratory for Modern Drug Delivery and High-Efficiency, School of Pharmaceutical Science and Technology, Tianjin University, Tianjin, China
e-mail: suyanfang@tju.edu.cn

H. Chen, M. Zhang (eds.), *Structure and Health Effects of Natural Products on Diabetes Mellitus*, https://doi.org/10.1007/978-981-15-8791-7_10

α-Glucosidase inhibitors is an important class of antidiabetic drugs, like acarbose, miglitol, and voglibose already used in clinic. α-Glucosidase inhibitors can prevent the sudden rise in postprandial hyperglycaemia by inhibiting the hydrolysis of oligosaccharides into absorbable monosaccharides [2]. However, acarbose, miglitol, and voglibose have different adverse reactions, including abdominal distension, abdominal pain, diarrhoea, and so on. Therefore, it is expected to find new safer glucosidase inhibitors from medicinal plants for the prevention and treatment of diabetes. In this chapter, Plant-derived compounds with α-glucosidase inhibitory activity in the literature since 1980 are summarized.

2 Flavonoids

88 Flavonoids (Table 10.1) with α-glucosidase inhibitory effects have been reported, including 30 flavone/flavonols (Fig. 10.1), 12 dihydroflavone/dihydroflavonols (Fig. 10.2), 12 isoflavones (Fig. 10.3), 11 flavanes and isoflavanes (Fig. 10.4), 16 xanthones (Fig.10.5), and other types of Flavonoids (Fig. 10.6).

3 N-Containing Compounds

29 N-containing compounds (Fig. 10.7, Table 10.2) have been reported to show α-glucosidase inhibitory activity. Of these compounds, 1-deoxynojirimycin (**12**, DNJ) is an analogue of glucose and a potent α-glucosidase inhibitor. 16 derivatives or structural analogues (**13-29**) of DNJ have been isolated from plants.

4 Terpenoids

12 Terpenoids (Fig. 10.8, Table 10.3) covering one iridoid, six diterpenoids, and five triterpenoids have been reported to show α-glucosidase inhibitory.

Miscellaneous

Besides flavonoids, N-containing compounds, terpenoids, some other classes of plant-derived compounds (Table 10.4), like long chain fatty acids, are also reported to show α-glucosidase inhibitory activity.

Table 10.1 Flavonoids with α-glucosidase inhibitory activity

No.	Names of compounds [ref]	Inhibitory effects/IC_{50}	Plant sources
1	Norartocarpetin [3]	13.4 μM	*Morus alba*
2	Vitexin [4]	12.9 μM	*Ficus deltoidea*
3	Isorhamnetin[5]	-	*Corchorus olitorius*
4	Isovitexin [6, 7]	34.37 μM	*Achillea fragrantissima* .
		32 μM	*Passiflora bogotensis*
5	Isovitexin-4'-methyl ether [6]	187 μM	*Achillea fragrantissima*
6	Isoorientin [7]	29 μM	*Passiflora bogotensis*
7	Isovitexin-2″-*O*-Rhamnoside [7]	35 μM	
8	Isoorientin-2″-*O*-Rhamnoside [7]	34 μM	
9	Apigenin-6-*C*-α-L-Rhamnopyranosyl-(1→2)-(6″-*O*-acetyl)-β-D-Glucopyranoside[7]	19 μM	
10	Luteolin-6-*C*-α-L-Rhamnopyranosyl-(1→2)-(6″-*O*-acetyl)-β-D-Glucopyranoside [7]	25 μM	
11	Acacetin-6-*C*-(6″-acetyl-β-D-Glucopyranoside)-8-*C*-α-L-Arabinpyranoside [6]	1.5 μg/mL	*Achillea fragrantissima*
12	Apigenin-8-*C*-α-L-Rhamnopyranosyl-(1→2)-β-D-Glucopyranoside [8]	560 μM	*Garcinia gracilis*
13	Hirta-flavonoside-B[9]	71 μM	*Euphorbia hirta*
14	Hirtacoumaro flavonoside [9]	22 μM	
15	Quercetrin [9]	151 μM	
16	Dimethoxy quercetrin[9]	182 μM	
17	Hirsutrin[10]	98.2 μM	*Zea mays*
18	3′-Methoxyhirsutrin [10]	245.5 μM	
19	Kaempferol [3]	33.7 μM	*Morus alba*
20	Quercetin[3]	19.6 μM	
21	5-Hydroxy-3-methoxy-flavone-7-*O*-[β-D-Apiosyl-(1→6)]-β-D-Glucoside [11]	43.4% Inhibition at 437μM/L	*Dendrobium devonianum*
22	Guaijaverin [12]	100 μM	*Myrcia multiflora*
23	Quercetin 3-*O*-β-D-Galactopyranoside[13]	450 μM	*Arcytophyllum thymifolium*
24	Rhamnetin[13]	380 μM	
25	Kaempferol 3-*O*-β-D-Glucopyranosyl-(1→2)-β-D-Galactopyranoside[13]	210 μM	
26	Isoquercitrin[14]	>1000 μM	*Alnus firma*
27	Quercetin-3-*O*-α-L-Arabinofuranoside[14]	>1000 μM	
28	Myricetin-3-*O*-β-D-Galactopyranoside[14]	>1000 μM	
29	Myricitrin [12]	420 μM	*Myrcia multiflora*
30	Desmanthin-1[12]	240 μM	
31 32	(2R)/(2S)-Euchrenone a_7[3]	6.28 μM	*Morus alba*

(continued)

Table 10.1 (continued)

No.	Names of compounds [ref]	Inhibitory effects/IC_{50}	Plant sources
33	Liquiritigenin [15]	3.61 mM	*Dalbergia odorifera*
34	(2S)-7-Prenyloxyeriodictyol [13]	28.1 μM	*Arcytophyllum thymifolium*
35	Pinocembrin[14]	>1000 μM	*Alnus firma*
36	Myrciacitrins I [12]	600 μM	*Myrcia multiflora*
37	Quercetin-3, 6, 7-trimethyl ether [4]	40 μM	*Achillea fragrantissima*
38	Kaempferol-3-*O*-Arabinofuranosyl-7-*O*-Rhamnopyranoside [16]	27.70 μM	*Geranium thunbergii*
39	Kaempferitrin[16]	>86.51 μM	
40	Alnustinol[14]	>1000 μM	*Alnus firma*
41	Catechin-(7, 8-*b*, *c*)-4α-(3, 4-dihydroxyphenyl)-2 (3H)-pyranone [16]	25 μM	*Eucommia ulmoides*
42	Catechin-(7, 8-*b*, *c*)-4β-(3, 4-dihydroxyphenyl)-2 (3H)-pyranone [16]	25 μM	
43	Daidzein [17]	23.01 μM	*Pueraria lobata*
44	Ononin [18]	422.89 μM	
45	Puerarin [18]	524.08 μM	
46	Formononetin [15]	0.51 mM	*Dalbergia odorifera*
47	Tectorigenin[15]	3.52 mM	
48	Biochanin A [15]	0.35 mM	*Pueraria thunbergiana*
49	Calycosin [15]	12.70 mM	
50	Genistein [15]	0.15 mM	
51	Puerarin[15]	>30 mM	
52	Daidzin [15]	>30 mM	
53	Formononetin-7-*O*-β-Glucopyranoside [15]	>30 mM	
54	Genistin [15]	>30 mM	
55	Crepidatuol B [19]	18.9 μM	*Dendrobium loddigesii*
56	Epicatechin-(4β,8)-epicatechin gallate (B2-3′-*O*-gallate) [20]	0.71 μM	*Rhodiola crenulata*
57	(2R)/(2S)-7-methoxyl-8-hydroxyethyl-2', 4'-dihydroxylflavane [3]	75.4 μM	*Morus alba*
58	(2S)-7-methoxyl-8-hydroxyethyl-4'-hydroxylflavane-2'-*O*-β-D-Glucopyranoside [3]	12.7 μM	
59	Epicatechin gallate (ECG) [20]	4.77 μM	*Rhodiola crenulata*
60	(+)-Catechin [14]	>1000 μM	*Alnus firma*
61	(-)-Epicatechin [14]	>1000 μM	
62	(3R)-Calussequinone [15]	29.38 mM	*Dalbergia odorifera*
63	Tuberosin [15]	4.68 mM	*Pueraria thunbergiana*

(continued)

Table 10.1 (continued)

No.	Names of compounds [ref]	Inhibitory effects/IC_{50}	Plant sources
64	Mucronulatol [15]	12.53 mM	*Dalbergia odorifera*
65	(3R)-5'-methoxyvestitol [15]	>30 mM	
66	Mangostanol [21]	30.5 μM	*Garcinia mangostana*
67	Allanxanthone E [21]	31.7 μM	
68	Mangostenone F [21]	21.0 μM	
69	α-Mangostin [21]	5.0 μM	
70	γ-Mangostin [21]	1.5 μM	
71	Mangostingone [21]	9.8 μM	
72	Cudraxanthone [21]	58.5 μM	
73	1, 5, 8-Trihydroxy-3-methoxy-2- (3-methylbut-2-enyl) xanthone [21]	50.5 μM	
74	Mangosenone [21]	45.0 μM	
75	Garcinone D [21]	29.9 μM	
76	8-Deoxygartanin [21]	21.5 μM	
77	Gartanin [21]	10.8 μM	
78	Smeathxanthone A [21]	6.9 μM	
79	β-Mangostin [21]	14.4 μM	
80	9-Hydroxycalabaxanthone [21]	34.2 μM	
81	Oxoethylmangostine [15]	63.5 μM	
82	Cyanidin-3-glucoside[10]	220.8 μM	*Zea mays*
83	Pelargonidin-3-glucoside [10]	198.2 μM	
84	Peonidin-3-glucoside [10]	260.3 μM	
85	Peonidin 3-*O*-[2-*O*-(6-*O*-*E*-feruloyl-β-D-Glucopyranosyl)-6-*O*-*E*-caffeoyl-β-D-Glucopyranoside]-5-*O*-β-D-Glucopyranoside [21]	200 μM	*Ipomoea batatas cv. Ayamurasaki*
86	Iobatflavate [18]	1.79 μM	*Pueraria lobata*
87	3S, 4R-Tuberosin [18]	378.89 μM	
88	Medicarpin [15]	2.93 mM	*Dalbergia odorifera*

The Inhibitory effects were not quantitatively described.

5 Isolation and Identification of Naturally Occurring α-Glucosidase Inhibitors Derived from Plants

Inspired by the application of the plants in the treatment of diabetes or by the α-glucosidase inhibitory activity of crude extracts of the medicinal plants, their chemical study was conducted to seek the ingredients with α-glucosidase inhibitory activity.

Flavonoids: Among the natural compounds with α-glucosidase inhibitory activity reported so far, flavonoids are the most abundant. The extraction and isolation of flavonoids depend on their classes, and polarity is primary consideration. Less polar

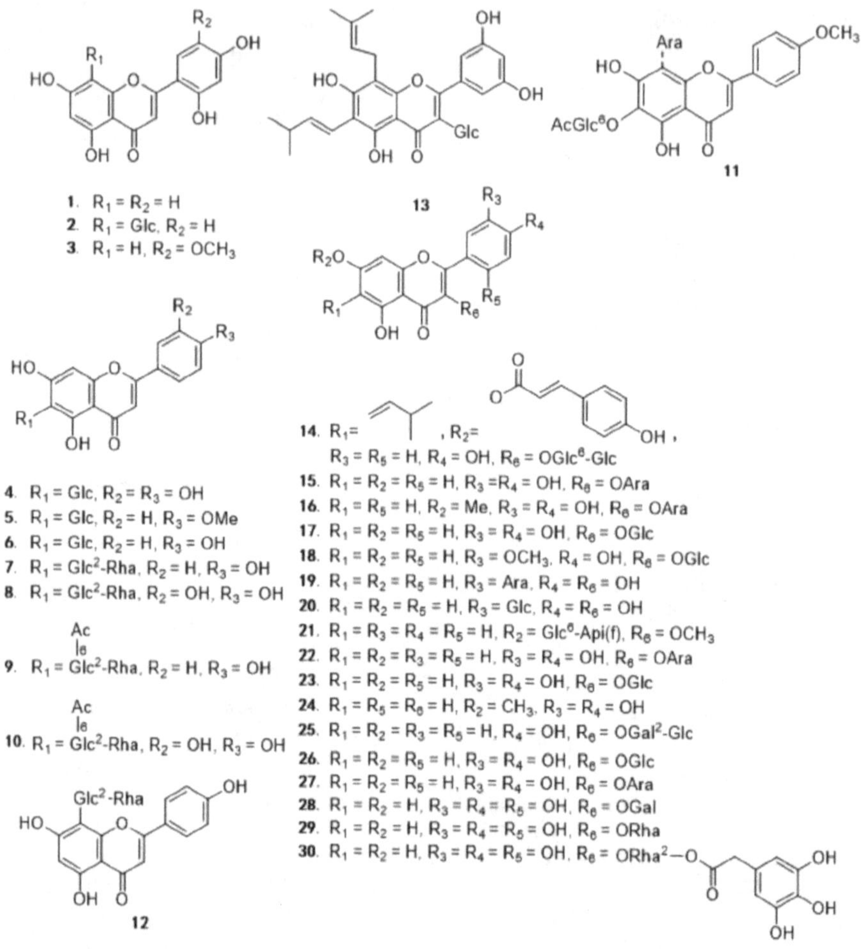

Fig. 10.1 Structures of flavones and flavonols with α-glucosidase inhibitory activity

flavonoids can be extracted with dichloromethane or ethyl acetate, while more polar flavonoid aglycones and glycosides can be extracted with *n*-butanol, methanol, ethanol, or ethanol–water mixtures. Open-column chromatography is extensively used in the further separation or purification of flavonoids, and support materials include polyamide, silica gel, and Sephadex LH-20. Reverse phase HPLC is also extensively used. NMR spectroscopy is an extremely powerful analytical technique for the identification of previously undescribed flavonoids, and HPLC-MS is also used in the recognition of known flavonoids with α-glucosidase inhibitory activity. For example, compounds **11**, **4**, and **5** (see Fig. 10.1 or Table 10.1) were obtained by the following process: The aerial parts of *Achillea fragrantissima* were extracted by cold percolation with 95% ethanol. The ethanolic extract was concentrated under reduced pressure to give a brown residue. The ethanolic extract was suspended in

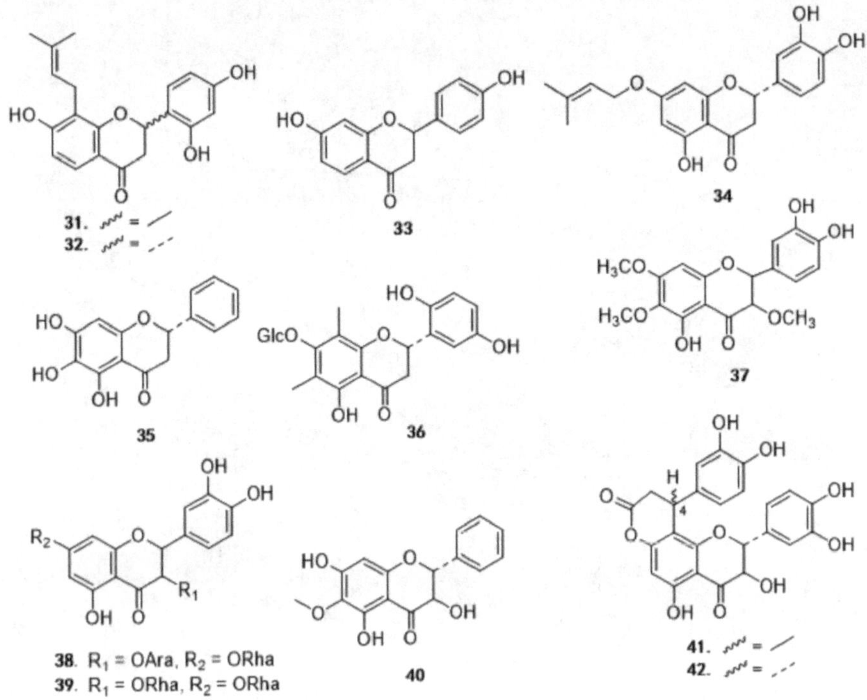

Fig. 10.2 Structures of dihydroflavone/dihydroflavonols with α-glucosidase inhibitory activity

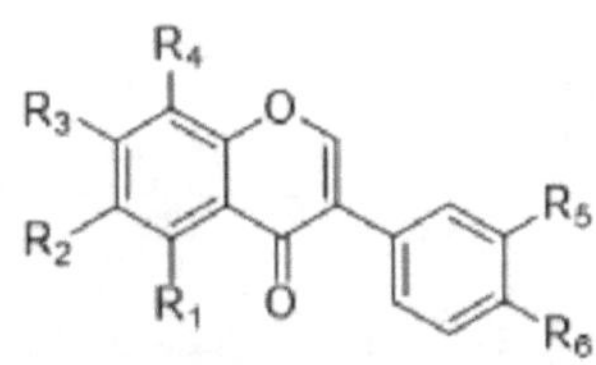

43. $R_1 = R_2 = R_4 = R_5 = H$, $R_3 = R_6 = OH$
44. $R_1 = R_2 = R_4 = R_5 = H$, $R_3 = Glc$, $R_6 = CH_3$
45. $R_1 = R_2 = R_4 = R_5 = H$, $R_3 = R_6 = OH$, $R_4 = Glc$
46. $R_1 = R_2 = R_4 = R_5 = H$, $R_3 = OH$, $R_6 = OCH_3$
47. $R_1 = R_3 = OH$, $R_2 = OCH_3$, $R_4 = R_5 = H$, $R_6 = OH$
48. $R_1 = R_3 = OH$, $R_2 = R_4 = R_5 = H$, $R_6 = OCH_3$
49. $R_1 = R_2 = R_4 = H$, $R_3 = R_5 = OH$, $R_6 = OCH_3$
50. $R_1 = R_3 = OH$, $R_2 = R_4 = R_5 = H$, $R_6 = OH$
51. $R_1 = R_2 = R_5 = H$, $R_3 = OH$, $R_4 = Glc$, $R_6 = OCH_3$
52. $R_1 = R_2 = R_4 = R_5 = H$, $R_3 = OGlc$, $R_6 = OH$
53. $R_1 = R_2 = R_4 = R_5 = H$, $R_3 = OGlc$, $R_6 = OCH_3$
54. $R_1 = OH$, $R_2 = R_4 = R_5 = H$, $R_3 = OGlc$, $R_6 = OCH_3$

Fig. 10.3 Structures of isoflavones with α-glucosidase inhibitory activity

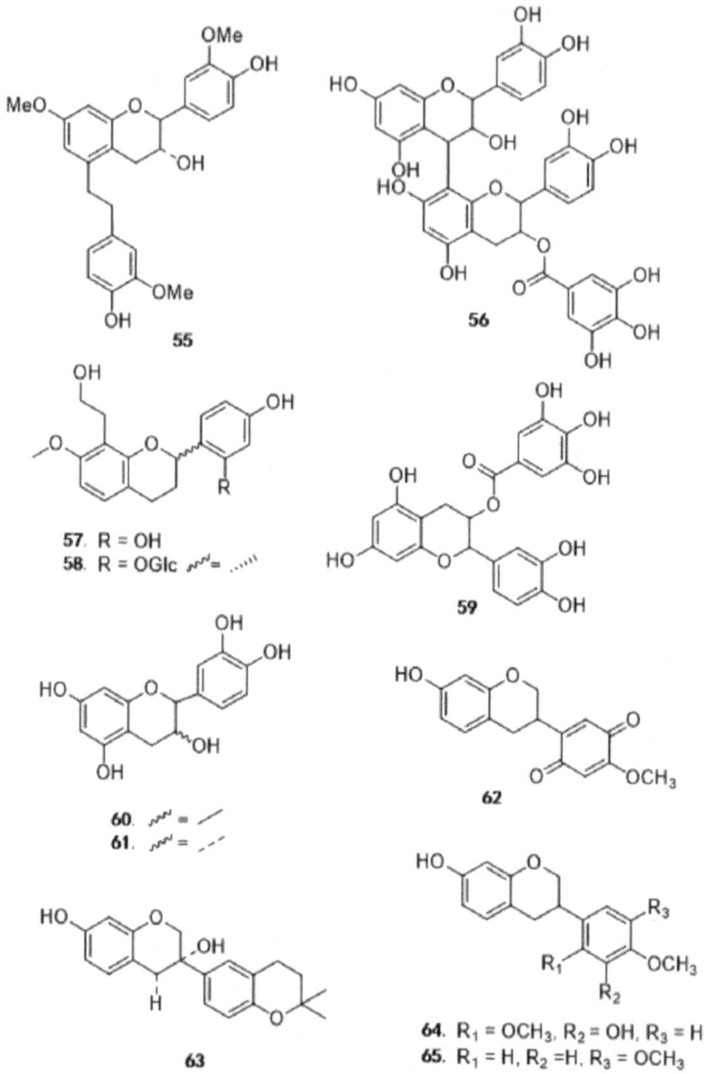

Fig. 10.4 Structures of flavanes and isoflavanes with α-glucosidase inhibitory activity

distilled water and subjected to liquid–liquid fractionation using chloroform and *n*-butanol-saturated with water. Fractions were separately concentrated under reduced pressure to yield the chloroform and *n*-butanol fraction, respectively, in addition to a residual water fraction. The *n*-butanol fraction was chromatographed over a VLC Silica gel column and Sephadex LH-20 columns to yield the above compounds [6].

N-containing compounds: Besides eleven aromatic amides, N-containing natural compounds with α-glucosidase inhibitory activity reported in the literature are members of polyhydroxylated piperidine or pyrrolidine alkaloids, or

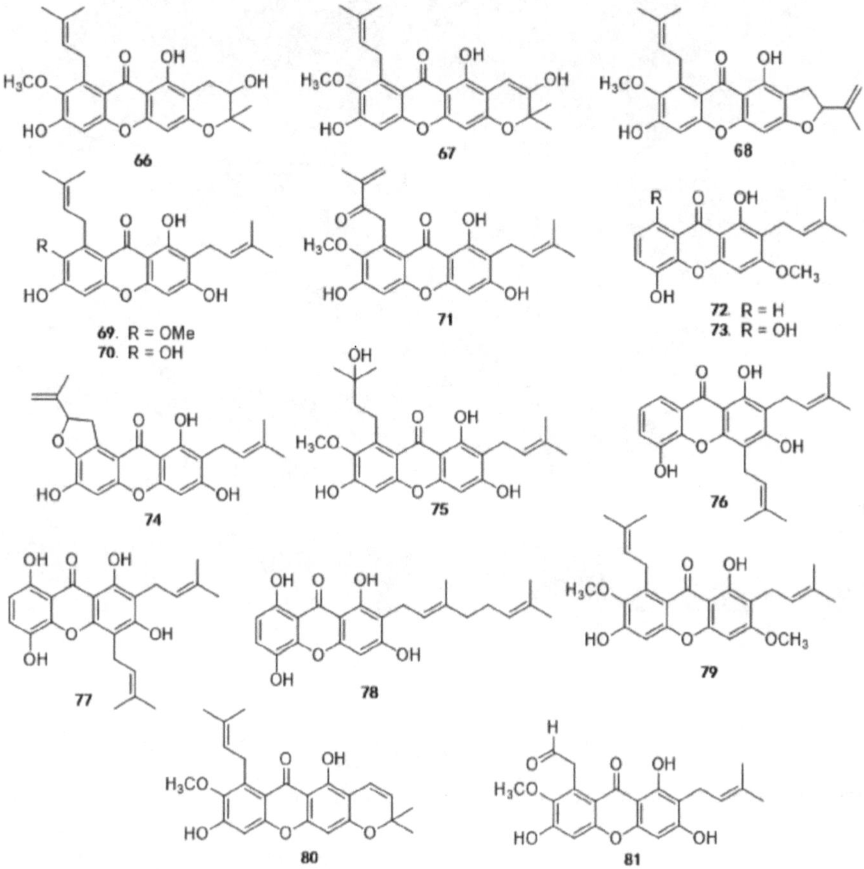

Fig. 10.5 Structures of xanthones with α-glucosidase inhibitory activity

polyhydroxy-nor-tropane alkaloids (see Table 10.2 or Fig. 10.7). As polyhydroxyl alkaloids with high-polarity, they are reported to be extracted with aqueous methanol and purified by various ion-exchange resins in the literature, e.g., the extraction and purification of eleven polyhydroxylated piperidine or pyrrolidine alkaloids (N-containing furanose and pyranose analogues) from the 50% aqueous MeOH extract of the bulbs of *Hyacinthus orientalis* by ion-exchange columns (Amberlite IR-120B and Amberlite CG-50) [29]. Identification of these compounds depends heavily on a variety of NMR (including extensive 1D and 2D NMR techniques) and MS spectroscopic analysis.

Terpenoids: Except an iridoid glucoside (compound **1**, in Fig. 10.8 or Table 10.3), all other terpenoids are common diterpenoid or triterpenoid aglycones with low polarity. Therefore, these compounds can be extracted with low-polarity solvents like petroleum ether, hexane, chloroform, dichloromethane, or ethyl acetate, and then purified by normal phase silica gel column chromatography. Their identification

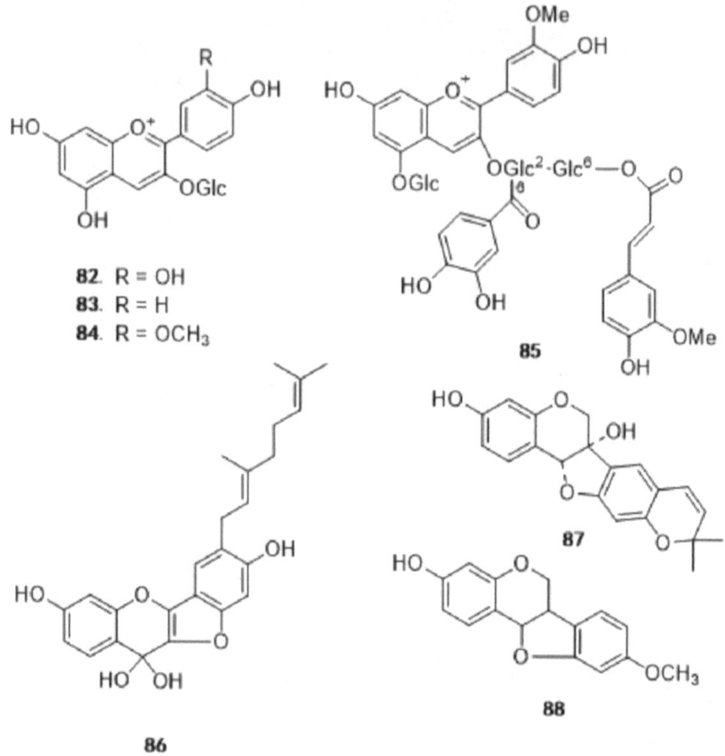

Fig. 10.6 Structures of other types of flavonoids with α-glucosidase inhibitory activity

can be accomplished by comparing their NMR spectroscopic data with those reported in the literature.

6 Future Perspectives

Literature summaries indicate that medicinal plants are important sources to discover α-glucosidase inhibitors. However, low content and time-consuming extraction limit the research and development of plant-derived α-glucosidase inhibitors. Furthermore, chemical modification or synthesis and SAR study of natural α-glucosidase inhibitors should be investigated, and standard extracts can be developed for the future prevention and treatment of type 2 diabetes.

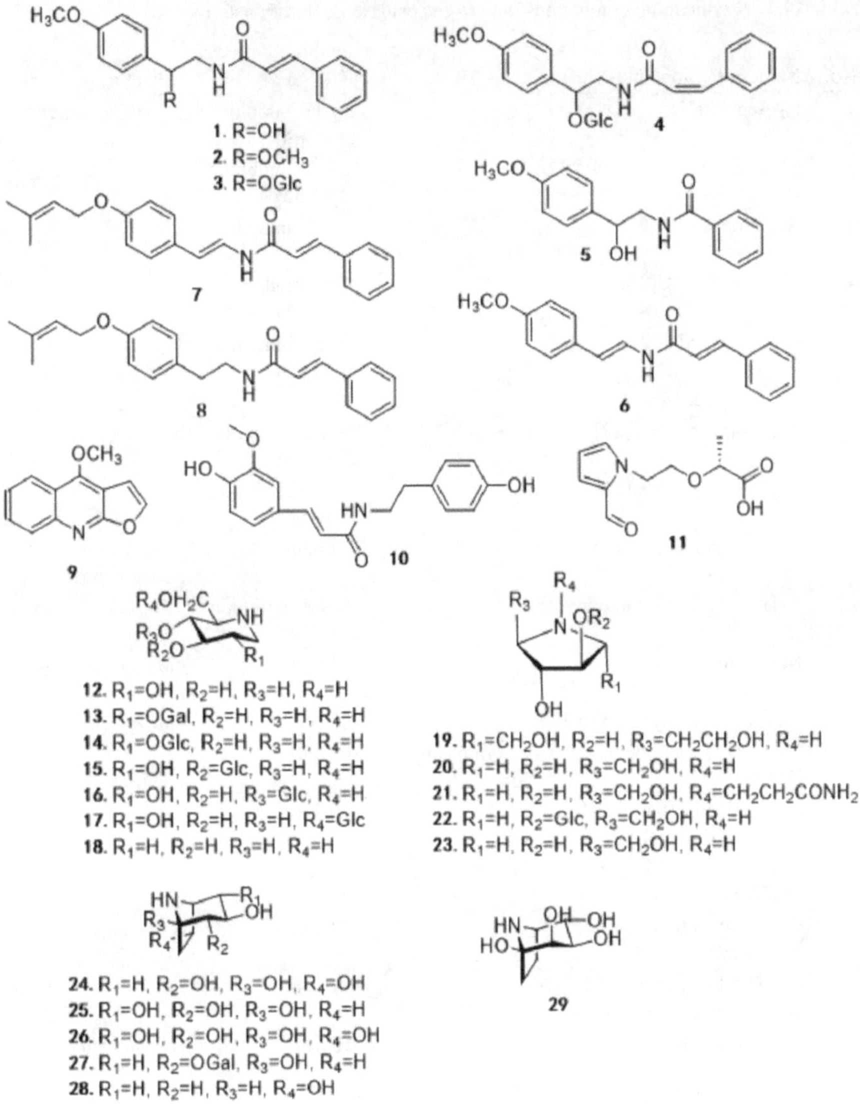

Fig. 10.7 Structures of N-containing compounds with α-glucosidase inhibitory activity

Table 10.2 N-containing compounds with α-glucosidase inhibitory activity

No.	Names of compounds [ref]	Inhibitory effects/ IC_{50}	Plant sources
1	(-)-Aegeline[22]	36.0% inhibition at 10 μM	*Aegle marmelos*
2	(-)-*O*-methylether aegeline[22]	12.6% inhibition at 10 μM	
3	Aegelinosides A[22]	17.6% inhibition at 10 μM	
4	Aegelinosides B[22]	8.8% inhibition at 10 μM	
5	(-)-Tembamide[22]	13.5% inhibition at 10 μM	
6	Anhydroaegeline[22]	35.8 μM	
7	Anhydromarmeline[22]	30.1% inhibition at 10 μM	
8	Dehydromarmeline[22]	19.7% inhibition at 10 μM	
9	Dictamnine[23]	2160.99 μM	*Dictamnus dasycarpus*
10	*N*-Trans-feruloyltyramine[24]	99.7% inhibition at 0.25g/L	*Clerodendranthus spicatus*
11	Norimester[25]	4.22 μM	*Morinda citrifolia Lin*
12	1-Deoxynojirimycin(DNJ)[26]	0.2μM	*Morus alba*
13	2-*O*-α-*D*-Galactopyranosyl-1-deoxynojirimycin [26]	0.39μM	
14	*O*-α-D-Glucopyranosyl-1-deoxynojirimycin[26]	<0.2μM	
15	*O*-β-D-Glucopyranosyl-1-deoxynojirimycin [26]	<0.2μM	
16	*O*-β-D-Glucopyranosyl-1-deoxynojirimycin [26]	<0.2μM	
17	*O*-β-***D***-Glucopyranosyl-1-deoxynojirimycin [27]	940μM	
18	Fagomine [28]	1.3μM	*Xanthocercis zambesiaca*
19	2, 5-Imino-2, 5, 6-trideoxy-D-manno-heptitol (6-deoxy-homoDMDP)[29]	2.2 μM	*Hyacinthus orientalis*
20	1, 4-Dideoxy-1, 4-imino-D-Arabinitol[27]	0.15 μM	*M. alba*
21	(2R, 3R, 4R)-2-Hydroxymethyl-3, 4-Dihydroxypyrrolidine-*N*-propionamide[27]	12.5 μM	
22	1, 4-Dideoxy-1, 4-imino-(2-*O*-β-D-Glucopyranosyl) -D-Arabinitol[26]	1-5 μM	*M. alba*
23	1, 4-Dideoxy-1, 4-imino-D-Ribitol[26]	9.0 μM	
24	Calystegine B1[30]	10.0% inhibition at 125 μM	*Lycium chinense*

(continued)

Table 10.2 (continued)

No.	Names of compounds [ref]	Inhibitory effects/ IC_{50}	Plant sources
25	Calystegine B2[30]	6.4% inhibition at 125 μM	*L. chinense*
26	Calystegine C1[30]	7.2% inhibition at 125 μM	*L. chinense*
27	*O*-α-D-Galactopyranosyl-calystegine B2[26]	less than 50% inhibition at 1000 mM	*M. alba*
28	3β, 6β-Dihydroxynortropane[26]	<0.2μM	
29	Calystegine B3[30]	6.2% inhibition at 125 μM	*L. chinense*

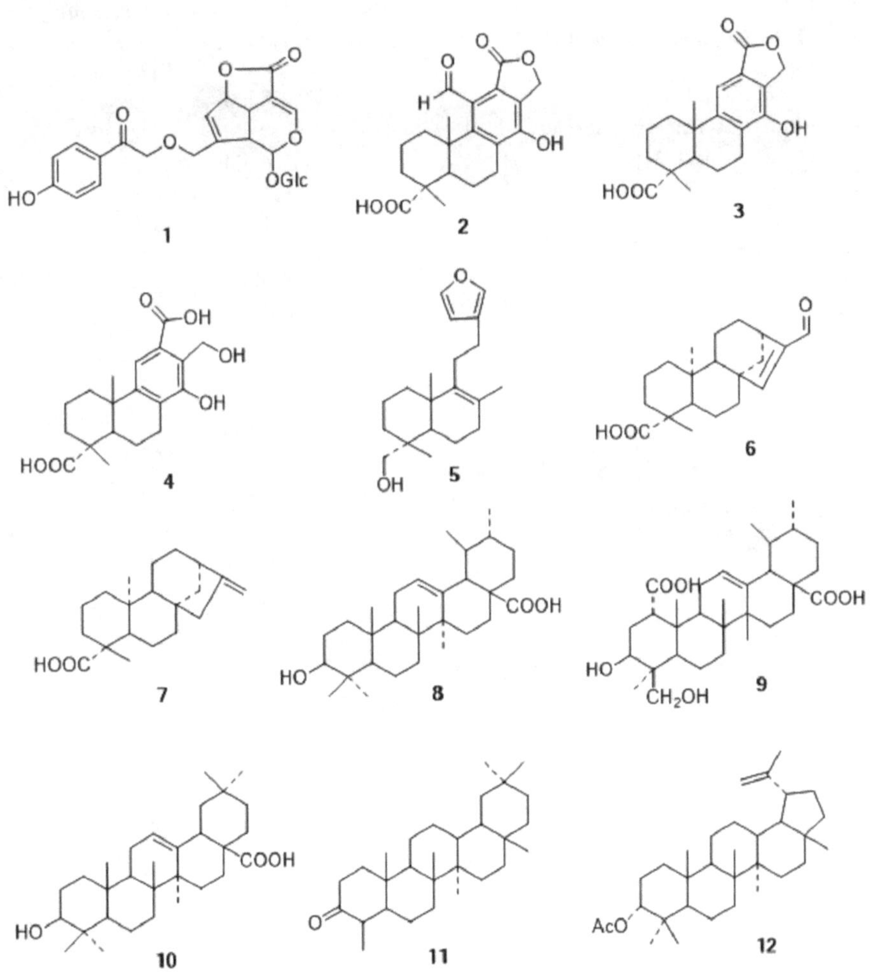

Fig. 10.8 Structures of terpenoids with α-glucosidase inhibitory activity

Table 10.3 Terpenoids with α-glucosidase inhibitory activity

No.	Names of compounds [ref]	Inhibitory effects/IC_{50}	Plant sources
1	10-*O*-*p*-Hydroxybenzoyl deacetyl asperuloside [13]	250 μM	*Arcytophyllum thymifolium*
2	14-Hydroxyabieta-8, 11, 13-triene-11-carbaldehyde-18-oic-12-carboxy-13-(1-hydroxy-1-methylethyl)-lactone [31]	0.379 mM	*Phlomis tuberosa*
3	14-Hydroxyabieta-8,11,13-triene-17-oic-12-carboxy-13-(1-hydroxy-1-methylethyl)-lactone [32]	0.624 mM	
4	14, 16-Dihydroxyabieta-8, 11, 13-triene-15, 17-dioic acid [31]	0.721 mM	
5	Phlomisol (15, 16-eposy-8, 13(16), 14-labdatrien-19-ol [31]	0.067 mM	
6	Ent-kaur-15-en-17-al-19-oic acid [32]	463.61 μM	*Acanthopanax trifoliatus*
7	Ent-16-Kauren-19-oic acid [33]	0.112 mM	*Sphagneticola trilobata*
8	Ursolic acid [34]	7.226 μM	*Luculia pinciana*
9	1α, 3β, 24-Trihydroxy-ursolic acid [34]	37.82 μM	
10	Oleanolic acid [34]	6.306 μM	
11	Friedelin [35]	20.1% inhibition at 100 μM	*Ficus drupacea*
12	Epilupeol acetate [35]	16.1% inhibition at 100 μM	

Table 10.4 Terpenoids with α-glucosidase inhibitory activity

No.	Names of compounds [ref]	Inhibitory effects/ IC_{50}	Plant sources
1	3-Caffeyl-5-feruloylquinic acid [36]	289.96 μM	*Artemisia selengensis*
2	Ferulic acid [10]	1.29 mM	*Zea mays*
3	Protocatechuic acid [10]	0.55 mM	
4	Vanillic acid [10]	1.22 mM	
5	2, 4, 6-Trihydroxybenzoic acid [10]	0.66 mM	
6	*p*-Hydroxycinnamic acid [10]	0.55 mM	
7	Scopoletin [34]	0.182 mM	*Luculia pinciana*
8	5-Methoxy-7-hydroxy-9, 10-dihydro-1, 4-phenanthrenequinone [34]	126.88 μM	*Dendrobium formosum*
9	5′-Hydroxy-3′-dehydrotoddanin [37]	170.0 μM	*Arcytophyllum nitidum*
10	Cneorumcoumarin B [37]	179.5 μM	
11	Hedyotiscone B [37]	215.3 μM	
12	Fraxetin [37]	210.9 μM	
13	9-(4′-Hydroxyphenyl)-2-methoxyphenalen-1-one [13]	0.013 mM	*Musa* spp. (Baxijiao)
14	*N-p*-coumaroyl serotonin [38]	47.2 μM	*Carthamus tinctorius*
15	*N*-Feruloyl serotonin [38]	99.8 μM	
16	Resveratrol [39]	71% inhibition at 0.2 mM	*Rumex bucephalophorus*
17	Rumexoid [40]	43% inhibition at 0.2 mM	
18	Oleic acid [40]	0.069 mM	-
19	Linoleic acid [40]	0.075 mM	
20	Eicosapentaenoic acid [40]	0.10 mM	
21	α-Linolenic acid [40]	0.16 mM	
22	Arachidonic acid [40]	0.29 mM	
23	Palmitic acid [40]	0.58 mM	
24	β-Sitosterol [41]	277.7 μM	-
25	Daucosterol [42]	0.429 mM	-

References

1. https://www.who.int/health-topics/diabetes
2. Kumar D, Ghosh R, Pal BC (2013) α-Glucosidase inhibitory terpenoids from *Potentilla fulgens* and their quantitative estimation by validated HPLC method. J Funct Food 5(3):1135–1141
3. Yang Z, Wang Y, Wang Y, Zhang YJFC (2012) Bioassay-guided screening and isolation of α-glucosidase and tyrosinase inhibitors from leaves of Morus alba. Food Chem 131(2):617–625
4. Choo CY, Sulong NY, Man F, Wong TW (2012) Vitexin and isovitexin from the Leaves of *Ficus deltoidea* with in-vivo α-glucosidase inhibition. J Ethnopharmacol 142(3):776–781
5. Oboh G, Ademiluyi AO, Akinyemi AJ, Henle T, Saliu JA, Schwarzenbolz U (2012) Inhibitory effect of polyphenol-rich extracts of jute leaf (*Corchorus olitorius*) on key enzyme linked to

type 2 diabetes (α-amylase and α-glucosidase) and hypertension (angiotensin I converting) in vitro. J Funct Food 4(2):450–458
6. Ezzat SM, Salama MM (2014) A new α-glucosidase inhibitor from *Achillea fragrantissima* (Forssk.) Sch. Bip. growing in Egypt. Nat Prod Res 28(11):812–818. https://doi.org/10.1080/14786419.2014.891203
7. Cardenas PA, Freddy AR, Costa GM (2015) Isolation of *C*-glycosylflavonoids with alpha-glucosidase inhibitory activity from *Passiflora bogotensis* Benth by gradient high-speed counter-current chromatography. J Chromatogr B 990:104–110
8. Supasuteekul C, Nonthitipong W, Tadtong S, Likhitwitayawuid K, Tengamnuay P, Sritularak B (2016) Antioxidant, DNA damage protective, neuroprotective, and α-glucosidase inhibitory activities of a flavonoid glycoside from leaves of *Garcinia gracilis*. Rev Bras Farm 26 (3):312–320
9. Sheliya MA, Rayhana B, Ali A, Pillai KK, Aeri V, Sharma M, Mir SR (2015) Inhibition of α-glucosidase by new prenylated flavonoids from *euphorbia hirta* L. herb. 176:1–8
10. Nile SH, Park SW (2014) Antioxidant, alpha-Glucosidase and xanthine oxidase inhibitory activity of bioactive compounds from maize (*Zea mays* L.). Chem Biol Drug Des 83 (1):119–125. https://doi.org/10.1111/cbdd.12205
11. Sun J, Zhang F, Yang M, Zhang J, Chen L, Zhan R, Li L, Chen YJNPR (2014) Isolation of α-glucosidase inhibitors including a new flavonol glycoside from. *Dendrobium devonianum*. 28 (21):1900–1905
12. Yoshikawa M, Shimada H, Nishida N, Li Y, Toguchida I, Yamahara J, Matsuda H (1998) Antidiabetic principles of natural medicines. II. Aldose reductase and alpha-glucosidase inhibitors from Brazilian natural medicine, the leaves of *Myrcia multiflora* DC. (Myrtaceae): structures of myrciacitrins I and II and myrciaphenones A and B. Chem Pharm Bull 46(1):113–119
13. Milella L, Milazzo S, Leo MD, Saltos MBV, Faraone I, Tuccinardi T (2016) Alpha-Glucosidase and alpha-Amylase Inhibitors from *Arcytophyllum thymifolium*. J Nat Prod 79(8):2104–2112
14. Yu YB, Miyashiro H, Nakamura N, Hattori M, Park JC (2007) Effects of triterpenoids and flavonoids isolated from alnus firma on HIV-1 viral enzymes. Arch Pharm Res 30(7):820–826
15. Choi CW, Choi YH, Cha MR, Yoo DS, Shi YR (2010) Yeast α-glucosidase inhibition by isoflavones from plants of *Leguminosae* as an in vitro alternative to acarbose. J Agric Food Chem 58(18):9988–9993
16. Choi SJ, Jin KK, Jang JM, Shin KH, Lim SS (2012) Rapid identification of the α-glucosidase inhibitory compounds from Thunberg's Geranium (*Geranium thunbergii Sieb*. et Zucc.). Food Sci Biotechnol 21(4):987–996
17. Watanabe J, Kawabata J, Kurihara H, Biochem RNJBB (1997) Isolation and identification of α-glucosidase inhibitors from Tochu-cha (Eucommia ulmoides). Biosci Biotechnol Biochem 61 (1):177–178
18. Wang X-L, Le W, Yu M, Xiao J, Duan D-Z (2017) Constituents with potent alpha-glucosidase inhibitory activity from *Pueraria lobat*a (Willd.) ohwi. Bioorgan Med Chem Lett 27 (9):1993–1998
19. Yu L, Ming K, Hu GP, Wu RB, Wang J, Lan L, Lin YCJM (2014) Loddigesiinols G–J: α-Glucosidase Inhibitors from Dendrobium loddigesii. Molecules 19(6):8544–8555
20. Chu Y-H, Wu S-H, Hsieh J-F (2014) Isolation and characterization of α-glucosidase inhibitory constituents from *Rhodiola crenulata*. Food Res Int 57:8–14
21. Ryu HW, Cho JK, Curtis-Long MJ, Yuk HJ, Kim YS, Jung S, Kim YS, Lee BW, Park KHJP (2011) α-Glucosidase inhibition and antihyperglycemic activity of prenylated xanthones from *Garcinia mangostana*. Phytochemistry 72(17):2148–2154
22. Phuwapraisirisan P, Puksasook T, Jong-Aramruang J, Letters MC (2008) Phenylethyl cinnamides: a new series of α-glucosidase inhibitors from the leaves of *Aegle marmelos*. Bioorg Med Chem Lett 18(18):4956–4958
23. Yang XAF, Lin J, Tang J (2019) Chemical constituents extracted from *Dictamnus dasycarpus* and their α-glucosidase inhibitory activity. J China Pharm Univ 50(01):44–48

24. Chen H-Q, Zhang R-R, Mei W-L, Cai C-H (2019) A new eudesmane type sesquiterpene from *cultivated Clerodendranthus* spicatus in Hainan. Zhongguo Zhong Yao Za Zhi 44(1):95–99
25. Zhang B, Zhou X-M, Zhao T, Wang J-L (2019) Chemical constituents from *fermented Noni* juice. J Nat Prod 44(18):4015–4020
26. Asano N, Oseki K, Tomioka E, Kizu H, Matsui KJ (1994) N-containing sugars from *Morus alba* and their glycosidase inhibitory activities. Carbohydr Res 259(2):243–255
27. Asano N, Yamashita T, Yasuda K, Ikeda K, Kang SR (2001) Polyhydroxylated Alkaloids Isolated from Mulberry Trees (*Morus alba* L.) and Silkworms (*Bombyx mori* L.). J Agric Food Chem 49(9):4208–4213
28. Kato A, Asano N, Kizu H, Matsui K, Nash RJ (1997) Fagomine isomers and glycosides from *Xanthocercis zambesiaca*. J Nat Prod 60(3):312–314
29. Asano N, Kato A, Miyauchi M, Kizu H, Kameda Y, Watson AA, Nash RJ, Fleet GW (1998) Nitrogen-containing furanose and pyranose analogues from hyacinthus orientalis. J Nat Prod 61 (5):625–628
30. Huo LM, Zheng XH, Chen F, Liu YJ, Zhou GX (2017) Chemical constituents of the aqueous extract from Lycii Cortex. Journal of Jinan University (Natural Science and Medicine Edition) 38(05):443–449
31. Yang Y, Gu L, Xiao Y, Liu Q, Hu H, Wang Z, Chen K (2015) Rapid Identification of α-Glucosidase Inhibitors from *Phlomis tuberosa* by Sepbox Chromatography and Thin-Layer Chromatography Bioautography. Plos One 10(2):e0116922
32. Li CY, Li DL, Sun YK, Zhang TW, Dong YQ, Zhang KJFI (2016) α-Glucosidase Inhibition of Extracts and Constituents of *Acanthopanax trifoliatus*(Linn.) Merr. Molecules 17 (6):6269–6276
33. Jia Y, Ren H, Zhou Z, Xu X, Tan J (2015) Qualitative and quantitative analysis of diterpenoid compounds in *Sphagneticola trilobata* (L.) Pruski (南美蟛蜞菊中二萜化学成分的定性与定量分析研究). J Trop Subtrop Bot 23(6):703–709
34. Kang W, Zhang L, Song Y (2009) α-Glucosidase inhibitors from *Luculia pinciana*. Zhongguo Zhong Yao Za Zhi 34(4):406–409
35. Phan VK, Chau VM (2012) Chemical constituents of *Ficus drupacea* leaves and their α-glucosidase inhibitory activities. Bull Chem Kor Soc 34(1):263–266
36. Xie X(2018) Study on α-glucosidase inhibitors screening, purification and inhibition mechanism in Artemisia selengensis Turcz. Nanchang University.
37. Inthongkaew P, Chatsumpun N, Supasuteekul C, Kitisripanya T, Putalun W, Likhitwitayawuid K, Sritularak B (2017) α-Glucosidase and pancreatic lipase inhibitory activities and glucose uptake stimulatory effect of phenolic compounds from *Dendrobium formosum*. Rev Bras Farm 27:480–487
38. Sheng Z, Dai H, Pan S, Wang H, Hu Y, Ma W (2014) Isolation and Characterization of an α-Glucosidase Inhibitor from *Musa spp.* (Baxijiao) Flowers. Molecules 19:10563–10573. https://doi.org/10.3390/molecules190710563
39. Takahashi T, Miyazawa M (2012) Potent α-glucosidase inhibitors from safflower (Carthamus tinctorius L.) seed. Phytother Res 26(5):722–726
40. Zohar K, Itzhak B, Moshe A, Flaishman L (2006) Antioxidant activity and inhibition of α-glucosidase by trans-resveratrol, piceid, and a novel trans-stilbene from the roots of *Israeli Rumex bucephalophorus* L. J Agric Food Chem 54(4):1243–1247
41. Liu B, Kongstad KT, Wiese S, Jager AK, Staerk D (2016) Edible seaweed as future functional food: Identification of alpha-glucosidase inhibitors by combined use of high-resolution alpha-glucosidase inhibition profiling and HPLC-HRMS-SPE-NMR. Food Chem 203:16–22
42. Tabussum A, Riaz N, Saleem M, Ashraf M (2013) a-Glucosidase inhibitory constituents from *Chrozophora plicata*. Phytochem Lett 6(4):614–619

Chapter 11
Rhinacanthin-C and Its Potential to Control Diabetes Mellitus

Pharkphoom Panichayupakaranant, Muhammad Ajmal Shah, and Thongtham Suksawat

Abstract Diabetes mellitus (DM) is the seventh major cause of mortality worldwide. Ninety percent of diabetic patients suffer from type-2 DM, which still requires an economical, effective, and safe oral antihyperglycemic treatment. Natural products, especially from plant resources, provide a variety of bioactive compounds with different therapeutic effects. This chapter summarizes the research related to the antidiabetic potential of rhinacanthin-C, a major bioactive naphthoquinone ester of *Rhinacanthus nasutus* leaves. Moreover, its physicochemical properties, natural and plant tissue culture resources, quantitative HPLC analysis for the standard markers, rhinacanthin-C, -D, and -N, and some toxicity assessments have been described. Rhinacanthin-C exerts its antidiabetic effect by various mechanisms, i.e., noncompetitive inhibition of α-glucosidase with a synergistic effect to acarbose; enhanced glucose uptake in muscle and adipose cells; regeneration of insulin-secreting beta cells; and anti-adipogenic effect. It also exerts protective effects against diabetes and its complications, such as a nephroprotective effect against diabetic nephropathy, by the scavenging of free radicals and inhibition of advanced glycation end products (AGEPs). The method for preparation of rhinacanthins enriched *R. nasutus* leaf extracts (RRE) has been described using a simple green extraction and fractionation method. RRE exhibits the equivalent biological activities to rhinacanthin-C. In contrast, its cost of production is much lower than pure rhinacanthin-C. Therefore, RRE is highly recommended as an alternative of rhinacanthin-C for pharmaceutical applications in terms of its efficacy, safety, and low cost. However, further studies, including acute and chronic toxicities, pharmacokinetic study as well as clinical trials are required for ensuring its use for the treatment of DM.

P. Panichayupakaranant (✉) · T. Suksawat
Department of Pharmacognosy and Pharmaceutical Botany, Faculty of Pharmaceutical Sciences, Prince of Songkla University, Hat-Yai, Songkhla, Thailand
e-mail: pharkphoom.p@psu.ac.th

M. A. Shah
Department of Pharmacognosy, Faculty of Pharmaceutical Sciences, Government College University, Faisalabad, Pakistan

H. Chen, M. Zhang (eds.), *Structure and Health Effects of Natural Products on Diabetes Mellitus*, https://doi.org/10.1007/978-981-15-8791-7_11

Keywords Rhinacanthin-C · Rhinacanthin · Naphthoquinone · *Rhinacanthus nasutus* · Antidiabetic · Hypoglycemic

1 Introduction

Diabetes mellitus (DM) can be described as a metabolic disability with multiple etiologies, characterized by hyperglycemia, and metabolic alteration in carbohydrates, proteins, and lipids. DM is a serious global health concern. In 2018, DM affected more than 8.5% of the global population. It has been predicted that by the year 2040, there will be at least 640 million diabetic patients. In 2012, over 1.5 million died directly as a result of DM, and another 2.2 million deaths were caused by high blood glucose [1]. DM is a complicated chronic disorder that requires consecutive medical care. Apart from glycemic control, multi-complication risk management approaches are critical [2]. DM is a multifactorial chronic hyperglycemia, characterized by endocrinologically altering the body's metabolic compounds, i.e. carbohydrates, proteins, and lipids [3]. It begins with a complete or partial decrease in insulin production and/or insulin interaction with the receptor [4]. Pathogenesis of DM occurs from either irregular insulin secretion from the pancreas or impairment of insulin action, resulting in persistent high blood glucose and glucose intolerance. There are two types of diabetes, including type-1 (insulin-dependent) and type-2 (noninsulin-dependent) DM classified accordingly to their pathogenesis [5]. Patients losing insulin-secreting beta cells are commonly diagnosed as type-1 diabetic patients, whereas type-2 DM patients occur either by a decrease in insulin synthesis or by weak involvement of insulin with its insulin receptors. Although both types have different pathophysiology, the symptom of hyperglycemia is similar.

Over 90% of DM patients are considered to be type-2 DM [6]. Current analysis of data expresses that nearly 451 million people have type-2 DM [7], and it is predicted that this number will escalate to 693 million by the year 2045 [8]. Obesity and physical inactivity are the main causes of type-2 DM. Hyperlipidemia and hyperglycemia are consequences of obesity leading to the progression of type-2 DM and complication of cardiovascular diseases [9]. Moreover, the release of free fatty acids and pro-inflammatory cytokines from adipocytes in obese patients leads to over-secretion of insulin from beta cells and reduces insulin receptors resulting in insulin being desensitized [10]. Oxidative stress generated by overproduction of free radicals is implicated in chronic DM patients and their complications. The oxidative stress generations, which deteriorate DM include advanced protein glycation, transcription factor triggering, and protein kinase-C activation [11]. Non-enzymatic glycation occurs from reactions between amino groups of proteins and carbonyl groups of reducing sugar, which subsequently rearrange into stable compounds, advanced glycation end products (AGEPs). Generations of AGEPs stimulate positive feedback of cells damaging oxidative stress [12]. DM complication is one of the serious problems in most DM patients. Neuropathy, nephropathy, retinopathy, and cardiovascular issues are severe complications that critically increase the risk of DM mortality [13].

While type-1 DM mainly requires insulin treatment, antidiabetic, and glucose-lowering drugs are necessary for type-2 DM [2]. Various mechanisms of those antidiabetic drugs apply therapeutic effects, including stimulation of insulin secretion, increase of peripheral glucose absorption, and delay and reduction of hepatic gluconeogenesis [14]. Sulfonylurea, meglitinides, biguanides, thiazolidinediones, and α-glucosidase inhibitors are the most common synthetic drugs used in type-2 DM patients [2]. However, some of these drugs are high cost and cause various short- and long-term side effects. Metformin, the most widely used antidiabetic agent that belongs to the biguanide group may cause B12 deficiency. Glipizide belongs to sulfonylureas and can cause weight gain. Thiazolidinedione is an antidiabetic drug that may increase the risk of congestive heart failure due to the fluid retention effect. Acarbose, a high cost α-glucosidase inhibitor which can cause short term gastrointestinal side effects and is associated with cardiac issues in long-term users. Nonetheless, the main critical side effects of taking synthetic conventional antidiabetic agents are hypoglycemia episodes, weight gain, cardiovascular effects, and renal complications [15]. Beyond the side effects of these antidiabetic drugs, drug–drug interaction effects from combination treatment among antidiabetic agents are concerning [16]. Hence, finding the perfect combination treatment for diabetic patient without side effect issues remains a challenge.

Alternative treatments for DM are increasingly presented to patients. Natural products have a variety of sources of pharmacological active molecules [17]. Many treatments for diabetic patients involve the use of natural compounds in both developed and developing countries [17]. The plant kingdom comprises a greater proportion of undiscovered reservoirs of bioactive compounds. Though these natural compounds may act as drugs or precursors of synthetic derivatives, biological active compounds are to be involved in a profound understanding of physiological processes. Conventional hypoglycemic plants may provide a worthwhile originator of recent oral antihyperglycemic agents for establishment of novel medicinal entities, or as a basic nutritional additive to present therapies. Recently, the introduction of new technologies to promote research on antidiabetic phytomedicines has increased. Over 400 medicinal plants have been reported for antidiabetic properties. However, only around 100 plants have been confirmed for antidiabetic mechanisms [4]. Medicinal plants that contain alkaloids, terpenoids, polysaccharides, carotenoids, flavonoids, and naphthoquinones are more likely to have the antidiabetic potential [18]. For instance, cinnamaldehyde, from *Cinnamomum zeylanicum*; davidigenin, a flavonoid from *Artemisia dracunculus*, vindoline, an alkaloid from *Catharanthus roseus*, costunolide, a terpenoid from *Costus speciosus*, neomyrtillin, a glycoside from *Vaccinium myrtillus*, and rhinacanthin-C, a naphthoquinone ester from *Rhinacanthus nasutus* have all shown antidiabetic activity [19]. Diverse therapeutic potential and multi-mechanism are ideal mechanism aspects in antidiabetic agents. As conventional antidiabetic agents, plants also have the potential to provide various mechanisms, i.e., improving the performance of pancreatic tissue resulting in increased insulin secretion, improving metabolic profiles including blood glucose level and lipid profile, interfering glucose absorption from the gastrointestinal tract, decreasing body weight, and possessing antiglycation activity to prevent complications from the oxidative stress [20].

Naphthoquinones are biologically active compounds isolated from various phytogenic families, including Acanthaceae, Balsaminaceae, and Plumbaginaceae. They have been reported to possess many biological effects, such as antimicrobial, anti-inflammatory, antioxidant, anticancer, antidiabetic, anti-obesity, antiglycation, and neuroprotective activities [21]. *Rhinacanthus nasutus* (Acanthaceae family), a unique medicinal plant, has been traditionally used to treat various diseases, including DM. The major phytochemical compounds in *R. nasutus* are 1,4-naphthoquinone esters, called rhinacanthins. However, rhinacanthin-C is the most active marker for antidiabetic activity in *R. nasutus* [22]. Therefore, this chapter seeks to summarize the antidiabetic potential of rhinacanthin-C. Moreover, its botanical source, physicochemical properties, mechanism of action, and toxicity are described as necessary.

2 Rhinacanthin-C and Its Botanical Source

Rhinacanthin-C or 3-(3-hydroxy-1,4-dioxo-1,4-dihydronaphthalen-2-yl)-2,2-dimethylpropyl(2*E*,6*E*)-2,6-dimethylocta-2,6-dienoate or [3-(3-hydroxy-1,4-dioxo-2-naphthyl)-2,2-dimethyl-propyl](2*E*,6*E*)-2,6-dimethylocta-2,6-dienoate is a naphthoquinone ester (Fig. 11.1) with the molecular formula of $C_{25}H_{30}O_5$ and a molecular weight of 410.5. Rhinacanthin-C has been isolated from *Rhinacanthus nasutus* (L.) Kurz (Acanthaceae family) (Fig. 11.2), and the name "rhinacanthin" is derived from the plant genus name from which it was first isolated, "Rhinacanthus," and "C" is derived from the alphabet ordering of naphthoquinone discovery from *R. nasutus* [23]. Rhinacanthin-C was purified from the leaf extract of *R. nasutus* [24]. It is a reddish-yellow oily liquid and is soluble in alcohol, chloroform, hexane, and ethyl acetate [25].

Rhinacanthus nasutus (L.) Kurz or *R. communis* Nees. (Thai name: Thong Phan Chang; English name: Snake jasmine; Chinese name: 白鹤灵芝) is a medicinal plant native to Thailand and Southeast Asia. *R. nasutus* and its flowers were shown in Fig. 11.2.

The plant is well known in traditional medicines for the treatment of various diseases including DM [26]. Most of the traditional therapeutic uses of *R. nasutus* have been scientifically validated and it has been reported that the majority of the

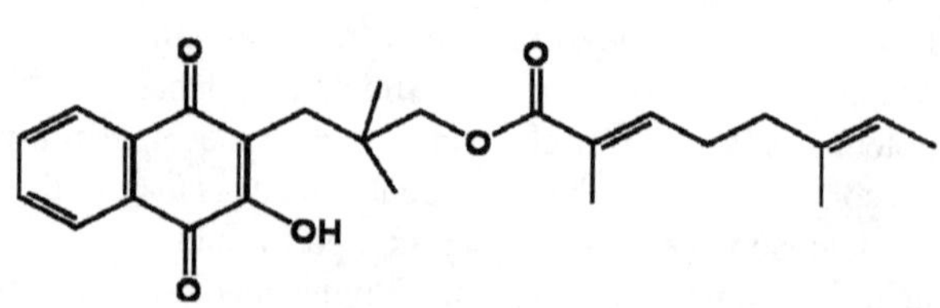

Fig. 11.1 Chemical structure of rhinacanthin-C

Fig. 11.2 *Rhinacanthus nasutus* and its flowers

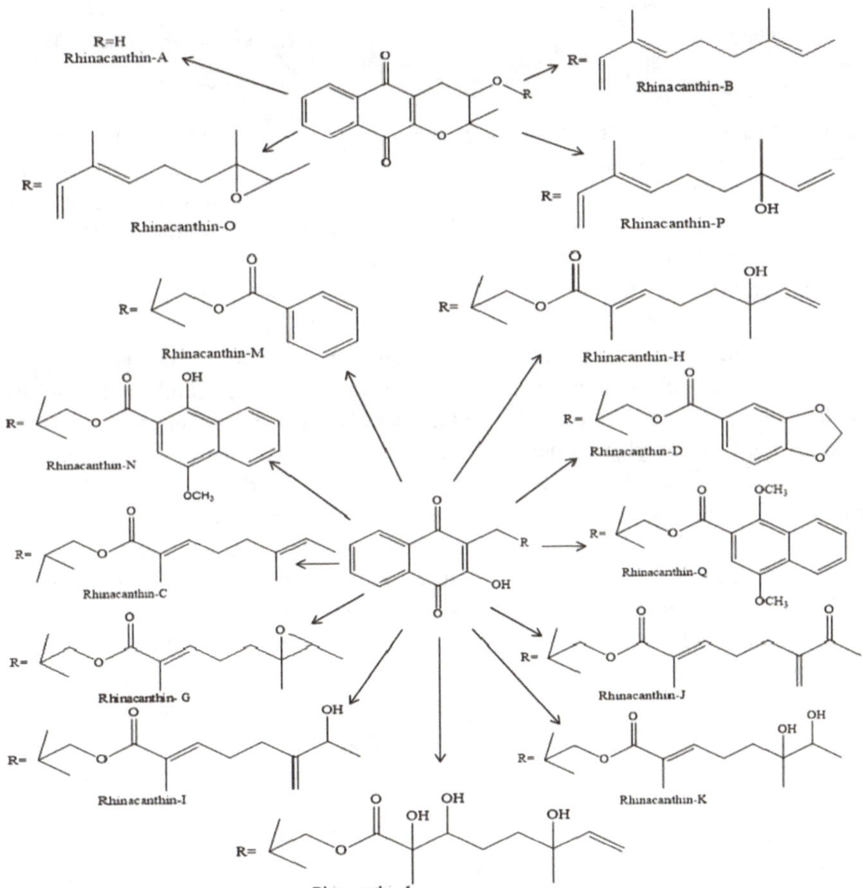

Fig. 11.3 Chemical structures of naphthoquinone esters found in *R. nasutus*

Table 11.1 Biological activities of *R. nasutus* extracts and its naphthoquinones esters

Phytochemicals	Biological activities	References
Rhinacanthin-C, rhinacanthins rich extract	Analgesic	[27]
Rhinacanthin-C, -D, -N	Antiallergic	[28]
Rhinacanthin-C, -N, -Q, rhinacanthins rich extract	Antibacterial	[25, 29]
Rhinacanthin-C, -D, -N, rhinacanthins rich extract	Antifungal	[30]
Rhinacanthin-C, -D, -N, rhinacanthins rich extract	Anti-inflammatory	[27, 31]
Rhinacanthin-C, rhinacanthins-rich extract,	Antioxidant	[32, 33]
Rhinacanthin-A, -B, -C, -G, -H, -I, -K, -M, -Q	Antiplatelet	[34, 35]
Ethanol and aqueous extracts, rhinacanthin-C, -N, -Q, liposomal formulations of rhinacanthin-C, -N, -Q	Antiproliferative	[23, 25, 36–38]
Rhinacanthin-M, -N, -Q	Antitumor activity	[23]
Rhinacanthin-C, -D, -N, -Q	Antiviral	[39, 40]
Methanol extract, rhinacanthin-A, -B, -C, -D, -G, -H, -I, -K, -M, -N, -Q	Cytotoxic	[24, 34]
Methanol extract, rhinacanthin-C	Antihyperlipidemic	[32, 41]
Aqueous and 80% ethanol extracts	Immunomodulatory	[42]
Rhinacanthin-C	Neuroprotective	[43, 44]

bioactivities are due to the naphthoquinone esters, namely rhinacanthins. Various naphthoquinone esters that have been found in *R. nasutus*, including rhinacanthin-A, -B, -C, -D, -G, -H, -I, -J, -K, -L, -M, -N, O, -P, and -Q and their reported biological activities are summarized in Fig. 11.3 and Table 11.1, respectively. The most bioactive constituent is rhinacanthin-C, which may have potential as a drug development candidate. However, it is not yet available commercially.

Recently, research has been conducted on the establishment of tissue cultures of *R. nasutus* in order to use them as an alternative source of the natural plant (Fig. 11.4). The cell suspension cultures of *R. nasutus* were established from the callus culture and maintained in liquid Gamborg's B5 (B5) medium supplemented with 2.0 mg/L 6-benzylaminopurine (BA) and 0.5 mg/L indole-3-butyric acid (IBA). However, the cell suspension cultures did not produce any rhinacanthin-C. Medium manipulation and elicitation techniques have been examined to induce rhinacanthin production in cell suspension cultures. Unfortunately, neither technique could induce rhinacanthin production in cell suspension cultures. This indicated that the dedifferentiation of *R. nasutus* cell suspension culture resulted in a loss of rhinacanthin production potential. However, the medium manipulation technique brought about an establishment of *R. nasutus* root cultures. Induction of *R. nasutus* root formation was accomplished on solid B5 medium, supplemented with 0.1 mg/L IBA. However, medium manipulation found that Murashige and Skoog (MS) medium supplemented with 3.0 mg/L IBA was the most suitable for increased root formation. Unfortunately, the root cultures produced only a low amount of rhinacanthin-C (0.003% w/w). However, when they were transferred into the semi-solid medium (4 g/L agar) of the same MS composition, the root cultures produced

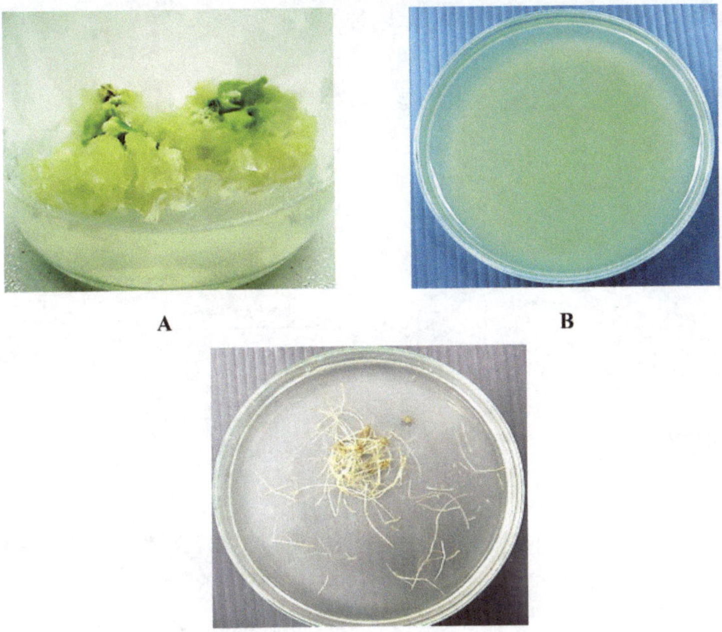

Fig. 11.4 Tissue cultures of *Rhinacanthus nasutus* (**a**) callus culture on solid B5 medium supplemented with 2.0 mg/L BA and 0.5 mg/L IBA; (**b**) cell suspension culture in liquid B5 medium supplemented with 2.0 mg/L BA and 0.5 mg/L IBA; (**c**) root culture in liquid MS medium supplemented with 3.0 mg/l IBA

more rhinacanthins, including rhinacanthin-C, -D, and -N, with the content of 0.34, 0.007, and 0.007% w/w, respectively [45].

3 Biogenetic Pathways of Rhinacanthin-C

Rhinacanthin-C is a naphthoquinone ester comprised of two core structures, namely the naphthoquinone unit and the isoprenyl unit. Plant naphthoquinones are synthesized differently among plant families, i.e., via 4-hydroxybenzoate (4HB)/mevalonic acid (MVA) pathway, *o*-succinylbenzoic acid (OSB) pathway, and acetate-polymalonate pathway [46]. Biosynthesis of rhinacanthin-C has not been fully discovered, yet literature review suggests that the biogenetic view of a naphthoquinone unit of rhinacanthin-C is involved in OSB pathway, starting from isomerization of chorismic acid (product from shikimic acid) to isochorismic acid. Two extra carbon atoms in naphthoquinone are provided from 2-oxoglutaric acid, which are added into isochorismic acid by the coenzyme thiamine diphosphate (TPP) via Michael-type reaction. Subsequently, elimination of pyruvic acid and

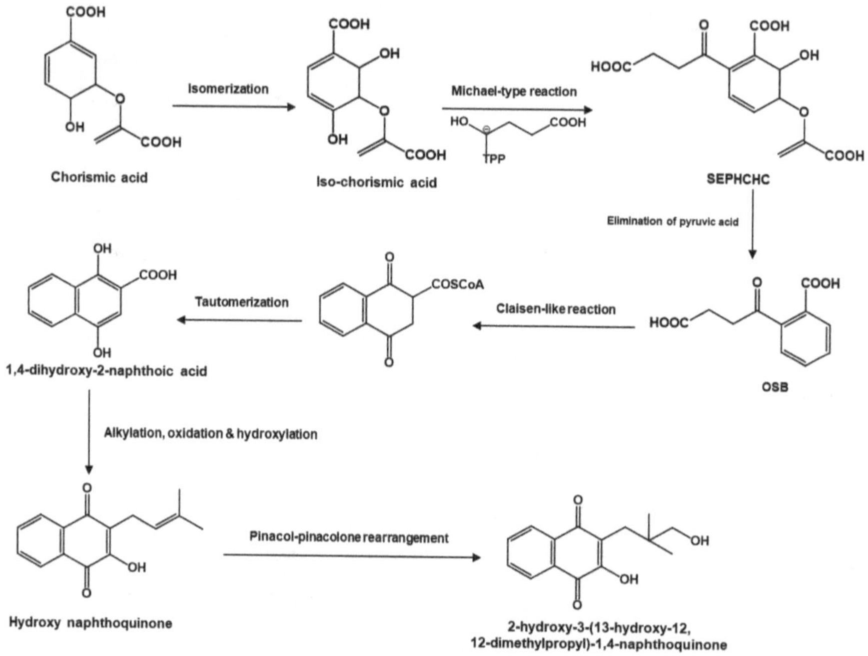

Fig. 11.5 Shikimate (OSB) pathway

dehydration are given the intermediate *o*-succinylbenzoic acid using *o*-succinyl benzoate synthase enzyme. Formation of coenzyme A ester with OSB leads to a Claisen-like condensation and allows ring formation. Tautomerization of dihydroxynaphthoic acid coenzyme A ester is more flavored given 1,4-dihydroxynaphthoic acid [47]. Transformation of 1,4-dihydroxynaphthoic acid is further catalyzed by alkylation, decarboxylation, oxidation, and hydroxylation into the isoprenylated hydroxynaphthoquinone. Spontaneously, an isoprenyl unit on the isoprenylated hydroxynaphthoquinone can be reorganized into epoxide species and cyclized via an ether linkage followed by hydrolysis and pinacol-pinacolone rearrangement given an aldehyde hydroxynaphthoquinone. Aldehyde hydroxynaphthoquinone can be reduced to 2-hydroxy-3-(13-hydroxy-12,12-dimethylpropyl)-1,4-naphthoquinone, the alcohol part of naphthoquinone ester [34]. At this step, the naphthoquinone unit from rhinacanthin-C is more likely to be synthesized through the OSB pathway because the end product from the OSB pathway is isoprenylated hydroxynaphthoquinone, which is further synthesized into 2-hydroxy-3-(13-hydroxy-12,12-dimethylpropyl)-1,4-naphthoquinone, the naphthoquinone unit that is a core structure of rhinacanthin-C (Fig. 11.5). Yet, other naphthoquinone biosynthesis pathways, i.e., 4HB/MVA and acetate-polymalonate pathways, are less like to give such a naphthoquinone unit of the core structure of rhinacanthin-C [46].

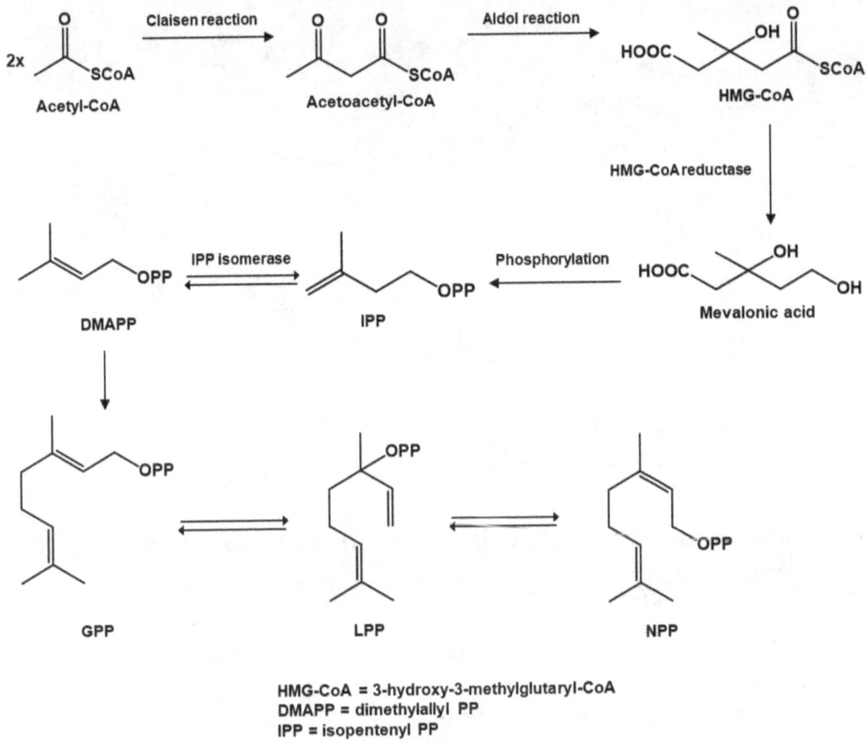

Fig. 11.6 Biosynthesis of geranyl pyrophosphate in the mevalonate pathway

Isoprenyl unit of rhinacanthin-C seems likely to be synthesized via mevalonate pathway, which is enzymatically catalyzed in the combination of dimethylallyl pyrophosphate (DMAPP) and isopentenyl pyrophosphate (IPP) using geranyl pyrophosphate synthase to produce geranyl pyrophosphate (GPP) (Fig. 11.6). Linalyl pyrophosphate (LPP) and neryl pyrophosphate (NPP) are the isomer forms of GPP that are relatively active to the addition reaction with other compounds, i.e., hydrocarbons, alcohols, aldehydes, or ester by reacting with acetyl-CoA [47]. Therefore, prenylation of monoterpene (GPP, Linalyl PP, and neryl PP) to the naphthoquinone unit (2-hydroxy-3-(13-hydroxy-12,12-dimethylpropyl)-1,4-naphthoquinone) via esterification is the final reaction step to obtain rhinacanthin-C.

Therefore, rhinacanthin-C may be synthesized through the shikimate (OSB) pathway coupled with the mevalonate pathway (Fig. 11.7). Nonetheless, a definite investigation of rhinacanthin-C biosynthesis pathway is still necessary. Since OSB pathway is localized in plastids and the mevalonate pathway is localized in the cytosol, then a biogenetic pathway of rhinacanthin-C seems to be located at chlorophylls in *R. nasutus* leaves [48]. However, after rhinacanthin-C is synthesized, it can be translocated and accumulated in the roots, because rhinacanthin-C is also markedly accumulated in the roots of *R. nasutus*.

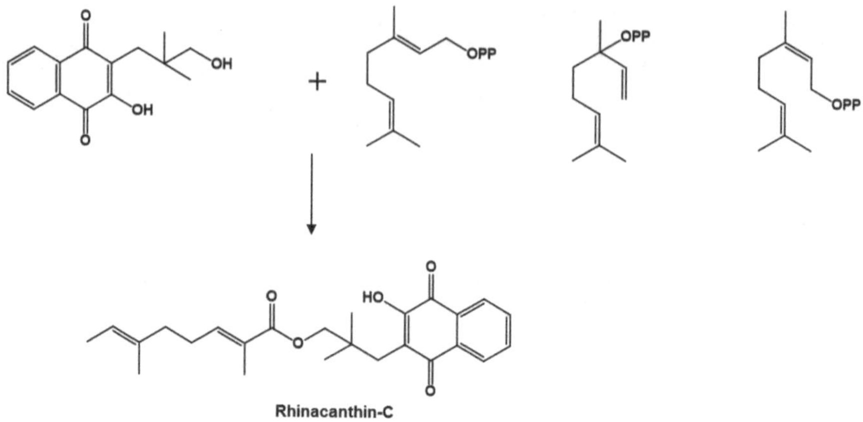

Fig. 11.7 Biogenetic pathway of rhinacanthin-C

4 Rhinacanthins Enriched *Rhinacanthus nasutus* Leaf Extract

Many conventional drugs and their precursors are derived from natural products, especially phytochemicals. However, there is a difference between administering a pure phytochemical and the same chemical in a form of plant extract. Whether phytochemical complexity is advantageous or not is still a matter of some debate. Recently, the concept of synergistic interaction between phytochemicals in plant extracts has become a topic of interest [49]. In the context of chemical complexity, it applies if the pharmacological effect of a chemical mixture is greater than the arithmetic sum of the effects of individual components. This is the basis of the preparation of an active constituent enriched herbal extract.

Currently, rhinacanthin-C is not available commercially. Moreover, the purification of rhinacanthin-C involves multistage process that requires a large amount of toxic solvents as well as being time and energy consuming, therefore increasing its cost of production. An alternative approach that presents the possibility of synergistic effect is to utilize rhinacanthins enriched rich extract (RRE), a semi-purified extract of *R. nasutus* leaves [50].

The extraction and fractionation processes are very important steps for the preparation of an active constituent enriched extract. A suitable method and solvent for extraction must be determined to increase the yield of active constituent in an herbal extract. Rhinacanthin-C, rhinacanthin-D, and rhinacanthin-N (Fig. 11.8), the major naphthoquinone esters found in *R. nasutus* leaves, possessed various pharmacological activities, such as antidiabetic, antifungal, antibacterial, antiviral, anti-inflammatory, anti-allergic, and cytotoxic activities (Table 11.1).

Some solvents were determined to increase the extraction of rhinacanthins from *R. nasutus* leaf powders. Although methanol gave the highest yield of the crude extract (18.4% w/w of the dried leaf), the obtained extract contained a rather low

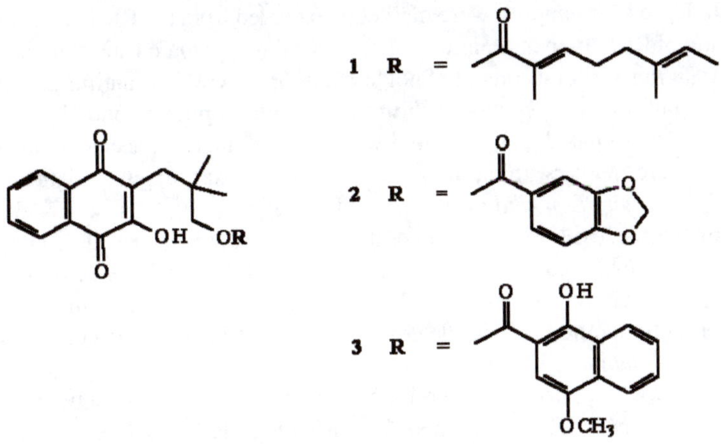

Fig. 11.8 Chemical structures of rhinacanthin-C (1), -D (2) and -N (3)

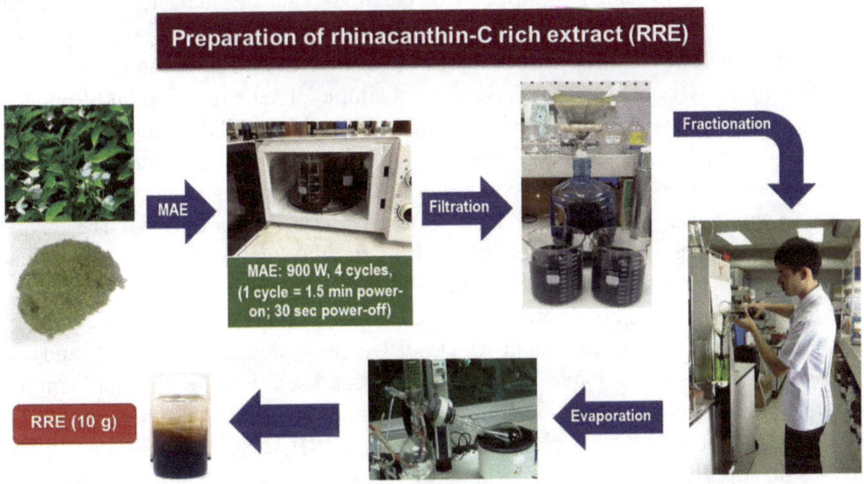

Fig. 11.9 Preparation of RRE using a green extraction

rhinacanthins content (6.7% w/w of dried extract). In contrast, ethyl acetate gave a rather low yield of crude extract (4.8% w/w of dried leaf) but contained the highest content of rhinacanthins (33.0% w/w of dried extract). This indicates that ethyl acetate is a selective and appropriate solvent for the extraction of rhinacanthins [30].

Rhinacanthins are anionic compounds that can be isolated using an anion exchange column chromatography. It has been reported that the Amberlite® IRA-67 column could improve rhinacanthin concentration to produce rhinacanthin enriched extracts from the crude ethyl acetate extracts of *R. nasutus* leaves. By this method, the content of total rhinacanthins in the rhinacanthins enriched *R. nasutus* extracts (RRE) was increased to 77.5% w/w. The other pigments, including chlorophylls and carotenoids, were markedly excluded from the RRE. To confirm an application of RRE as an antimicrobial agent, RRE was prepared and standardized to contain total rhinacanthin content of not less than 70% w/w. The antifungal activities of RRE against *Trichrophyton rubrum*, *T. mentagrophytes*, and *Microsporum gypseum* were evaluated and compared with those of the ethyl acetate crude extract and its standard markers, rhinacanthin-C, -D, and -N. Antifungal activities of RRE against *T. rubrum*, *T. mentagrophytes,* and *M. gypseum* (MICs of 7.5, 31.2 and 125 μg/mL, respectively) were stronger than those of the ethyl acetate crude extract (MICs of 31.2, 62.5 and 500 μg/mL, respectively). The antifungal activity of RRE was equal to that of rhinacanthin-C and D, but stronger than rhinacanthin-N. Moreover, antimicrobial activity determination of RRE and rhinacanthin-C against *Streptococcus mutans*, *Propionibacterium acnes*, *Staphylococcus aureus,* and *S. epidermidis* revealed that both exhibited antimicrobial activity against *S. mutans*, *P. acnes*, *S. aureus,* and *S. epidermidis* [29]. RRE exhibited potent bactericidal activity against Gram-positive anaerobic bacteria, *S. mutans* and *P. acnes* with MBCs of 4 and 32 μg/mL, respectively, and also showed moderate bactericidal activity against Gram-positive facultative anaerobic bacteria, *S. aureus* and *S. epidermidis* with the MBCs of 256 and 512 μg/mL, respectively. The antibacterial activity of RRE was also almost equal to that of rhinacanthin-C.

Currently, RRE has been obtained using a simple green extraction and fractionation method that is environmentally friendly (Fig. 11.9). The method used only green solvents, i.e., ethanol and water. A microwave-assisted extraction, which required low energy and was less time consuming has been used and a step of solvent evaporation prior to the fractionation process has been omitted. The developed green process in preparation of RRE is highly suitable for pharmaceutical applications in term of safety and low cost.

Moreover, RRE has been shown to have various biological activities equivalent to rhinacanthin-C in the treatment of DM (See more details in topics 6 and 7). Therefore, RRE is a suitable *R. nasutus* extract used for the development of pharmaceutical preparations, i.e., topical antimicrobial agents and nutraceuticals. According to the satisfactory biological activity of RRE, the extracts used in further studies and applications should be standardized to a total rhinacanthin content of no less than 70% w/w or no less than 60% rhinacanthin-C w/w.

5 Quantitative HPLC Analysis of Rhinacanthins

A reversed-phase HPLC method has been established for the simultaneous determination of rhinacanthin-C, -D, and -N in *R. nasutus* leaf extract [30]. The method involves the use of a TSK-gel ODS-80Ts column (5 μm, 4.6 × 150 mm) isocratically eluted with a mixture of methanol and 5% aqueous acetic acid (80:20, v/v). All three rhinacanthins were eluted within 20 minutes with good resolution (Fig. 11.10). The parameters of linearity, repeatability, accuracy, and specificity of the method were evaluated to validate the method. This method exhibited good specificity when

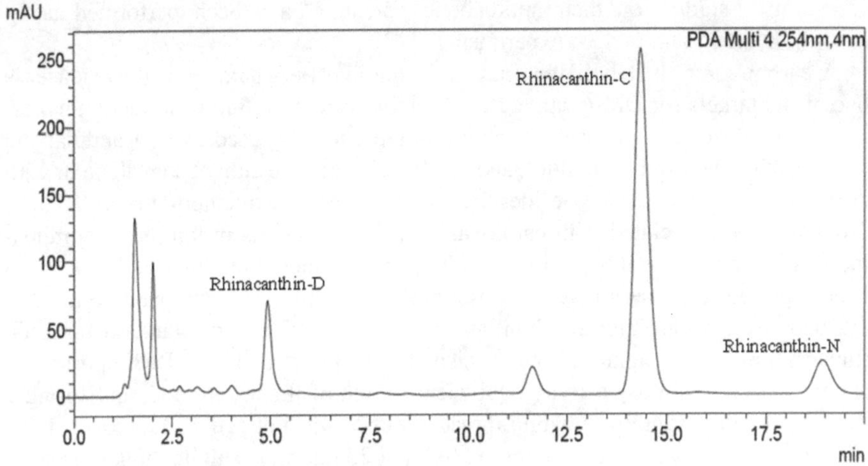

Fig. 11.10 HPLC chromatogram of ethyl acetate extract of *R. nasutus* leaves

determining the sample peaks using their absorbances produced by a photodiode array detector and comparing with the peaks of the authentic compounds. The limit of detection and quantification of all rhinacanthins were 0.75 and 3.0 μg/mL, respectively. A high degree of accuracy (the recoveries of the method were 94.3–100.9%) as well as precision (R.S.D. values less than 5%) were also achieved with good linearity (correlation coefficient ≥ 0.9999). Based on the HPLC analysis of *R. nasutus* leaf extract, rhinacanthin-C was the major naphthoquinone ester (1.9% w/w), while rhinacanthin-D (0.16% w/w) and rhinacanthin-N (0.07% w/w) were the minor ones.

6 Role of Rhinacanthin-C in Diabetes Mellitus

DM is a complex metabolic disorder that requires efficient multidisciplinary treatment protocols for its management [41]. Rhinacanthin-C is a major naphthoquinone ester of *R. nasutus* that has been recently reported for its hypoglycemic, hypolipidemic, and pancreatic protective properties in streptozotocin-nicotinamide induced diabetic rats [22]. Rhinacanthin-C considerably produced antihyperglycemic, antihyperlipidemic, and pancreatic protective effects by reducing the synthesis of inflammatory and apoptosis cytokines, i.e., I kappa B kinase β (Ikkβ), tumor necrosis factor-alpha (TNFα), and caspase-3 in diabetic rats. The insulin secretion was also enhanced by the activation of GLUT2 in the pancreas. Currently, rhinacanthins enriched *R. nasutus* leaf extract (RRE) has been prepared using a green extraction method and standardized to contain a total rhinacanthin content of no less than 70% w/w, with 62% w/w of rhinacanthin-C [51]. This extract has been determined for use as an alternative for rhinacanthin-C. Several comparative studies on their antidiabetic potential have been performed using in vitro, in silico and in vivo experiments.

A carbohydrate metabolizing enzyme in the intestine, namely α-glucosidase, is one of the targets for DM treatments, especially for controlling postprandial hyperglycemia. Various α-glucosidase inhibitors are currently used as oral antidiabetic drugs, such as voglibose, miglitol, and acarbose, which are either taken alone or with other hypoglycemic agents. Besides their high cost and gastroenteric irritation, these drugs may be associated with cardiovascular complications in long-term administration [9]. Some natural occurring naphthoquinones, including rhinacanthin-C, have been reported to possess α-glucosidase inhibitory activity. A comparative in vitro evaluation on α-glucosidase inhibitory activity of RRE and its standard markers, rhinacanthin-C, -D and -N have also been reported [51]. RRE possessed α-glucosidase inhibitory activity equivalent to that of rhinacanthin-C, and stronger than that of rhinacanthin-D, while rhinacanthin-N was not active. Moreover, RRE (IC_{50} of 25 μg/mL) and rhinacanthin-C (IC_{50} of 23 μg/mL) exhibited much stronger inhibitory activity than that of a standard drug, acarbose (IC_{50} of 395 μg/mL). Based on kinetic studies, both RRE and rhinacanthin-C were noncompetitive α-glucosidase inhibitors, while acarbose was a competitive inhibitor. Interestingly, RRE and rhinacanthin-C showed a synergistic inhibitory effect with acarbose at low concentrations. In addition, the mechanism of α-glucosidase inhibitor of rhinacanthin-C has been determined by molecular docking. A ligand-protein interaction profile has confirmed the formation of various hydrogen and hydrophobic bonds between rhinacanthin-C and α-glucosidase. The observed affinity in molecular modeling studies between the ligand and enzyme has explained the high inhibitory effect of rhinacanthin-C against α-glucosidase.

The compounds that can effectively control hyperglycemia with anti-adipogenic potential are ideal antidiabetic agents. Therefore, an enhanced glucose uptake in adipose and muscle tissue with anti-adipogenic activity becomes a prominent strategy in controlling obesity associated with DM. Enhanced glucose uptake and anti-adipogenic activities of RRE compared with its standard markers in 3T3-L1 and L6 cells have been reported [33]. RRE (20 μg/mL), rhinacanthin-N (20 μg/mL), and insulin (0.58 μg/mL) exhibited an equivalent effect on enhanced glucose uptake in 3T3-L1 adipocytes but were higher than rhinacanthin-C (20 μg/mL) and rhinacanthin-D (20 μg/mL). In contrast, RRE (2.5 μg/mL) exhibited potent and equivalent enhanced glucose uptake in L6 myotubes when compared to rhinacanthin-C (2.5 μg/mL), insulin (2.90 μg/mL), and metformin (219.5 μg/mL), but higher activity than rhinacanthin-D (2.5 μg/mL) and rhinacanthin-N (2.5 μg/mL). The strong glucose uptake stimulatory effect of RRE might be due to the possible synergism among the rhinacanthins. The mechanism of enhanced glucose uptake by rhinacanthins may be via an insulin-independent tyrosine kinase pathway that is previously reported for shikonin, a 1,4-naphthoquinone of *Lithospermum erythrorhizon* [52].

RRE (20 μg/mL) and rhinacanthin-C (20 μg/mL) also had equivalent potent anti-adipogenic activity in 3T3-L1 adipocytes, but higher than rhinacanthin-D and rhinacanthin-N, at the same concentration. The anti-adipogenic potential of rhinacanthins may be correlated with the previous report of shikonin that inhibited adipogenesis via inhibiting FABP4 and LPL gene expression. It has been reported that 1,4-naphthoquinones exhibit their anti-adipogenic effects by both upstream (SREBP1C) and downstream (PPARγ and C/EBPα) regulations [53].

Antihyperglycemic and antihyperlipidemic activities of RRE (15 mg/kg equivalent to rhinacanthin-C content) in comparison to rhinacanthin-C (15 mg/kg) and glibenclamide (600 μg/kg) in nicotinamide-streptozotocin induced diabetic rats for 28 days have been evaluated. Insulin is an important hormone that regulates the metabolism of carbohydrates, proteins, and lipids by promoting their synthesis, degrading, and releasing into the blood circulatory system. Insulin also acts as a stimulator for glucose, amino acid, and fatty acid uptake and enhances the expression of certain metabolic enzymes [54]. A combination of streptozotocin and nicotinamide have been used for partial destruction of the pancreas to develop diabetes in the rat. Due to partial destruction of the pancreas, insulin levels become lower and this affects the overall metabolic process. RRE, rhinacanthin-C, and glibenclamide significantly reduced the fasting blood glucose (FBG), glycated hemoglobin (HbA1c), and food/water intake, while increasing the insulin level and body weight in diabetic rats without affecting the normal rats. During hyperglycemia, glycated hemoglobin (HbA1c) is increased when the excess sugar reacts with the protein. HbA1c is a laboratory marker of diabetes and the associated risk of diabetic complications due to the formation of advanced glycation products [55]. RRE, rhinacanthin-C, and glibenclamide also normalized the serum lipid, liver, and kidney biomarkers in diabetic rats without affecting the normal rats. Based on a histopathology study of the pancreas, RRE, rhinacanthin-C, and glibenclamide evidently restored the islets of Langerhans in diabetic rats. Pancreatic protective effects of RRE can be correlated to the previous study on rhinacanthin-C [22]. Rhinacanthin-C is a potent antioxidant, anti-inflammatory, and anti-apoptotic agent that can be correlated to the pancreatic protective effect [27]. These results have confirmed the equivalent antidiabetic potential among RRE and rhinacanthin-C.

7 Role of Rhinacanthin-C in the Complications of Diabetes Mellitus

Oxidative stress is a cause of several disorders, including DM and its complications such as neuropathy, retinopathy, nephropathy, and cardiovascular problems that ultimately increase the mortality risk. The excessive production of free radicals can be controlled by intake of antioxidants [3]. A non-enzymatic glycation reaction, i.e., nucleophilic addition reaction between proteins and the reducing sugars in chronic hyperglycemic conditions triggers DM complications [56]. Hyperglycemia and oxidative stress are the major factors that accelerate the formation of advanced glycation end products (AGEPs) that further damage cells and increase diabetic complications [57].

Several compounds have been used to inhibit AGEPs formation, e.g., aminoguanidine but their toxicity and adverse effects limit the use of these agents [58]. Recently, *R. nasutus* leaf extracts have been reported to exhibit antiglycation activity. Moreover, RRE, rhinacanthin-C, -D, and -N have been evaluated for superoxide scavenging and antiglycation effects. Among these, RRE exhibited the most potent superoxide scavenging activity [59]. Based on the antiglycation assay, RRE (IC_{50} of 39.7μg/mL) showed nearly equivalent antiglycation activity to that of rhinacanthin-C (IC_{50} of 37.3 μg/mL), but higher than that of rhinacanthin-D (IC_{50} of 50.4 μg/mL) and rhinacanthin-N (IC_{50} of 89.5 μg/mL). Furthermore, the structure-activity relationship has been explained by a molecular docking experiment. The in silico studies indicated that rhinacanthins interacted with specific residues of albumin that protected the protein from non-enzymatic glycation [33]. The potent superoxide scavenging and antiglycation activity of RRE and rhinacanthin-C support therapeutic application in the complications of DM.

The major trigger of type 2 DM is insulin resistance, which is associated with the liberation of inflammatory mediators and fatty acids from adipocytes, which in turn activate insulin hypersecretion and reduce receptor response to insulin [60]. Along with other side effects, insulin and oral hypoglycemic drugs usually cause weight gain, an exaggerating factor in the induction and development of type 2 DM [13]. *R. nasutus* leaf extracts have been reported for anti-obesity potential [61].

The major chronic microvascular complication of DM is diabetic nephropathy and it is considered that more than 40% of diabetic patients are nearly nephropathic. Furthermore, diabetic nephropathy is the biggest key factor involved in the development of end-stage kidney damage worldwide [62]. Various factors, including oxidative stress, reactive oxygen species (ROS) inflammation, increased polyol pathway, and increased formation of AGEPs have been linked with the initiation and development of diabetic nephropathy. The continuous synthesis of ROS leads to oxidative stress, which has been triggered by continuous hyperglycemia and participates in diabetic nephropathy development. In addition, higher excretion of albumin and improper functioning of the glomerulus observed in diabetic nephropathy has been associated with the larger formation of ROS oxidative stress. Moreover, the oxidant/antioxidant system imbalance in the renal cells causes the exhaustion of the endogenous anti-radical enzymes resulting in the synthesis of inflammatory mediators, therefore assisting the chronicity of DM linked kidney cell damage. It is believed that the treatment that can accelerate the anti-radical defense system and inhibit the synthesis of multiple inflammatory mediators can be fruitful in the cure or prevention of kidney failure caused by DM [63].

Recently, rhinacanthin-C and RRE have been evaluated for their nephroprotective activity contrary to diabetic nephropathy. Significant reductions in the kidney index, pro-inflammatory mediators, and renal oxidative stress markers have been observed in both rhinacanthin-C and RRE treated diabetic rats. In addition, rhinacanthin-C and RRE enhanced renal levels of glutathione, catalase, superoxide dismutase levels and reduced diabetic induced renal cell injury.

Conclusively, rhinacanthin-C and RRE provided nephroprotective effect against diabetic nephropathy via the inhibition of oxidative stress and inflammation. Therefore, both rhinacanthin-C and RRE could be considered as alternative agents for the prevention of diabetic nephropathy [64].

8 Pharmacokinetics and Toxicity of Rhinacanthin-C

The pharmacokinetic and toxicity profiles of rhinacanthin-C were reported using computational studies [41]. Based on Lipinski's rule, rhinacanthin-C exhibited druggable behavior. ADMET calculations revealed that rhinacanthin-C can easily pass through the blood-brain barrier. This can be linked with the previous report on the role of rhinacanthin-C in modulation and attenuation of brain apoptosis in subarachnoid hemorrhage [43]. The oral bioavailability and intestinal absorption of rhinacanthin-C have also been reported in both in vitro and in silico studies [65]. P-glycoprotein transports the drug from cells to the intestine, thus causing hindrance in the drug bioavailability [66]. The in silico studies showed that rhinacanthin-C is a potent inhibitor of P-glycoprotien, thus presenting good bioavailability. The hERG gene has recently been explored as an important target in preclinical drug development [67]. On the basis of the in silico studies, rhinacanthin-C has low affinity for the inhibition of hERG, thus increasing the druggable behavior of rhinacanthin-C. Moreover, the ADMET results showed the affinity of rhinacanthin-C as a substrate of CYP450 3A4, and its terminal methyl group and the aromatic part appear to be susceptible to oxidation [41].

Although the toxic evidence of rhinacanthin-C has not been clearly demonstrated, there is an in vivo toxicity study of *R. nasutus* extracts in mice. The subcutaneous injection and feeding 10 g/kg of *R. nasutus* ethanol and water extracts through a gastric tube found no toxicity. In addition, an in vitro toxicity study of the chloroform extract of *R. nasutus* root in the Raji cell line found no significant toxicity [68]. Other toxicity information of rhinacanthin-C from in silico calculation has obtained a toxicity profile as non-toxic, non-lethal, and range in category III of acute oral toxicity ($LD_{50} = 500–5000$ mg/kg), which is more than ten times higher than therapeutic dose of rhinacanthin-C [41].

9 Conclusions

Rhinacanthin-C, a major bioactive naphthoquinone ester of *R. nasutus* is responsible for the antidiabetic potential of *R. nasutus* extracts. Rhinacanthin-C exerts its antidiabetic effect by various mechanisms, i.e., noncompetitive inhibition of α-glucosidase with a synergistic effect to acarbose; enhanced glucose uptake in muscles and adipose cells; regeneration of insulin-secreting beta cells; and anti-adipogenic effect. It also exerts protective effects against diabetes and its complications, such as nephroprotective effect against diabetic nephropathy, by scavenging free radicals and inhibiting AGEPs. RRE prepared by a simple green extraction and fractionation method also exhibits equivalent biological activities to rhinacanthin-C. In contrast, its cost of production is much lower than pure rhinacanthin-C. Therefore, RRE is highly recommended as an alternative of rhinacanthin-C for pharmaceutical applications in terms of its efficacy, safety, and low cost. However, further studies, including acute and chronic toxicities, pharmacokinetic study as well as clinical trials are required for ensuring its use for the treatment of DM.

References

1. Molina GA, Esparza R, López-Miranda JL, Hernández-Martínez AR, España-Sánchez BL, Elizalde-Peña EA, Estevez M (2019) Green synthesis of Ag nanoflowers using *Kalanchoe Daigremontiana* extract for enhanced photocatalytic and antibacterial activities. Colloids Surf B: Biointerfaces 180:141–149
2. Amer Diabet A (2020) Diabetes care in the hospital: Standards of medical care in diabetes-2020. Diabetes Care 43:S193–S202. https://doi.org/10.2337/dc20-S015
3. Ullah A, Khan A, Khan I (2016) Diabetes mellitus and oxidative stress-A concise review. Saudi Pharm J 24(5):547–553. https://doi.org/10.1016/j.jsps.2015.03.013
4. Perez RM, Zavala MA, Perez S, Perez C (1998) Antidiabetic effect of compounds isolated from plants. Phytomedicine 5(1):55–75. https://doi.org/10.1016/s0944-7113(98)80060-3
5. Maritim AC, Sanders RA, Watkins JB (2003) Diabetes, oxidative stress, and antioxidants: a review. J Biochem Mol Toxicol 17(1):24–38. https://doi.org/10.1002/jbt.10058
6. Kahn SE, Hull RL, Utzschneider KM (2006) Mechanisms linking obesity to insulin resistance and type 2 diabetes. Nature 444(7121):840–846. https://doi.org/10.1038/nature05482
7. Cho NH, Shaw JE, Karuranga S, Huang Y, Fernandes JDR, Ohlrogge AW, Malanda B (2018) IDF diabetes atlas: global estimates of diabetes prevalence for 2017 and projections for 2045. Diabetes Res Clin Pract 138:271–281. https://doi.org/10.1016/j.diabres.2018.02.023
8. Shah MA, Keach JE, Panichayupakaranant P (2018) Antidiabetic naphthoquinones and their plant resources in Thailand. Chem Pharm Bull 66(5):483–492
9. Fisman EZ, Motro M, Tenenbaum A (2008) Non-insulin antidiabetic therapy in cardiac patients: current problems and future prospects. In: Fisman EZ, Tenenbaum A (eds) cardiovascular diabetology: clinical, metabolic and inflammatory facets. Adv Cardiol 45:154–170
10. Klein G, Kim J, Himmeldirk K, Cao Y, Chen X (2007) Antidiabetes and anti-obesity activity of *Lagerstroemia speciosa*. Evid Based Complement Alternat Med 4(4):401–407. https://doi.org/10.1093/ecam/nem013
11. Xi M, Hai C, Tang H, Chen M, Fang K, Liang X (2008) Antioxidant and antiglycation properties of total saponins extracted from traditional Chinese medicine used to treat diabetes mellitus. Phytother Res 22(2):228–237. https://doi.org/10.1002/ptr.2297
12. S-i Y, Maeda S, Matsui T, Ueda S, Fukami K, Okuda S (2012) Role of advanced glycation end products (AGEs) and oxidative stress in vascular complications in diabetes. Biochim Biophys Acta-Gen Sub 1820(5):663–671. https://doi.org/10.1016/j.bbagen.2011.03.014
13. Phung OJ, Scholle JM, Talwar M, Coleman CI (2010) Effect of noninsulin antidiabetic drugs added to metformin therapy on glycemic control, weight gain, and hypoglycemia in type 2 diabetes. Jama-J Am Med Assoc 303(14):1410–1418. https://doi.org/10.1001/jama.2010.405
14. Gupta P, Bala M, Gupta S, Dua A, Dabur R, Injeti E, Mittal A (2016) Efficacy and risk profile of anti-diabetic therapies: conventional vs traditional drugs-A mechanistic revisit to understand their mode of action. Pharmacol Res 113:636–674. https://doi.org/10.1016/j.phrs.2016.09.029

15. McClintock HFV, Morales KH, Small DS, Bogner HR (2016) Patterns of adherence to oral hypoglycemic agents and glucose control among primary care patients with type 2 diabetes. Behav Med 42(2):63–71. https://doi.org/10.1080/08964289.2014.904767
16. Carver C (2006) Insulin treatment and the problem of weight gain in type 2 diabetes. Diabetes Educ 32(6):910–917. https://doi.org/10.1177/0145721706294259
17. Kooti W, Farokhipour M, Asadzadeh Z, Ashtary-Larky D, Asadi-Samani M (2016) The role of medicinal plants in the treatment of diabetes: a systematic review. Electron Physician 8 (1):1832–1842. https://doi.org/10.19082/1832
18. Kajbaf F, De Broe ME, Lalau J-D (2016) Therapeutic concentrations of metformin: a systematic review. Clin Pharmacokinet 55(4):439–459. https://doi.org/10.1007/s40262-015-0323-x
19. Hung H-Y, Qian K, Morris-Natschke SL, Hsu C-S, Lee K-H (2012) Recent discovery of plant-derived anti-diabetic natural products. Nat Prod Rep 29(5):580–606. https://doi.org/10.1039/c2np00074a
20. Leiherer A, Muendlein A, Drexel H (2013) Phytochemicals and their impact on adipose tissue inflammation and diabetes. Vasc Pharmacol 58(1–2):3–20. https://doi.org/10.1016/j.vph.2012.09.002
21. Panichayupakaranant P, Ahmad MI (2016) Plumbagin and its role in chronic diseases. Adv Exp Med Biol 929:229–246
22. Adam SH, Giribabu N, Rao PV, Sayem ASM, Arya A, Panichayupakaranant P, Korla PK, Salleh N (2016) Rhinacanthin C ameliorates hyperglycaemia, hyperlipidemia and pancreatic destruction in streptozotocin-nicotinamide induced adult male diabetic rats. Eur J Pharmacol 771:173–190. https://doi.org/10.1016/j.ejphar.2015.12.028
23. Bukke S, Raghu PS, Sailaja G, Kedam TR (2011) The study on morphological, phytochemical and pharmacological aspects of *Rhinacanthus nasutus* (L) kurz (A review). J Appl Pharm Sci 1 (8):26–32
24. Wu TS, Tien HJ, Yeh MY, Lee KH (1988) Isolation and cytotoxicity of rhinacanthin-A and -B, two; naphthoquinones, from *Rhinacanthus nasutus*. Phytochemistry 27(12):3787–3788
25. Siripong P, Kanokmedakul K, Piyaviriyagul S, Yahuafai J, Chanpai R, Ruchirawat S, Oku N (2006) Antiproliferative naphthoquinone esters from *Rhinacanthus nasutus* Kurz. roots on various cancer cells. J Tradit Med 23:166–172
26. Brimson JM, Tencomnao T (2014) Medicinal herbs and antioxidants: potential of *Rhinacanthus nasutus* for disease treatment? Phytochem Rev 13(3):643–651. https://doi.org/10.1007/s11101-013-9324-2
27. Bhusal N, Panichayupakaranant P, Reanmongkol W (2014) *In vivo* analgesic and anti-inflammatory activities of a standardized *Rhinacanthus nasutus* leaf extract in comparison with its major active constituent rhinacanthin-C. Songklanakarin J Sci Technol 36(3):325–331
28. Tewtrakul S, Tansakul P, Panichayupakaranant P (2009) Anti-allergic principles of *Rhinacanthus nasutus* leaves. Phytomedicine 16(10):929–934. https://doi.org/10.1016/j.phymed.2009.03.010
29. Puttarak P, Charoonratana T, Panichayupakaranant P (2010) Antimicrobial activity and stability of rhinacanthins-rich *Rhinacanthus nasutus* extract. Phytomedicine 17(5):323–327. https://doi.org/10.1016/j.phymed.2009.08.014
30. Panichayupakaranant P, Charoonratana T, Sirikatitham A (2009) RP-HPLC analysis of rhinacanthins in *Rhinacanthus nasutus*: validation and application for the preparation of rhinacanthin high-yielding extract. J Chromatogr Sci 47(8):705–708. https://doi.org/10.1093/chromsci/47.8.705
31. Tewtrakul S, Tansakul P, Panichayupakaranant P (2009) Effects of rhinacanthins from *Rhinacanthus nasutus* on nitric oxide, prostaglandin E-2 and tumor necrosis factor-alpha releases using RAW264.7 macrophage cells. Phytomedicine 16(6–7):581–585. https://doi.org/10.1016/j.phymed.2008.12.022
32. Rao PV, Madhavi K, Naidu MD (2011) Hypolipidemic properties of *Rhinacanthus nasutus* in streptozotocin induced diabetic rats. J Pharmacol Toxicol 6(6):589–595

33. Shah MA, Muhammad H, Mehmood Y, Khalil R, Ul-Haq Z, Panichayupakaranant P (2017) Superoxide scavenging and antiglycation activity of rhinacanthins-rich extract obtained from the leaves of *Rhinacanthus nasutus*. Pharmacogn Mag 13(52):652–658. https://doi.org/10.4103/pm.pm_196_17
34. Wu TS, Hsu HC, Wu PL, Leu YL, Chan YY, Chern CY, Yeh MY, Tien HJ (1998) Naphthoquinone esters from the root of *Rhinacanthus nasutus*. Chem Pharm Bull 46 (3):413–418
35. Wu TS, Hsu HC, Wu PL, Teng CM, Wu YC (1998) Rhinacanthin-Q, a naphthoquinone from *Rhinacanthus nasutus* and its biological activity. Phytochemistry 49(7):2001–2003. https://doi.org/10.1016/s0031-9422(98)00425-7
36. Gotoh A, Sakaeda T, Kimura T, Shirakawa T, Wada Y, Wada A, Kimachi T, Takemoto Y, Iida A, Iwakawa S, Hirai M, Tomita H, Okamura N, Nakamura T, Okumura K (2004) Antiproliferative activity of *Rhinacanthus nasutus* (L.) Kurz extracts and the active moiety, rhinacanthin C. Biol Pharm Bull 27(7):1070–1074. https://doi.org/10.1248/bpb.27.1070
37. Siriwatanametanon N, Fiebich BL, Efferth T, Prieto JM, Heinrich M (2010) Traditionally used Thai medicinal plants: *In vitro* anti-inflammatory, anticancer and antioxidant activities. J Ethnopharmacol 130(2):196–207. https://doi.org/10.1016/j.jep.2010.04.036
38. Boueroy P, Saensa-Ard S, Siripong P, Kanthawong S, Hahnvajanawong C (2018) Rhinacanthin-C extracted from *Rhinacanthus nasutus* (L.) inhibits cholangiocarcinoma cell migration and invasion by decreasing MMP-2, uPA, FAK and MAPK pathways. Asian Pac J Cancer Pre 19(12):3605–3613
39. Sendl A, Chen JL, Jolad SD, Stoddart C, Rozhon E, Kernan M, Nanakorn W, Balick M (1996) Two new naphthoquinones with antiviral activity from *Rhinacanthus nasutus*. J Nat Prod 59 (8):808–811. https://doi.org/10.1021/np9601871
40. Tran Minh N, Nguyen Thi Thanh P, Nguyen Minh K, Park S, Kwak HJ, Nguyen Xuan N, Bui Thi Thu T, Bui Huu T, Song J-H, Ko H-J, Kim SH (2019) A new naphthoquinone analogue and antiviral constituents from the root of *Rhinacanthus nasutus*. Nat Prod Res 33(3):360–366. https://doi.org/10.1080/14786419.2018.1452004
41. Shah MA, Reanmongkol W, Radenahmad N, Khalil R, Ul-Haq Z, Panichayupakaranant P (2019) Anti-hyperglycemic and anti-hyperlipidemic effects of rhinacanthins-rich extract from *Rhinacanthus nasutus* leaves in nicotinamide-streptozotocin induced diabetic rats. Biomed Pharmacother 113:108702. https://doi.org/10.1016/j.biopha.2019.108702
42. Punturee K, Wild CP, Kasinrerk W, Vinitketkumnuen U (2005) Immunomodulatory activities of *Centella asiatica* and *Rhinacanthus nasutus* extracts. Asian Pac J Cancer Pre 6(3):396–400
43. Chang C-Z, Wu S-C, Kwan A-L, Lin C-L (2016) Rhinacanthin-C, a fat-soluble extract from *Rhinacanthus nasutus*, modulates high-mobility group box 1-related neuro-inflammation and subarachnoid hemorrhage-induced brain apoptosis in a rat model. World Neurosurg 86:349–360. https://doi.org/10.1016/j.wneu.2015.08.071
44. Chuang KA, Li MH, Lin NH, Chang CH, Wen SF (2017) Rhinacanthin C alleviates amyloid- β fibrils' toxicity on neurons and attenuates neuroinflammation triggered by LPS, amyloid-β, and interferon-γ in glial cells. Oxidative Med Cel Longev 14:1–18
45. Panichayupakaranant P, Meerungrueang W (2010) Effect of medium composition and light on root and rhinacanthin formation in *Rhinacanthus nasutus* cultures. Pharm Biol 48 (11):1192–1197. https://doi.org/10.3109/13880200903578747
46. Widhalm JR, Rhodes D (2016) Biosynthesis and molecular actions of specialized 1,4-naphthoquinone natural products produced by horticultural plants. Hortic Res 3. https://doi.org/10.1038/hortres.2016.46
47. Dewick PM (2009) Medicinal natural products: a biosynthetic approach 3rd Edition. Kew Bull 53(2):499
48. Koricheva J, Larsson S, Haukioja E, Keinanen M (1998) Regulation of woody plant secondary metabolism by resource availability: hypothesis testing by means of meta-analysis. Oikos 83 (2):212–226. https://doi.org/10.2307/3546833

49. Wills RBH, Bone K, Morgan M (2000) Herbal products: Active constituents, modes of action and quality control. Nutr Res Rev 13(1):47–77. https://doi.org/10.1079/095442200108729007
50. Viljoen AM, Vermaak I, Chen W (2013) Quality control of phytomedicines - Is the herbal medicine industry facing a crisis? Planta Med 79(13):1121–1121
51. Shah MA, Khalil R, Ul-Haq Z, Panichayupakaranant P (2017) α-Glucosidase inhibitory effect of rhinacanthins-rich extract from *Rhinacanthus nasutus* leaf and synergistic effect in combination with acarbose. J Funct Foods 36:325–331
52. Kamei R, Kitagawa Y, Kadokura M, Hattori F, Hazeki O, Ebina Y, Nishihara T, Oikawa S (2002) Shikonin stimulates glucose uptake in 3T3-L1 adipocytes via an insulin-independent tyrosine kinase pathway. Biochem Biophys Res Commun 292(3):642–651. https://doi.org/10.1006/bbrc.2002.6714
53. Lee H, Kang R, Yoon Y (2010) Shikonin inhibits fat accumulation in 3T3-L1 adipocytes. Phytother Res 24(3):344–351. https://doi.org/10.1002/ptr.2942
54. Saltiel AR, Kahn CR (2001) Insulin signalling and the regulation of glucose and lipid metabolism. Nature 414(6865):799–806. https://doi.org/10.1038/414799a
55. Jeffcoate SL (2004) Diabetes control and complications: The role of glycated haemoglobin, 25 years on. Diabetic Med 21(7):657–665. https://doi.org/10.1046/j.1464-5491.2003.01065.x
56. Abdallah HM, El-Bassossy HM, Mohamed GA, El-Halawany AM, Alshali KZ, Banjar ZM (2017) Mangostanaxanthones III and IV: advanced glycation end-product inhibitors from the pericarp of *Garcinia mangostana*. J Nat Med 71(1):216–226. https://doi.org/10.1007/s11418-016-1051-8
57. Hays NP, Galassetti PR, Coker RH (2008) Prevention and treatment of type 2 diabetes: current role of lifestyle, natural product, and pharmacological interventions. Pharmacol Ther 118 (2):181–191. https://doi.org/10.1016/j.pharmthera.2008.02.003
58. Peng X, Zheng Z, Cheng K-W, Shan F, Ren G-X, Chen F, Wang M (2008) Inhibitory effect of mung bean extract and its constituents vitexin and isovitexin on the formation of advanced glycation endproducts. Food Chem 106(2):475–481. https://doi.org/10.1016/j.foodchem.2007.06.016
59. Sompong W, Adisakwattana S (2015) Inhibitory effect of herbal medicines and their trapping abilities against methylglyoxal-derived advanced glycation end-products. BMC Complement Altern Med 15:394. https://doi.org/10.1186/s12906-015-0897-8
60. Dandona P, Aljada A, Bandyopadhyay A (2004) Inflammation: the link between insulin resistance, obesity and diabetes. Trends Immunol 25(1):4–7. https://doi.org/10.1016/j.it.2003.10.013
61. Sompong W, Muangngam N, Kongpatpharnich A, Manacharoenlarp C, Amorworasin C, Suantawee T, Thilavech T, Adisakwattana S (2016) The inhibitory activity of herbal medicines on the keys enzymes and steps related to carbohydrate and lipid digestion. BMC Complement Altern Med 16(1):439. https://doi.org/10.1186/s12906-016-1424-2
62. Tanios BY, Ziyadeh FN (2012) Emerging therapies for diabetic nephropathy patients: beyond blockade of the renin-angiotensin system. Nephron Extra 2(1):278–282. https://doi.org/10.1159/000343312
63. Abou-Hany HO, Atef H, Said E, Elkashef HA, Salem HA (2018) Crocin mediated amelioration of oxidative burden and inflammatory cascade suppresses diabetic nephropathy progression in diabetic rats. Chem-Biol Interact 284:90–100. https://doi.org/10.1016/j.cbi.2018.02.001
64. Zhao L-L, Makinde EA, Shah MA, Olatunji OJ, Panichayupakaranant P (2019) Rhinacanthins-rich extract and rhinacanthin C ameliorate oxidative stress and inflammation in streptozotocin-nicotinamide-induced diabetic nephropathy. J Food Biochem 43(4):e12812. https://doi.org/10.1111/jfbc.12812
65. Fricker G, Miller DS (2002) Relevance of multidrug resistance proteins for intestinal drug absorption *in vitro* and *in vivo*. Pharmacol Toxicol 90(1):5–13. https://doi.org/10.1034/j.1600-0773.2002.900103.x
66. Fromm MF (2000) P-glycoprotein: a defense mechanism limiting oral bioavailability and CNS accumulation of drugs. Int J Clin Pharmacol Ther 38(2):69–74

67. Nogawa H, Kawai T (2014) hERG trafficking inhibition in drug-induced lethal cardiac arrhythmia. Eur J Pharmacol 741:336–339. https://doi.org/10.1016/j.ejphar.2014.06.044
68. Murakami A, Kondo A, Nakamura Y, Ohigashi H, Koshimizu K (1993) Possible antitumor promoting properties of edible plants from Thailand, and identification of an active constituent, cardamonin, of *Boesenbergia pandurata*. J Agric Chem Soc Jpn 57(11):1971–1973

Chapter 12
Regeneration of Beta Cells by Inhibition of pro-Apoptotic Proteins through Phytocompound in STZ Induced Diabetic Albino Wistar Rats: In Vivo and In Silico Approach

R. Jasmine and A. Sherlin Rosita

Abstract Type 1 diabetes mellitus (T1DM) is due to severe insulin deficiency by loss of insulin-producing β-cells. Hence, induction of β-cell proliferation is a promising therapeutic strategy to regenerate β-cell mass. Both extrinsic (receptor-mediated) and intrinsic (mitochondria-driven) pathway constitutes the apoptotic pathway. Extrinsic pathway driven by external factors and mitochondria-driven intrinsic pathway are known to operate at a balance and regulate pro-apoptotic and anti-apoptotic proteins. Hence, there is a future hope for therapeutic application to slow or even prevent β-cell apoptosis in T1DM by inhibiting pro-apoptotic proteins employing natural phytocompounds. Among the several plants screened, ***Elephantopus scaber*** was found to have a remarkable anti-diabetic activity. Terpenoid, the bioactive phytocompound isolated from the mentioned plant was found to increase insulin and restore glucose homeostasis. Experimental analysis in STZ diabetic rats demonstrated an increase in insulin by regeneration of beta cells. However, the underlying mechanism of beta cell regeneration was analysed by inhibition of pro-apoptotic proteins. Thus, restoring pancreatic beta cell function by inhibiting pro-apoptotic proteins may be an effective strategical approach for the prevention and treatment of diabetes, which may open doors for establishing phytocompounds as operative anti-diabetics. Restoring insulin secretion by regenerating beta cells in STZ induced diabetic rats can be traced to the inhibition of pro-apoptotic proteins by phytocompound through in silico study.

Keywords Apoptosis · Beta cell · Diabetes · Phytocompounds · In silico

R. Jasmine (✉) · A. Sherlin Rosita
Department of Biotechnology and Bioinformatics, Bishop Heber College, Tiruchirappalli, Tamilnadu, India

H. Chen, M. Zhang (eds.), *Structure and Health Effects of Natural Products on Diabetes Mellitus*, https://doi.org/10.1007/978-981-15-8791-7_12

1 Introduction

Diabetes mellitus (DM), a major global health problem is an epidemic [1]. As of WHO report, an estimated 425 million people had diabetes worldwide, and trends suggest that rates will continue to rise. Diabetes at least doubles a person's risk of early death. In 2017, diabetes resulted in approximately 3.2 to 5.0 million deaths. It is a major metabolic disorder characterized by chronic hyperglycaemia as a result of a relative or absolute lack of insulin or the actions of insulin [2]. The condition affects the metabolism of carbohydrates [3], protein, fat [4], water and electrolytes leading to structural changes in a range of cells especially those of the vascular system, subsequently leading to long-term complications of diabetes.

1.1 Blood Glucose Homeostasis

One of the most well-regulated homeostatic mechanisms in vertebrates is the maintenance of stable levels of glucose in blood [5]. Glucose is the transport carbohydrate in animals, and its concentration in the blood affects every cell in the body [6]. The range between 80 and 120 mg/dL of blood glucose is considered to be the normal range. However, both increase or decrease can cause complications and become lethal [7]. The concentration of glucose in the blood is monitored by the glucose receptor cells of the pancreas. Insulin secreted by the beta cells in the islets of Langerhans of the pancreas aids in regulation of glucose metabolism. Elevation of blood glucose in both type I and type II diabetes leads to several biochemical disturbances, which in turn impairs carbohydrates, fat and protein metabolism. The alpha and beta cells of the pancreas secrete glucagon and insulin, respectively, and these two hormones are antagonistic, and have opposite effects on blood glucose. Under normal physiological conditions, normal glycaemic index is maintained by the balance between glucagon and insulin.

According to WHO, two types of diabetes as insulin-dependent diabetes mellitus (IDDM) and non-insulin-dependent diabetes mellitus (NIDDM) have been recognized [8]. In type I diabetes, beta cells of islets of Langerhans are impaired or destroyed, whereas obesity leading to insulin resistance may contribute to type II diabetes [9].

1.2 Complications of Diabetes Mellitus

Diabetes mellitus is a metabolic disorder accompanied by several complications, which may be acute, sub-acute and chronic. These complications may range from polyuria, polyphagia, polydipsia to severe hypertension, neuropathy, nephropathy, myocardial infarction, atherosclerosis, retinopathy, foot ulcers and urinary tract infections.

1.3 Types

Though increase in blood sugar levels lead to diabetes mellitus, yet the underlying reason differentiates diabetes to two types. Insulin deficiency due to impairment or autoimmune destruction of beta cells [10] leads to type I diabetes mellitus, which accounts for less than 10% of the population, especially targeting the young [11]. Majority of the population suffer from type II diabetes due to insulin resistance.

1.4 Apoptotic Pathways

There are two main apoptotic pathways, the extrinsic (receptor-mediated) and the intrinsic (mitochondrial-driven). Intrinsic pathway is driven by internal factors, while the extrinsic pathway is activated upon ligation of the cell surface death receptors(s), which in turn activates a downstream effector mechanism orchestrated by the caspase family of cysteine proteases (Fig. 12.1) [12]. Caspase-3 is the key enzyme in apoptosis and is found to be the unifying point of the extrinsic and intrinsic pathways (Fig. 12.1). Hence researchers have suggested the possibility of caspase 3 inhibitors in preventing islet apoptosis, thereby leading to a successful islet transplantation[13].Fas receptor, also known as CD95, a death receptor seen on the surface of several cells is known to trigger apoptotic signals, thereby inducing caspases leading to apoptosis [14]. Apoptosis is known to downregulate the caspase pathway [15] and upregulate cellular damage and is governed by both pro- and anti-aopototic members of the BCL-2 protein family.

1.5 Proposed Mechanisms for Protective/Regenerative Effect of Phytochemicals

The protective effect of *E. scaber* on streptozotocin-induced β-cells destruction, increasing islets size and beta cell number and restoring beta cell morphology was already reported. Although the mentioned evidence can fully support the beta cell regeneration, yet the exact molecular mechanisms involved have not been satisfactorily proved. However, decreasing apoptosis, reducing oxidative stress of beta cells and immunomodulation were the suggested mechanisms of action, which needed scientific proof.

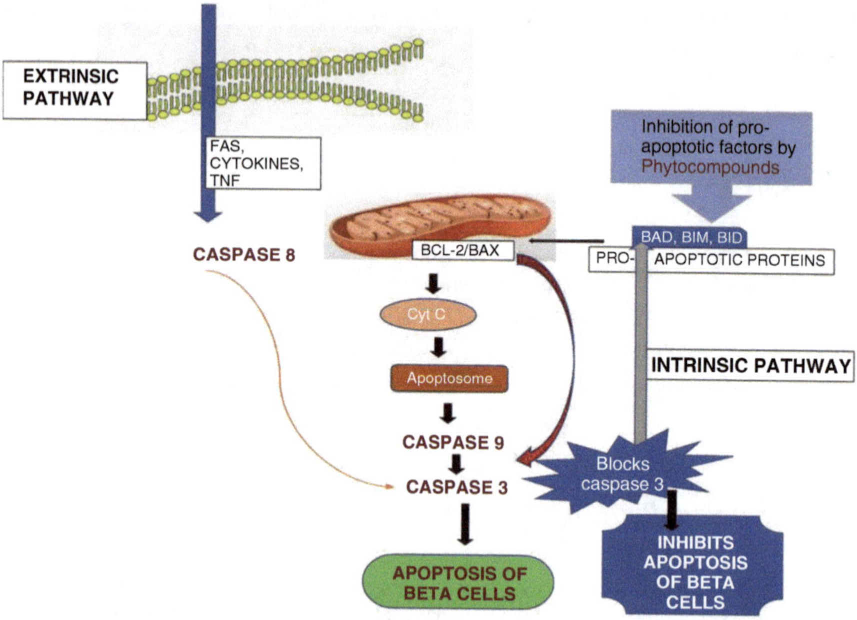

Fig. 12.1 Schematic representation of the extrinsic (receptor-mediated) and intrinsic (mitochondria-driven) beta cell apoptotic pathways as opposed to the inhibition of pro-apoptotic proteins by phytocompounds (blue) leading to inactivation of caspase 3 followed by regeneration of beta cells

Glucose homeostasis depends on the efficient production of insulin by pancreas, which is attributed to several factors as beta cell replication rates, size and numbers of islets of Langerhans, morphology of beta cells and islet neogenesis [16]. Also the balance between beta cell replication rates and death rate determines the efficiency of the functioning of the islets of Langerhans. Decreasing apoptosis results in the enhancement of beta cell viability and increase in insulin production, which has been our focus.

The existence and functioning of both the intrinsic and extrinsic apoptotic pathways are essential for efficient growth and differentiation of cells. Both the pathways are activated by internal and external signals, leading to the stimulation of caspases 9 and 8, respectively. The two pathways converge and activate stimulate effector caspase 3, which in turn leads to DNA fragmentation and cell death [17, 18].

Onset of hyperglycaemia is attributed towards the autoimmune destruction of beta cells in type I insulin-dependent diabetic patients. Hence, prevention of beta cell destruction and enhancing its regeneration are the focus towards therapeutic management of type I diabetes. In this regard, development of phytochemical products with beta cell regenerative property can be a promising option for the patients who have lost their mass of functional islet cells. Among the hundreds of plants that have been investigated for diabetes, a compound from *E. scaber* has shown the regenerative property, which has been highlighted in this chapter. For most of the herbs, however, the number of studies supporting their beneficial effects on pancreas is not enough. The exact mechanism responsible for the protective/regenerative effects of phytochemicals on pancreatic islets is yet to be elucidated. However, inhibition of pro-apoptotic proteins of phytochemicals may in part mediate their protective action against pancreatic beta cell apoptosis, as suggested by our in silico study. Though the molecular mechanism involved in the regeneration of beta cells, yet this approach may offer a future hope for type I diabetic patients, who could be treated at the earliest stages of diabetes to prevent the destruction of beta cells and also stimulate their regeneration in destroyed conditions. Future clinical trials on this topic are essentially warranted.

2 Experimental Procedures

2.1 Chosen Plant

2.1.1 Collection of Plant Materials and Extract Preparation

Elephantopus scaber (whole plants) (ES) (Fig. 12.2) was collected from Palode, Thiruvananthapuram district, Kerala, India during rainy season. *E. scaber* roots had been washed and air-dried and powdered and were extracted using acetone (the solvent was chosen based on preliminary screening-results not shown). About 500 g *E.scaber* roots yielded 190 g powder.

Fig. 12.2 Elephantopus scaber roots

2.1.2 Bioassay Guided Fractionation of *E. scaber*

The fractionation of the plant extract using column chromatography, yielded fractions, which on purification resulted in pure single bioactive compound. On subject to spectral study, the structure of the compound was elucidated [19].

2.2 *Animals*

Animals chosen for the study included male adult albino rats (Wistar strain), weighing about 100–150 g. They were procured from Tamil Nadu Veterinary and Animal Sciences University, Madhavaram, Chennai, Tamil Nadu, India. The lab where the animals were housed was maintained at a temperature of 22 + 2°C and 45 + 5% humidity with a 12 h day and 12 h night cycle with free access to food and water.

2.2.1 Induction of Diabetes

The rats under study were made diabetic by intraperitoneal injection of streptozotocin (STZ). After the observation period of 7 days, rats with a fasting blood glucose ranging between 280 and 350 mg/dL were labelled as diabetic and included for the study.

2.2.2 Protocol

The fasting blood glucose in diabetic rats was determined and measured after 1 week of induction of diabetes by STZ [20]. Rats with blood glucose >300 mg/dL were chosen for the study. They were grouped into five groups (as shown below), with five rats in each group.

Group 1

Normal rats treated with vehicle alone (carboxy methyl cellulose (CMC) 0.1%; 1mL/kg body weight);

Group 2

STZ treated diabetic control- diabetic control group was given distilled water every day through oral intubations [20];

Group 3

STZ induced diabetic rats treated with terpenoid from ES;

Group 4
STZ induced diabetic rats treated with 0.6 mg/kg glibenclamide [21]
Group 5
STZ induced diabetic rats treated with 0.3 IU/kg humulin [22]

2.2.3 Histological Studies

Pancreas obtained from different experimental groups of animals were washed and sliced into pieces and fixed for 3 days in Bouin-Hollande fixative. After series of washing in 70% alcohol, the tissues were cleared using xylene and then were infiltered with molten paraffin. Rotary microtome was used to obtain 3–5 μm thick section of the pancreatic tissues. Philips 200 (Holland) transmission electron microscope (TEM) was used to obtain electron micrographs at 2000–70,000× magnifications.

2.3 *In Silico Analysis to Study the Role of Compounds on pro-Apoptotic Proteins*

2.3.1 Protein Preparation

The three dimensional crystal structures of Pro-apoptotic proteins Bad, Bak and 177 AU4 Bim (with 1G5JID: 1BXLID: and 1PQ1ID: respectively) were obtained from RCSB 178 protein Data Bank (http://www.rcsb.org/). The water molecules were removed using Discovery Studio 4.1 Visualizer. The polar and nonpolar hydrogens were added. Gasteiger charges were computed and files were saved in . pdbqt format with the help of AutoDock tools.

2.3.2 Ligand Preparation

The ligand terpenoid was drawn and prepared for docking studies using ChemSketch. The structure of the ligands obtained from ChemSketch is given in Fig. 12.1. The ligands were converted to pdb file by using open babel 2.4. The structure was optimized by Argus Lab. Kollman united partial atomic charges were added. The druggability and pharmacokinetic assessment of the compound were done using pkCSM [23].

2.4 Molecular Docking

The molecular docking analysis was performed using AutoDock tools 4.0. It employs to dock the small molecule terpenoid with the macromolecules (Pro-apoptotic proteins Bad, Bak, and Bim). This software predicts how a substrate or a drug is docked into the receptor [24]. The torsions were fixed for the ligand. The grid box of $60 \times 60 \times 60$ with 3.75 Å was set where the ligand can interact. The rigid grid box was attained using Autogrid. It was followed by AutoDock with Lamarckian Genetic algorithm to obtain the best docking conformation. The docking was performed in two trials, and average binding energy was calculated. The ligand pose with better binding affinity was selected by analysing the docked poses. The pose with best binding affinity was visualized using Discovery Studio 4.1 Visualizer and PoseViewer.

3 Results

3.1 Anti-Diabetic Activity of Fractions/Compounds

Based on promising results obtained with the crude extract of *E. scaber*, the acetone extract of ES was fractionated. The bioactive compound from chosen plant was further investigated for anti-diabetic activity and whether it could increase β-cell regeneration in the STZ induced mouse model of type 1 diabetes. Administration of terpenoid started causing a substantial reduction of non-fasting blood glucose levels within 2 h of intraperitoneal injection and the results of the administration of the various fractions and compound on the blood glucose levels and insulin levels of STZ-diabetic rats have been reported earlier [19].

3.2 Histological Analysis—EM Study

Morphometric analysis of pancreas sections showed destruction of β-cells destroyed in STZ induced diabetic rats, whereas the beta cell morphology and numbers in terpenoid administered rats resembled more like the drug treated group. This indicated that the beta cells were proliferating at a higher rate (Fig. 12.3). Moreover, β-cell area and insulin content were increased in terpenoid treated group, which was compared with vehicle-treated diabetic mice.

Transmission electron micrographs (TEM) of β-cells of untreated rats revealed that the β-cells contained abundant secretion granules (Fig. 12.3). Each granule had a large clear space around it. Cytoplasmic organelles were also present. TEM of β-cells of diabetic control rats revealed the nucleus to be dark and shrunken and to possess an irregular outline (Fig. 12.3) The cytoplasmic matrix was densely

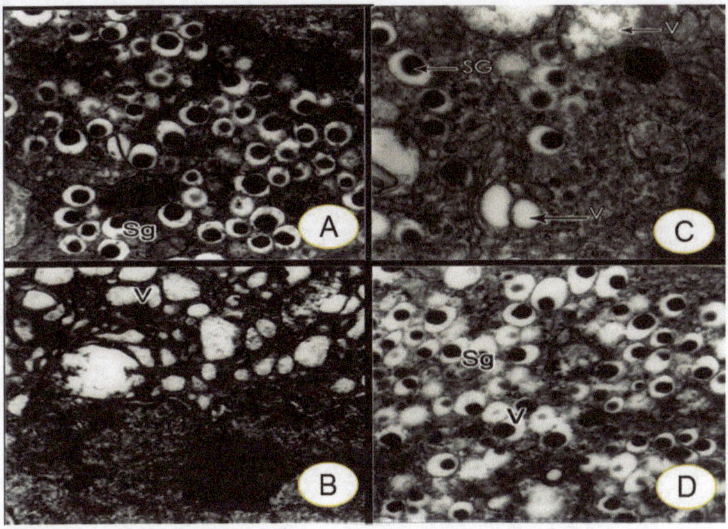

Fig. 12.3 Electron Microscopic photos (×15000) of beta cells of islets of Langerhans from rats A-Normal; B-STZ induced Diabetic; C-Drug treated; D-Terpenoid treated

vacuolated. Secretion granules were totally absent. TEM of compound treated diabetic rats showed the cytoplasm of the β-cell to contain numerous secretion granules and a few vacuoles and capillaries were prominent.

3.3 *Non-toxicity*

To assess the safety of the compounds administered to the rats, the kidney and liver functions were analysed. There was no marked change in the biomarkers as urea, creatinine for kidney functioning and SGOT, SGPT and alkaline phosphatase for liver functioning, respectively, indicating that the mice were not generally affected by the compounds treatment (data not shown). No haemolysis was observed in human erythrocytes which also suggest the non-toxicity of the plant.

3.4 *In Silico Analysis—Docking*

The general pharmacokinetic descriptors of the optimized ligand which were predicted using pkCSM are tabulated in Table 12.1. Notably, the ligand terpenoid shows better BBB permeability and it is also confirmed that the compound terpenoid has good intestinal oral absorption. The metabolic nature is confirmed to be safe as it is predicted to be inhibiting any of the cytochrome P2, P3 enzymes. In computer-

Table 12.1 The important predicted properties of ADME by pkCSM

ADME	Property	Values/Category	Unit
Absorption	Water solubility	–6.392	log mol/L
	Caco2 permeability	1.187	log Papp in 10
	Intestinal absorption	97.339	%
	Skin permeability	–2.936	log Kp
	P-glycoprotein substrate	No	
	P-glycoprotein I inhibitor	Yes	
	P- glycoprotein II inhibitor	Yes	
Distribution	VDss (Human)	0.306	log L/kg
	Fraction unbound (Human)	0	Fu
	BBB permeability	–0.221	log BB
	CNS permeability	–2.023	log PS
Metabolism	CYP1A2 inhibitor	No	
	CYP2C19 inhibitor	No	
	CYP2C9 inhibitor	No	
	CYP2D6 inhibitor	No	
	CYP3A4 inhibitor	No	
Excretion	Total clearance	0.436	log ml/min/kg
	Renal OCT2 substrate	No	

assisted drug design, molecular docking plays a major role in screening the most effective compounds economically with fewer efforts [25]. AutoDock 4.0 facilitates to extract the best docked poses by determining its mechanistic values such as lowest binding affinity, least binding energy, and ligand efficiency and inhibitory constant (ki). Docking studies were performed on the compound terpenoid with pro-apoptotic proteins. AutoDock results were analysed and the accuracy was calculated based on the lowest binding free energy and the hydrogen bonds interactions between the pro-apoptotic proteins and the ligand. Table 12.2 depicts the results of molecular mechanistic values that were obtained from molecular docking studies. From the docked results of the three pro-apoptotic factors (Bim, Bak, and Bad), we observed the least Binding energy with the compound terpenoid as −9.58, −7.9 and −9.75 kcal/mol, respectively.

3.5 *Pro-apoptotic Proteins*

The Bcl-2 family consists of anti-apoptotic and pro-apoptotic proteins. Anti-apoptotic proteins, such as Bcl-2, Bcl-XL, Bcl-w, Mcl-1, and Bfl-1 (Bcl-2A1) contain four Bcl-2 homology (BH) domains, while the pro-apoptotic members are divided into proteins with three BH domains BH1-BH3 (Bax, Bak, and Bok), and proteins with only a BH3 domain (e.g., Bim, Bad, Bik, Bmf, Bid, Noxa and Puma).

Table 12.2 Molecular docking results of Terpenoid bonding with apoptotic factors Bad, Bak, and Bim

Protein	Ligand	Binding energy (kcal/mol)	Ligand efficiency	Intermolecular energy (kcal/mol)	Inhibitory constant (K_i) (nM)	Interaction	
						Hydrogen bonding	Hydrophobic bonding
BIM	Terpenoid	−9.58	−0.33	−10.18	195.05	SER 4 SER 2	VAL 80 ASP 189 TRP 188
BAK		−7.9	−0.27	−8.49	112.12	HIS 113 ILE 114 SER122	GLY 572 THR118 LEU112
BAD		9.75	−0.33	−10.05	218.22	TRP 173	THR 176 GLU 128 HIS 181 PHE 127

Multi-domain pro-apoptotic proteins Bax and Bak are absolutely required for apoptosis [26].

3.6 Phytocompound as Inhibitors of pro-Apoptotic Proteins

In this study, we identified a terpenoid from *E. scaber* as a selective inhibitor of beta cell apoptosis, by suppressing the function of pro-apoptotic proteins. We further demonstrated that the terpenoid administration to experimental drug induced diabetic rats, promotes β-cells replication and regeneration in STZ induced diabetic mice. Utilizing small molecule inducers of β-cell proliferation is one of the most tangible methods to restore functional β-cell mass. We have pinpointed that inhibition of pro-apoptotic proteins promotes β-cell proliferation.

4 Discussion

Apoptotic mechanism is characterized by shrinking cells, condensed chromatin, fragmentation of DNA and production of apoptotic bodies [27]. Apoptosis plays a vital role in the re-designing of the pancreas after birth and leads to the development of final mass of islet cells, including beta cells [28]. Pancreatic β-cell apoptosis is also a pathological feature that is common to both Type 1 diabetes mellitus (T1DM) and T2DM. Excess apoptosis is the underlying cause for β-cell loss for both T1DM and T2DM. The role of apoptosis in the life and death of beta cells is seen to be vital. Anti-apoptotic member of the Bcl family, including Bcl-2 and Bcl-xL, blunts intrinsic death signalling by blocking the recruitment of pro-apoptotic members to the mitochondria [29]. The cumulative data support that high glucose modulates the balance of pro-apoptotic caspase family and anti-apoptotic Bcl proteins towards apoptosis, thus leading to β-cell death.

Treatment of type I diabetic patients in the past included endogenous activation of beta cells or by exogenous methods including transplantation of islets from cadavers and also grafting of new beta cells produced in vitro through cell engineering. Recently, several approaches towards production of human insulin-secreting beta cells have emerged, including in vivo activation of existing beta cells by using several chemical or natural compounds as inhibitors of various pathways leading to regeneration of beta cells.

Restoring functional beta cell mass is an important therapeutic goal for both type 1 and type 2 diabetes. Here, we have shown that a natural terpenoid can act as inhibitor of pro-apoptotic proteins and such inhibitor strongly and selectively could increase human beta cell proliferation in vitro and in vivo. Remarkably, the terpenoid was found to enhance glucose dependent insulin secretion. Thus, understanding

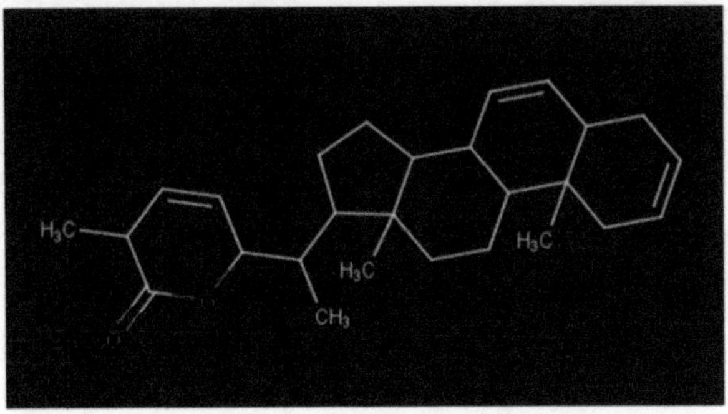

Fig. 12.4 The 2D structure of Terpenoid

the mechanisms involved in increasing beta cell mass is an area of major research interest in diabetes. Hence, we have shown that inhibition of Bad, Bid, Bax proteins by the terpenoid from *E. scaber* is an attractive approach to enhance beta cell proliferation. Further studies are required to understand the mechanism of this effect.

The 2D structure of Terpenoid was shown in Fig. 12.4. In the docking methodology, the most notable sites of interaction depends on Ligand efficiency and intermolecular energy, i.e., binding energy per atom of ligand to protein and between the non-bounded atoms to stabilize the ligand energetically and to improve drug efficacy [30]. Therefore from the results obtained from molecular docking analysis (Table 12.2) the observed ligand efficiency are −0.33, −0.27 and −0.33,which indicates that the compound terpenoid is effective with active pro-apoptotic factors (Bim, Bak and Bad) and by comparing its binding energy, the compound terpenoid is more effective on Bim and Bad. Moreover, the inhibitory constant (k_i) value of Bim and Bad 195.05 and 218.22 nM ensured the firm binding affinity of the compound terpenoid with the active site of pro-apoptotic proteins (Fig. 12.6). Furthermore, the 2D poses of the best hits of the compound terpenoid with pro-apoptotic proteins were generated using Accelrys Discovery Studio Visualiser 4.1. The presence of hydrogen bond interactions and hydrophobic interactions between the compound and pro-apoptotic factors (Bim, Bak, and Bad) was viewed using Pose viewer. There is also the probability of engineering either the drug or the target, or the interface between the drug and the target, to increase the binding affinity. An increase in the number of hydrophobic atoms in the active core of drug–target interface further increases the biological activity of the drug lead. In the present study the hydrogen and hydrophobic bonds were observed between Bad, Bim and Bak molecules and terpenoid. Since terpenoid has the strong hydrogen and hydrophobic bond with Bim, Bad and Bak it increases the binding affinity towards target and receptor complexes.

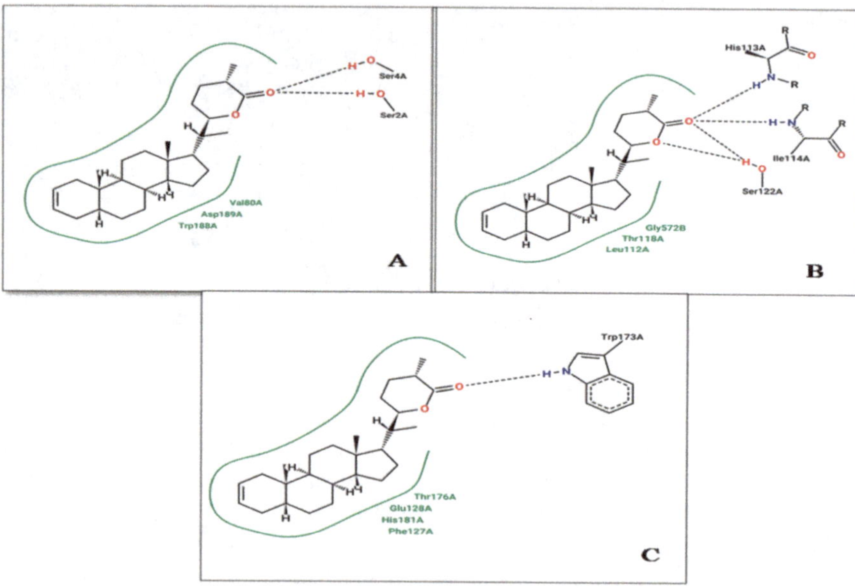

Fig. 12.5 Hydrogen bonding and hydrophobic bonding interaction of terpenoid with Pro-apoptotic proteins (**a**) Bim(1PQT), (**b**) Bak(1BXL) and (**c**) Bad(1G5J) by PoseView

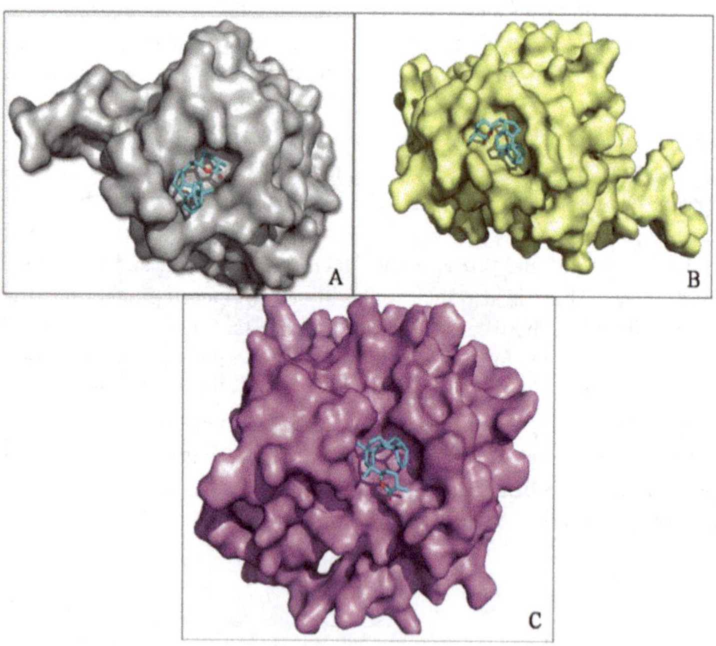

Fig. 12.6 Illustration of docked complex of terpenoid with Pro-apoptotic proteins (**a**) Bim(1PQT), (**b**) Bak(1BXL)

5 Conclusion

Preserving beta cells survival and function in early-stage diabetic patients and restoring beta cells mass/function in the advanced-stage patients are the logical solutions for the treatment of diabetes. Studies using animal models of diabetes support the idea that phytocompounds can preserve the function and survival of beta cells. The role of the terpenoid to inhibit pro-apoptotic proteins suggest its protective action of beta cells by inhibition of pro-apoptotic proteins.

References

1. Halimi S (2006) Diabetes in 2005 (Le diabete en 2005). Nephrol Ther 2(Suppl 1):S2–S7
2. Kumar PJ, Clark M (2002) Diabetes mellitus and other disorders of metabolism. Clin Med 2:1069–1121
3. Wu C, Khan SA, Peng L-J, Lange AJ (2006) Roles for fructose-2,6-bisphosphate in the control of fuel metabolism: beyond its allosteric effects on glycolytic and gluconeogenic enzymes. In: Weber G, Weber CEF, Cocco L (eds) Advances in enzyme regulation, vol 46, proceedings, vol 46. Advances in enzyme regulation, pp 72–88. https://doi.org/10.1016/j.advenzreg.2006.01.010
4. Kris-Etherton PM, Binkoski AE, Zhao G, Coval SM, Clemmer KF, Hecker KD, Jacques H, Etherton TD (2002) Dietary fat: assessing the evidence in support of a moderate-fat diet; the benchmark based on lipoprotein metabolism. Proc Nutr Soc 61(2):287–298. https://doi.org/10.1079/pns2002157
5. Kaji M, Nomura M, Tamura Y, Ito S (2007) Relationships between insulin resistance, blood glucose levels and gastric motility: an electrogastrography and external ultrasonography study. J Med Invest JMI 54(1–2):168–176. https://doi.org/10.2152/jmi.54.168
6. Ogata H, Tokuyama K, Nagasaka S, Ando A, Kusaka I, Sato N, Goto A, Ishibashi S, Kiyono K, Struzik ZR, Yamamoto Y (2007) Long-range correlated glucose fluctuations in diabetes. Methods Inf Med 46(2):222–226
7. Raccah D (2006) Importance of blood glucose management in the multifactorial approach of absolute cardiovascular risk in type 2 diabetes: the lessons from the Steno 2 study. Diabetes Metab 32(2):2S48–42S51. https://doi.org/10.1016/s1262-3636(06)70486-4
8. Dunne MJ, Cosgrove KE, Shepherd RM, Aynsley-Green A, Lindley KJ (2004) Hyperinsulinism in infancy: from basic science to clinical disease. Physiol Rev 84(1):239–275. https://doi.org/10.1152/physrev.00022.2003
9. Yajnik CS (2001) The insulin resistance epidemic in India: fetal origins, later lifestyle, or both? Nutr Rev 59:1):1–1):9
10. Daneman D (2006) Type 1 diabetes. Lancet 367(9513):847–858. https://doi.org/10.1016/s0140-6736(06)68341-4
11. Yoon J-W, Jun H-S (2005) Autoimmune destruction of pancreatic beta cells. Am J Ther 12 (6):580–591. https://doi.org/10.1097/01.mjt.0000178767.67857.63
12. Lee SC, Pervaiz S (2007) Apoptosis in the pathophysiology of diabetes mellitus. Int J Biochem Cell Biol 39(3):497–504. https://doi.org/10.1016/j.biocel.2006.09.007
13. Nakano M, Matsumoto I, Sawada T, Ansite J, Oberbroeckling J, Zhang HJ, Kirchhof N, Shearer J, Sutherland DER, Hering BJ (2004) Caspase-3 inhibitor prevents apoptosis of human islets immediately after isolation and improves islet graft function. Pancreas 29 (2):104–109. https://doi.org/10.1097/00006676-200408000-00004

14. Chandra J, Zhivotovsky B, Zaitsev S, Juntti-Berggren L, Berggren PO, Orrenius S (2001) Role of apoptosis in pancreatic beta-cell death in diabetes. Diabetes 50:S44–S47. https://doi.org/10.2337/diabetes.50.2007.S44
15. Gross A, McDonnell JM, Korsmeyer SJ (1999) BCL-2 family members and the mitochondria in apoptosis. Genes Dev 13(15):1899–1911. https://doi.org/10.1101/gad.13.15.1899
16. Montanya E, Tellez N (2009) Pancreatic remodeling: beta-cell apoptosis, proliferation and neogenesis, and the measurement of beta-cell mass and of individual beta-cell size. Method Mol Biol 560:137–158. https://doi.org/10.1007/978-1-59745-448-3_11
17. Forouzanfar F, Goli AA, Asadpour E, Ghorbani A, Sadeghnia HR (2013) Protective effect of Punica granatum L. against serum/glucose deprivation-induced PC12 cells injury. Evid Based Complement Alternat Med. https://doi.org/10.1155/2013/716730
18. Sharma V, Kalim S, Srivastava MK, Nanda S, Mishra S (2009) Oxidative stress and coxsackievirus infections as mediators of beta cell damage: a review. Sci Res Essays 4(2):42–58
19. Daisy P, Jasmine R, Ignacimuthu S, Murugan E (2009) A novel steroid from Elephantopus scaber L. an Ethnomedicinal plant with antidiabetic activity. Phytomedicine 16(2–3):252–257. https://doi.org/10.1016/j.phymed.2008.06.001
20. Pushparaj P, Tan CH, Tan BKH (2000) Effects of Averrhoa bilimbi leaf extract on blood glucose and lipids in streptozotocin-diabetic rats. J Ethnopharmacol 72(1–2):69–76. https://doi.org/10.1016/s0378-8741(00)00200-2
21. Dhanabal SP, Kokate CK, Ramanathan M, Kumar EP, Suresh B (2006) Hypoglycaemic activity of Pterocarpus marsupium Roxb. Phytother Res 20(1):4–8. https://doi.org/10.1002/ptr.1819
22. Schmidt RE, Dorsey DA, Beaudet LN, Plurad SB, Parvin CA, Miller MS (1999) Insulin-like growth factor I reverses experimental diabetic autonomic neuropathy. Am J Pathol 155 (5):1651–1660. https://doi.org/10.1016/s0002-9440(10)65480-6
23. Pires DEV, Blundell TL, Ascher DB (2015) pkCSM: predicting small-molecule pharmacokinetic and toxicity properties using graph-based signatures. J Med Chem 58(9):4066–4072. https://doi.org/10.1021/acs.jmedchem.5b00104
24. Morris GM, Lim-Wilby M (2008) Molecular docking. Method Mol Biol 443:365–382. https://doi.org/10.1007/978-1-59745-177-2_19
25. Lu Y, Wang Y, Xu Z, Yan X, Luo X, Jiang H, Zhu W (2009) C-X center dot center dot center dot H contacts in biomolecular systems: how they contribute to protein-ligand binding affinity. J Phys Chem B 113(37):12615–12621. https://doi.org/10.1021/jp906352e
26. Wei MC, Zong WX, Cheng EHY, Lindsten T, Panoutsakopoulou V, Ross AJ, Roth KA, MacCregor GR, Thompson CB, Korsmeyer SJ (2001) Proapoptotic BAX and BAK: a requisite gateway to mitochondrial dysfunction and death. Science 292(5517):727–730. https://doi.org/10.1126/science.1059108
27. Kerr JF, Wyllie AH, Currie AR (1972) Apoptosis: a basic biological phenomenon with wide-ranging implications in tissue kinetics. Br J Cancer 26(4):239–257. https://doi.org/10.1038/bjc.1972.33
28. Finegood DT, Scaglia L, Bonner-Weir S (1995) Dynamics of beta-cell mass in the growing rat pancreas. Estimation with a simple mathematical model. Diabetes 44(3):249–256. https://doi.org/10.2337/diabetes.44.3.249
29. Green DR (2005) Apoptotic pathways: ten minutes to dead. Cell 121(5):671–674. https://doi.org/10.1016/j.cell.2005.05.019
30. Qian S-B, Waldron L, Choudhary N, Klevit RE, Chazin WJ, Patterson C (2009) Engineering a ubiquitin ligase reveals conformational flexibility required for ubiquitin transfer. J Biol Chem 284(39):26797–26802. https://doi.org/10.1074/jbc.M109.032334

Chapter 13
Nitrogenous Compounds from Plant Origin in Management of Diabetes Mellitus

Ankit P. Laddha and Yogesh A. Kulkarni

Abstract Diabetes mellitus is a chronic metabolic disorder that occurs when the pancreas is no longer able to make insulin or the body is not able to make use of the insulin. Diabetes mellitus affects people from developed as well as developing countries. According to the International Diabetic Federation 2017 report, there is tremendous increase in number of diabetic people throughout the world from last few decades. The total number had reached 327 million in the year 2017. High prevalence, complex pathogenesis and rapidly developing complications result in an urge in the need of an effective treatment against diabetic progression.

Currently, insulin therapy, pharmacotherapy and diet therapy are available treatment approaches to control diabetes. Beside these, medicinal plants and their phytoconstituents play an important role in the management of diabetes mellitus. Many researchers are working on these phytoconstituents and using them as pharmacophores to develop new chemical entities as novel therapy against diabetes mellitus. Alkaloids, glycosides, flavonoids, terpenoids and tannins are the important phytochemicals reported in management and can be explored for discovery of new antidiabetic drugs.

Nitrogen-containing compounds like amino acids, amides, polyamines and alkaloids are found primarily in plants and are especially common in certain families of flowering plants. More than 3000 different types of alkaloids have been identified in a total of more than 4000 plant species. Different alkaloids like indoles, quinolones, isoquinolines, pyrrolidines, pyrrolizidines, pyridines, tropanes and polyamines like putrescine and cadaverine have been isolated from several medicinal plants and investigated for their antidiabetic activity in different animal models. Berberine, jambosine, vindoline, jatrorrhizine, magnoflorine, palmatine are some well-known alkaloids studied systematically for their role in diabetes mellitus.

The present chapter focuses on phytochemistry and pharmacological activity of nitrogen-containing compounds from plant origin in the management of diabetes mellitus.

A. P. Laddha · Y. A. Kulkarni (✉)
Shobhaben Pratapbhai Patel School of Pharmacy & Technology Management, SVKM's NMIMS, Mumbai, India

H. Chen, M. Zhang (eds.), *Structure and Health Effects of Natural Products on Diabetes Mellitus*, https://doi.org/10.1007/978-981-15-8791-7_13

Keywords Nitrogen-containing compounds · Alkaloids polyamines · Diabetes mellitus · Photochemistry

1 Introduction

Diabetes is a disease that occurs when blood glucose is too high in the body for prolonged duration of time. Glucose is a main source of energy that comes from food.

Insulin, a hormone synthesized from β-cells of pancreas helps to use the glucose received from food carbohydrates and convert it to energy. Insulin helps to keep blood glucose level from getting high (hyperglycaemia) or low (hypoglycaemia) [1]. According to World Health Organization (WHO) 2019 report, diabetes can be classified into type I (Insulin dependent) which develops because of destruction to β-pancreatic cells, type II (Non-insulin dependent) caused due to development of insulin resistance and hybrid forms which include immune medicated and ketosis prone diabetes. Other specific types include endocrine disorder, drug and chemical induced, infection related diabetes and gestational type which develops during pregnancy [2]. Abnormality in carbohydrate, fat and protein metabolism is common factor in diabetes and it is developed because of decreased action of insulin at receptor site. Decreased insulin action develops as a result of inadequate insulin secretion and altered tissue response to insulin. Abnormality in insulin secretion and insulin action is frequently observed or co-exist in same patient and is a common cause of hyperglycaemia [3]. Prolonged and uncontrolled hyperglycaemia is associated with long term damage, dysfunction and failure of vital organs like kidney, eye, nerve, heart and leads to the development of diabetic complications like diabetic nephropathy, diabetic retinopathy, diabetic neuropathy and diabetic cardiomyopathy [4].

According to International Diabetic Federation Atlas, 435 million people were reported to have diabetes till year 2017 and it was estimated to reach 629 million by the year 2045 [5]. Various plants and their phytoconstituents have been thoroughly studied by many researchers for their effect in treatment of diabetes. Many traditional systems of medicine provide evidence of plants and their role in diabetes. Various phytoconstituents have also been reported for their beneficial effect in diabetes. Nitrogen-containing compounds like alkaloids, amino acids, amines and anthranoids are important phytoconstituents of plants and have been proven for their antidiabetic properties.

2 Nitrogen-Containing Compounds from Plants

Nitrogen-containing compounds obtained from plants are mainly classified into alkaloids, amino acids, amines and anthranoids.

Alkaloids like barberin, catharanthine, vindoline, vindolinine, vinblastine and vincristine are commonly studied phytoconstituents and have been reported by many researchers for their antidiabetic property. These phytoconstituents are mainly present in various medicinal plants like *Tinospora cordifolia, Catharanthus roseus, Vinca rosa, Trigonella foenum, Ginkgo biloba, Allium sativum, Tribulus terrestris,* etc [6].

Amino acids and amines such as alliin, leucine, isoleucine, betaine, choline, glutamine and alanine obtained from *Allium sativum, Curcuma longa* and *Aloe vera* also possess antidiabetic property [7]. Piperidine, pyrrolidine, putrescine, spermidine and tryptamine are the predominant amine components of natural amides, which are mainly distributed in various plant families like piperaceae and rutaceae and have potent antidiabetic activity.

2.1 Alkaloids and Diabetes

Alkaloids attracted interest of researchers in the past because of their various traditional uses. Nowadays, these compounds have created a strong scientific and economic platform in medicine and pharmaceutical industries. Many scientists have attempted to defined alkaloid at the time of their discovery. In 1805, Friedrich Serturner discovered morphine which was isolated from the seeds of poppy. Many scientists studied its chemistry and reported pharmacological properties. In 1819, W. Meifsner found that alkaloids are alkali in nature and gave first definition to alkaloid which is 'alkali like compound'. During 1817 to 1821, many researchers followed the reported steps of Serturner and isolated may alkaloids like brucine, quinine and caffeine.

In mid of 1980s scientists took one step forward and developed alkaloids synthetically. Scientist Ball tried to develop synthetic quinine in laboratory [8]. After successive discovery on alkaloid, in 1990s Waller and Nowacki disclosed many characteristics of alkaloid [9]. Their research showed that alkaloids are naturally occurring compounds, which contain nitrogen atom in their structure and are weakly basic in nature. Some semi-synthetic compounds derived from alkaloids have same property [10].

Alkaloids mainly contain carbon, nitrogen and hydrogen in their structure. In addition to this, they also contain oxygen, sulphur, chlorine, bromine and phosphorus. Alkaloids are widely distributed throughout the globe with wide spread biological properties. They are secondary metabolites and derived by termination and transamination reaction of amino acid. Alkaloids are mainly classified depending upon amino acid which give nitrogen to their structure and are derived from various amino acids like L-lysine, L-ornithine, L-tyrosine, L-tryptophan, L-histidine, L-phenylalanine [11].

More than 3000 different alkaloids are discovered till now [12]. They are isolated from plant, animal and marine sources [13]. Alkaloids derived from plants are isolated from more than 4000 plant species and possess wide range of biological activities like antimalarial, CNS stimulant, anticancer, anti-asthmatic, antidiabetic , anti-hypertensive, antispasmodic, vasodilatory, analgesic, antiarrhythmic and antibacterial[14, 15]. Alkaloids possess α-amylase and α-glucosidase inhibitory activity which is considered as one of the important treatment approach by decreasing postprandial blood glucose in type-II diabetic patient. It also acts as dipeptidyl peptidase-4 (DDP-4) and protein tyrosine phosphatase 1B inhibitor, which is associated with glucose metabolism and insulin secretion. Besides this, alkaloids also possess GLUT-4 activator property which deals with transportation of glucose in response to insulin across the membrane. Some alkaloids are reported for their insulin secretion stimulator and β-pancreatic cell regenerator property [6].

2.1.1 Alkaloids as α-amylase and α-glucosidase Enzyme Inhibitors (Fig. 13.1)

α-amylase is an enzyme present in pancreatic juice and saliva which helps in breakdown of large and insoluble molecule like starch into soluble ones. Whereas α-glucosidase is an enzyme present in small intestinal brush border and helps in a last step of digestion which includes breakdown of starch and disaccharide which is prominently present in human diet. α-amylase and α-glucosidase inhibitors delay the breaking of carbohydrate and prevent sudden increase of postprandial glucose in blood[16]. A research report by Gao H and co-workers showed that methanolic extract of *Adhatoda vasica,* which is rich in vasicine and vasicinol, possesses high α-glucosidase enzyme inhibitory activity. Their IC50 values were found to be 125 and 250 μM, respectively [17]. Chelerythrine and sanguinarine, benzophenanthridine alkaloids are reported for α-amylase inhibitory activity. Study by Zajoncova L and co-workers showed that chelerythrine and sanguinarine possess porcine pancreas α-amylase inhibitory activity. Their activity was determined using biosensor method which utilizes a flow system equipped with a peroxide electrode [18]. Mahanimbin, isolated from the leaves of *Murraya koenigii* and vindogentianine isolated from the leaves of *Catharanthus roseus* also possess antidiabetic activity via α-amylase inhibition. Various molecular docking studies showed that alkaloids having α-glucosidase inhibitory activity nummularine-R, cyclopeptide alkaloids and vindoline possess striking interactions with α-glucosidase protein [19]. Branch of *Piper umbellatum* is rich in piperumbellactam, which is present in three form piperumbellactam A, piperumbellactam B, and piperumbellactam C. All isoforms are found to be moderately active against α-glucosidase enzyme [20].

2.1.2 Alkaloids as Dipeptidyl Peptidase-4 Inhibitors (DDP-4 Inhibitors) (Fig. 13.2)

Controlling glucose metabolism is also one of the important options for managing diabetes. DPP-4 enzyme plays a crucial role in glucose metabolism. Secretion of

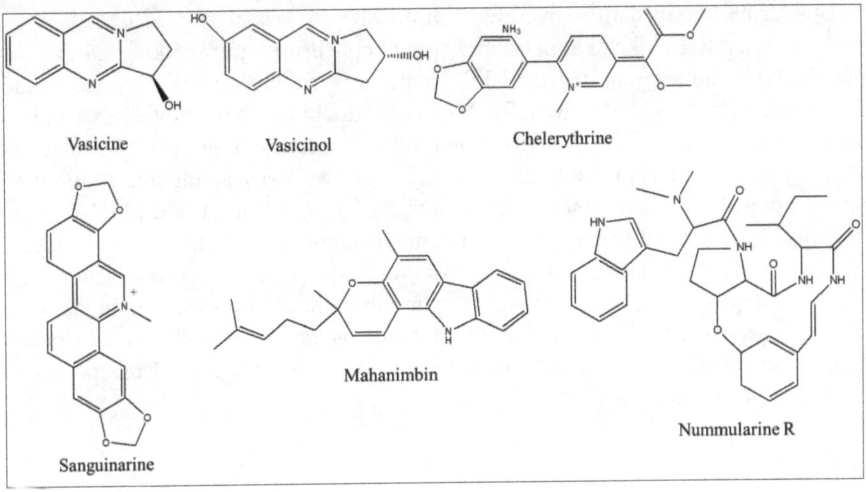

Fig. 13.1 Alkaloids having α-amylase and α-glucosidase inhibitory activity

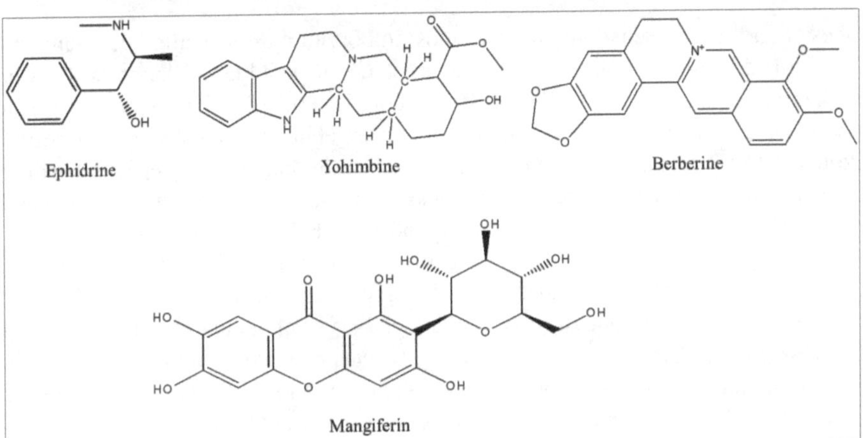

Fig. 13.2 Alkaloids as dipeptidyl peptidase-4 inhibitors (DDP-4 inhibitors)

DPP-4 degrades gut incretin hormone like glucagon like peptide-1 (GLP-1) and gastric inhibitory polypeptide (GIP). β-cell differentiation is induced by GLP-1 and GIP, thus resulting in increase in biosynthesis of insulin. In patient with type-II diabetes, levels of GLP-1 and GIP are reduced. Hence DDP-4 inhibitors are one of the important treatment options in the diabetes. Besides this, DDP-4 inhibitors also clinically reported for significant HbA1c reduction property after 1 h of treatment and widely preferred over other antidiabetic therapy because of low risk of hypoglycaemia and no effect on body weight. Various in-vivo and in-vitro study also showed its β-pancreatic cell regeneration property [21].

Ephedrine obtained from evergreen shrub *Ephedra sinica* and *E. alata* and its various derivatives like norephedrine, pseudoephedrine, norpseudoephedrine, *N*-methylephedrine are reported for DPP-4 inhibitory activity. Their IC_{50} was found in the range of 124–2800 μM [22]. Molecular docking study showed that indole alkaloids also possess DDP-4 enzyme inhibitory property. Out of 20 indole alkaloids tested, yohimbine has higher binding energy. It shows π–π stacking interaction with amino acid of DPP-4 receptor at its S1 and S2 hydrophobic pocket [23]. Another docking study proved that berberine, a benzylisoquinoline alkaloid present in *Berberis vulgaris*, *Berberis aristata* and *Mahonia aquifolium*, showed interaction with DPP-4 receptor, nitrogen atom of isoquinoline group of berberine binds with negatively charged acidic residue of glutamic acid and inhibits its expression [24]. Mangiferin a major phytochemical in *Mangifera indica* is also reported to have DPP-4 inhibitory activity [25].

2.1.3 Alkaloids as Protein Tyrosine Phosphatase 1B Inhibitors (PTP 1B) (Fig. 13.3)

Obesity and insulin sensitivity are two most important reasons behind development of type-II diabetes. Occurrence of drug resistance to insulin and leptin is a major challenge nowadays. Combating resistance could be an effective treatment option for diabetes. PTP1B is an important therapeutic target in type-II diabetes and obesity. Protein PTP1B acts as a negative regulator of insulin and leptin receptor signalling pathway and is most frequently present in skeletal muscle, adipose tissue, brain and liver. Imbalance between insulin receptor and PTP1B protein leads to the development of insulin resistance and impaired insulin signalling [26]. PTP1B inhibitors prevent this imbalance and impaired insulin signalling pathway.

Indole alkaloids are one of the important classes of alkaloids which are reported for their potent PTP 1B inhibitory activity. Leaves of *Catharanthus roseus* rich in vindoline and its various derivatives vindolicine, vindolinine and vindolidine are reported for their hypoglycaemic activity on pancreatic β-TC6 and myoblast C2C12 cells. Besides this, phidianidines A and B isolated form marine *Phidiana militaris* also possess PTP 1B inhibitory property. Pyridine alkaloids like fumosorinone, isolated from fungus *Isaria fumosorosea* are reported as potent PTP 1B inhibitor [27]. Protoberberine alkaloids which is an important group of isoquonoline alkaloids like berberine, coptisine and epiberberine and aporphine alkaloids, magnoflorine isolated from rhizoma *Coptidis* are also reported for PTP 1B inhibitory activity [28]. Naturally available canthinone alkaloid, picrasidine L and its derivative 3,4-dimethyl-canthin-5,6-dione, 4-ethyl-3-methyl-canthin-5,6-dione, eurycomine E, 5-methoxy-canthin-6-one are also found to inhibit activity of PTP 1B protein with the IC_{50} in between range of 19.8 and 28.89 μM. Among all these compounds picrasidine L was found to be the strongest inhibitor and promotes the activity in the insulin signalling pathway on human hepatocellular carcinoma (HepG2) cell line [29]. The norditerpenoid alkaloids obtained from the seeds of *Nigella sativa* like

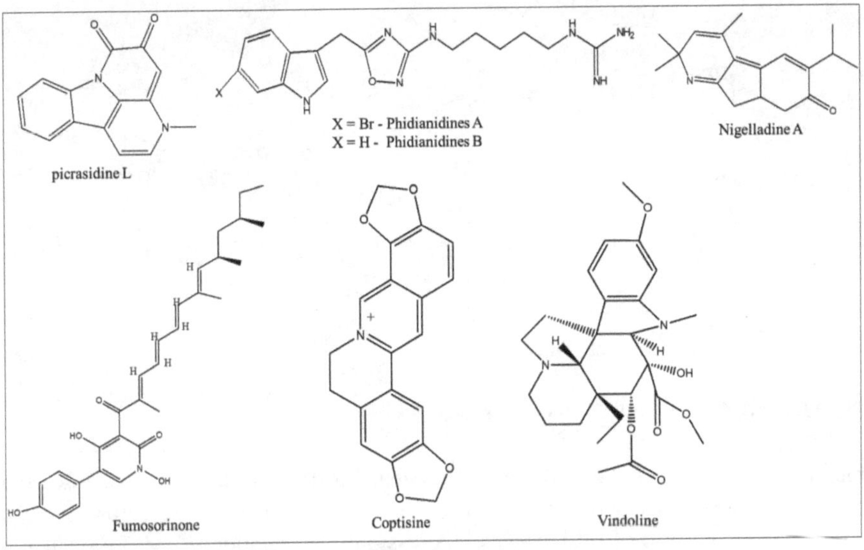

Fig. 13.3 Alkaloids as protein tyrosine phosphatase 1B inhibitors (PTP 1B)

nigelladine A, B and C and one pyrroloquinoline alkaloid, nigellaquinomine are also reported for their PTP 1B inhibitory activity [30].

2.1.4 Alkaloids as GLUT-4 Transporter Activator (Fig. 13.4)

Glucose transporter type 4 (GLUT-4), also known as solute carrier family 2 facilitated glucose transporter member 4, is a protein encoded in human with SLC2A4 gene. It is present in adipose tissues, skeletal and cardiac muscle and helps in insulin regulated glucose transport across the muscle against concentration gradient. Inside the cell, glucose is rapidly phosphorylated by glucokinase, hexokinase, and forms glucose-6-phosphate (G-6-P) which is converted into glycogen [31]. Alkaloids activate GLUT-4 and promote utilization of glucose by cells.

Neferine, dibenzyl isoquinoline alkaloids derived from the dietary *Nelumbo nucifera* (lotus seeds) have been reported for their effect on GLUT-4 expression. A study by Ping Zhao and co-workers showed that neferine upregulates expression of GLUT4. In addition to this neferine also increases intracellular Ca^{2+}, promotes glucose uptake and decreases insulin resistance. Their research proved that neferine increases GLUT4 expression via AMP-activated protein kinase (AMPK) and protein kinase C (PKC) pathway. Alkaloids present in *Murraya koenigii* leaves are well-known GLUT4 activators. Koenimbine, koenidine, *O*-methylmurrayamine A and mahanimbine found to stimulate glucose uptake and GLUT4 expression in skeletal muscle cells (L6- GLUT4 myc myotubes). In addition to this, all compounds have been reported for their glucose lowering property in streptozotocin induced diabetic

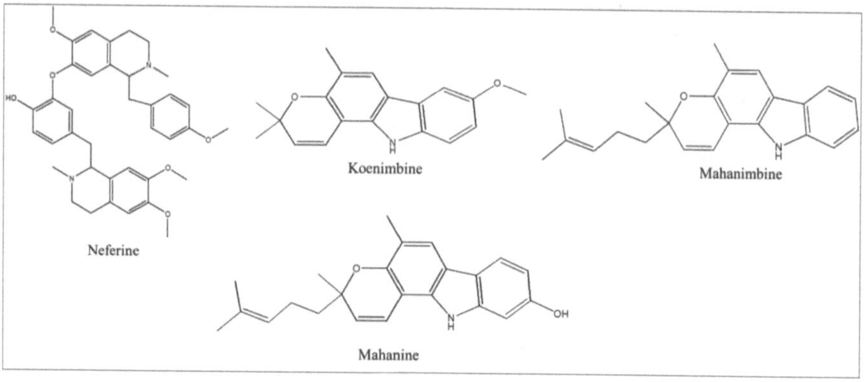

Fig. 13.4 Alkaloids as GLUT-4 transporter activator

rats. Koenidine at dose 30 mg/kg has also been reported for blood glucose lowering and insulin sensitivity activity in leptin receptor-deficient db/db mice for 15 days. The effect is because of GLUT4 overexpression by activation of AKT (Ser-473)-dependent signalling pathway [32]. Mahanine, another carbazole alkaloid also significantly enhanced GLUT4 translocation via AKT signalling pathway to the plasma membrane in L6 myotubes [33].

2.1.5 Alkaloids as Pancreatic β-cells Regenerators and Insulin Secretion Enhancer (Fig. 13.5)

Regeneration of pancreatic β cell is one of the important therapy in the management of type-I diabetes. Recent finding suggests that β-cell death is the final step in a series of mechanisms which are not limited to apoptotic damage related to immune mediated insulitis in type-1 diabetes. It also includes generation of reactive oxygen species, DNA alkylation, excess production of islet amyloid polypeptide with endoplasmic reticulum stress, protein complex formation and mitochondrial damage [34]. Development of oxidative stress, inflammation and apoptosis because of prolonged hyperglycaemia is one of the important reasons behind pancreatic β-cell death. Excessive production of proinflammatory cytokines like interlukin-1β, tumor necrosis factor (TNF)-α, and interferon (IFN)-γ, nuclear factor-kβ (NF-kβ) and endothelial growth factor like vascular endothelial growth factor (VEGF) and fibroblast growth factor (FGF-2) generates reactive oxygen species which damage pancreatic β-cell and worsen the condition of diabetes [35].

Many preclinical studies proved that alkaloidal compounds possess pancreatic β-cell regeneration property. Trigonelline, a bioactive alkaloid obtained from *Trigonella foenum-graecum* has been reported for their hypoglycaemic activity in alloxan induced diabetic rats. The study showed that decrease in blood glucose level in alloxan induced diabetic rats is because of regeneration of pancreatic β-cells due to trigonelline treatment at dose 75 mg/kg for 28 days [36]. Trigonelline is also

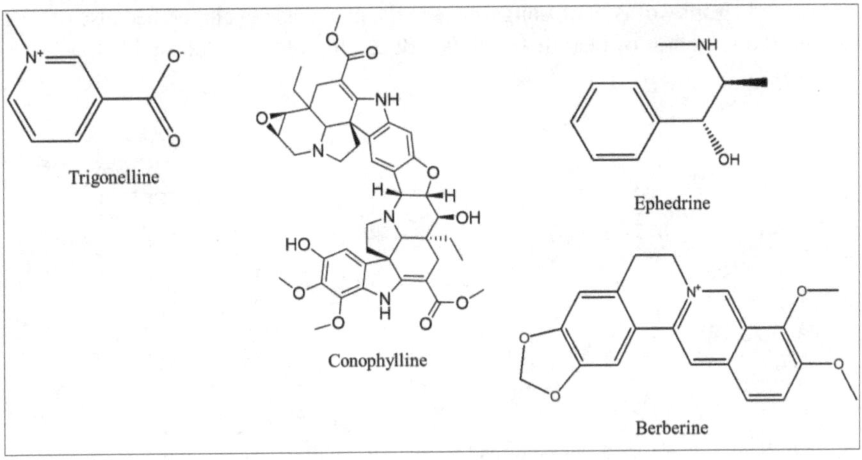

Fig. 13.5 Alkaloids as pancreatic β-cells regenerators and insulin secretion enhancer

effective in gestational diabetes. A study by Zhou J.-Y and team proved the effect of trigonelline during pregnancy in a mouse model of diabetes. Trigonelline treatment at dose 70 mg/kg to pregnant female STZ induced diabetic mice suppressed inflammation, regulated the secretion of adipocytokines, increased β-cell mass, replication and decreased β-cell apoptosis. It also prevented increased proinflammatory cytokines and reduced interleukin-10 level in placenta of diabetic pregnant mice [37]. Berberine is also scientifically proven for its glucose lowering activity. Berberine at dose 75, 150 and 300 mg/kg to STZ induced diabetic rats showed reduction in mitochondrial vacuolization and swelling, dilatation of the endoplasmic reticulum of pancreatic tissue after 28 days of treatment. Berberine also increased insulin secretion, improves anti-oxidant status and reduces lipid peroxidation in diabetic rats [38]. Non-obese diabetic mice treated with berberine at dose 50, 150 and 500 mg/kg significantly improved the decreased number of β-cell after 14 weeks of treatment. Beside this, it also decreased total cholesterol, triglyceride and lipid markers like low density lipoprotein (LDL) and improved high density lipoprotein (HDL) level [39]. Conophylline, a vinca alkaloid is studied using in-vivo and in-vitro technique for β-cell regeneration property. In-vitro study showed that conophylline (0.1 μg/mL) potentiated number of insulin-producing cells and insulin content in porcine cell culture treated with nicotinamide (10 mM) for 3 weeks. Whereas in-vivo study carried out in neonatal STZ induced diabetic rats showed that conophylline treatment for seven days increased the expression of neurogenin-3 (a protein which developed endocrine cell in the pancreas) by activating p38mitogen-activated protein kinase. Histopathological findings proved that conophylline increases number of β-cell mass and insulin content [40]. L-ephedrine is also reported for pancreatic β-cell regeneration property. A study carried out on STZ induced diabetic mice showed that L-ephedrine isolated from *Ephedra sinica herb at dose* 0.62 mg/kg for 14 days showed β-cell regeneration effect. Treatment regenerated atrophy of islet of Langerhans in the pancreas developed because of STZ restored the secretion of insulin and thus reduced high blood glucose [41].

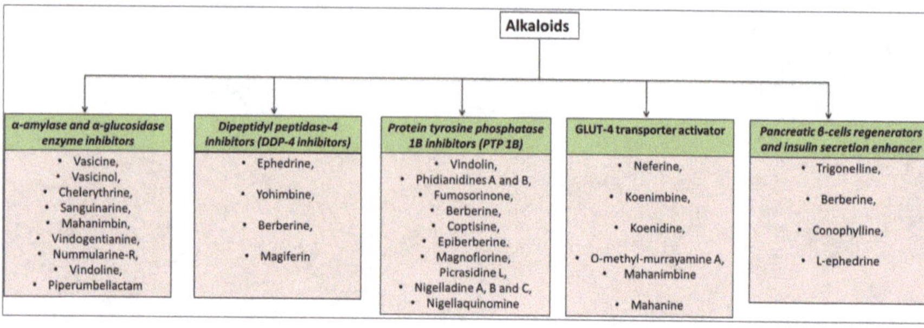

Fig. 13.6 Summary of various alkaloids and mechanisms in the management of diabetes

Table 13.1 Anti-hyperglycaemic effect of alkaloids

Sr. no	Drug	Source	Dose and duration	Animal model	References
1	Berberine and muscimol	*Helicteres isora* fruits	50 mg/kg, single dose	STZ (50 mg/kg) induced diabetic rats	[42]
2	Trigonelline	*Trigonella foenum-graecum* fruit	10 mg/12 h for 4 weeks	Alloxan-induce diabetic rabbits at dose 75 mg/kg at weekly intervals following 12 h fasting	[43]
3	Mahanimbine	*Murraya koenigii* leaves	50 and 100 mg/kg (*i. p.*) per week for 30 days	STZ induced diabetes at dose 45 mg/kg	[44]
4	2-Thionosparteine	*Lupinus angustifolius*	Single intraperitoneal injection at dose 8.6 mg/kg	Streptozotocin (15 mg/kg for 5 consecutive days	[45]
5	Fangchinoline	*Stephania tetrandra* Radix	0.3–3 mg/kg	Streptozotocin (STZ)-diabetic ddY mice	[46]
6	D-Fagomine	*Fagopyrum esculentum* fruit	0.96 g/kg feed	High-fat (HF) diet induced type-II diabetes	[47]
7	Cryptolepine	*Cryptolepis sanguinolenta* roots	10, 30, and 100 mg/kg for 15 days	Alloxan-induced diabetic rat at dose 140 mg/kg	[48]

2.1.6 Anti-hyperglycaemic Agent

Various alkaloids reported for anti-hyperglycaemic property have been listed in Table 13.1.

2.2 *Amino Acids and Amines in Diabetes (Fig. 13.7)*

Amino acids and amines especially dietary amines have been well reported for their antidiabetic potential. Many studies showed that amino acids and amines isolated from *Allium sativum, Curcuma longa* and *Aloe vera* are most widely used as anti-hyperglycaemic agents. Alliin, a cysteine amino acid derivative, is most commonly present in garlic and has been systematically studied for its hypoglycaemic activity. A study carried out by Baiqiang Zhai and co-worker on diet induced obese mice (C57BL/6J DIO mice) showed that allin treatment at dose 0.1 mg/mL for 8 weeks significantly enhanced glucose homeostasis, increased insulin sensitivity and improved the lipid profile [49]. A study showed that administration of garlic derived alliin to alloxan-induced diabetic rats at dose 200 and 500 mg/kg significantly decreased serum lipids and serum enzymes like liver glucose-6-phosphatase, intestinal HMG CoA reductase and liver hexokinase in addition to its hypoglycaemic effect [50]. Leucine, a dietary amino acid also improved glycaemic control. A study carried out in polygenic model with type-11 diabetes, NONcNZO10/LtJ (RCS10) (recombinant congenic strain developed at the Jackson laboratory to model human obesity-induced type-2 diabetes and metabolic syndrome) and monogenic model B6. Cg-A^y/J (A^y) for impaired central melanocortin receptor signalling, obesity and severe insulin resistance. Study showed that leucine supplementation provided for 8 months in drinking water at 1.5% concentration significantly reduced the level of HbA1c throughout the study in both the genetic model. Increased insulin sensitivity and decreased insulin level were observed after leucine supplementation. Improvement in mitochondrial oxidative stress and reduction in inflammatory markers were also observed in both the genetic diabetic models [51]. It has been reported that dietary leucine supplementation substantially decreases diet-induced hyperglycaemia, obesity and hypercholesterolemia in chow feed mice by increasing protein synthesis and decreasing protein catabolism via mTOR (mammalian target of rapamycin) pathway. Leucine and its derivative isoleucine also stimulate insulin-independent glucose uptake in skeletal muscle of rat by 73%. Oral administration of single dose isoleucine at dose 2.5 mL/100 g body weight to normal rats significantly reduced plasma blood glucose by altering glucose uptake via improving availability of ATP in the absence of increase in energy sensor, AMP-activated protein kinase activity in skeletal muscle [52]. Betaine, a modified amino acid consisting glycine with three methyl groups was tested in 12-week, parallel arm, randomized, double-masked, placebo-controlled trial. It showed significant reduction in blood glucose and improved metabolism in patients [53].

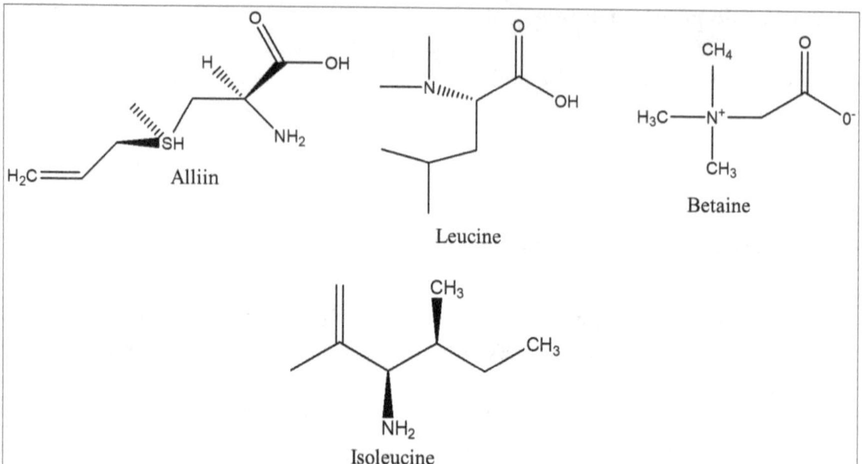

Fig. 13.7 Amino acids and amines in diabetes

3 Conclusion

Diabetes is a chronic metabolic disorder developed because of abnormality in insulin secretion, insulin action or both. Abnormal glucose metabolism and its uptake, down regulation of glucose transporters, and defect in insulin secretion from pancreatic β cell are the important reasons for development of type-I and type-II diabetes.

Nitrogen-containing compounds form plants have gained a lot of research interest and have been proven for their activity against diabetes. Alkaloids, amino acids and natural amines possess anti-oxidant and anti-inflammatory activities. They have also been reported for their activity in improving glucose metabolism and its uptake. Many nitrogen-containing compounds still need to be studied at the molecular level for their therapeutic effects. Systematic preclinical, clinical studies and formulation studies are required for nitrogen-containing compounds from plants for their establishment as antidiabetic drugs.

References

1. Laddha AP, Kulkarni YA (2019) Tannins and vascular complications of Diabetes: an update. Phytomedicine 56:229–245
2. Diabetes.. https://www.who.int/health-topics/diabetes. Accessed 7 Feb 2020
3. Association AD (2014) Diagnosis and classification of diabetes mellitus. Diabetes Care 37: S81–S90
4. Suryavanshi SV, Kulkarni YA (2017) NF-κβ: a potential target in the management of vascular complications of diabetes. Front Pharmacol 8:798
5. IDF (2019) IDF Diabetes Atlas 9th edition 2019, 9th edn. Int Diabetes Fed Diabetes Atlas, pp 1–176

6. Kumar A, Aswal S, Semwal RB, Chauhan A, Joshi SK, Semwal DK (2019) Role of plant-derived alkaloids against diabetes and diabetes-related complications: a mechanism-based approach. Phytochem Rev 18:1277–1298
7. Bharti SK, Krishnan S, Kumar A, Kumar A (2018) Antidiabetic phytoconstituents and their mode of action on metabolic pathways. Ther Adv Endocrinol Metab 9:81–100
8. Ball P (2008) History of science: quinine steps back in time. Nature 451:1065–1066
9. Aniszewski T (1994) From iodine to enzyme: a critical review of chemical and biological methods of lupine alkaloids analysis. Sci Legum 1:25–36
10. Scott T (1980) Alkaloid biology and metabolism in plants. Biochem Educ 8:30
11. Aniszewski T (2015) Definition, typology, and occurrence of alkaloids, 2nd edn. Alkaloids. https://doi.org/10.1016/b978-0-444-59433-4.00001-8
12. Kurek J (2019) Introductory chapter: alkaloids – their importance in nature and for human life. Alkaloids – their importance. Nat Hum Life. https://doi.org/10.5772/intechopen.85400
13. Manske RHF (1950) Chapter I sources of alkaloids and their isolation. Alkaloids Chem Physiol 1:1–14
14. Chaves KMS, Feitosa CM, da Araújo LS (2016) Alkaloids pharmacological activities – prospects for the development of phytopharmaceuticals for neurodegenerative diseases. Curr Pharm Biotechnol 17:629–635
15. Martynov A, Bomko T, Nosalskaya T, Kabluchko-Baruah T (2017) Structures and biological activity of cuprophyllins. https://doi.org/10.5281/zenodo.803831
16. Kazeem MI, Adamson JO, Ogunwande IA (2013) Modes of inhibition of α-amylase and α-glucosidase by aqueous extract of *Morinda lucida* Benth leaf. Biomed Res Int. https://doi.org/10.1155/2013/527570
17. Gao H, Huang YN, Gao B, Li P, Inagaki C, Kawabata J (2008) Inhibitory effect on α-glucosidase by *Adhatoda vasica* Nees. Food Chem 108:965–972
18. Zajoncová L, Kosina P, Vičar J, Ulrichová J, Peč P (2005) Study of the inhibition of α-amylase by the benzo[c]phenanthridine alkaloids sanguinarine and chelerythrine. J Enzyme Inhib Med Chem 20:261–267
19. Rahman N, Muhammad I, Nayab GE, Khan H, Aschner M, Filosa R, Daglia M (2019) Molecular docking of isolated alkaloids for possible α-Glucosidase inhibition. Biomol Ther 9:544
20. Tabopda TK, Ngoupayo J, Liu J et al (2008) Bioactive aristolactams from *Piper umbellatum*. Phytochemistry 69:1726–1731
21. Barnett A (2006) DPP-4 inhibitors and their potential role in the management of type 2 diabetes. Int J Clin Pract 60:1454–1470
22. Ojeda-Montes MJ, Ardid-Ruiz A, Tomás-Hernández S, Gimeno A, Cereto-Massagué A, Beltrán-Debón R, Mulero M, Garcia-Vallvé S, Pujadas G, Valls C (2017) Ephedrine as a lead compound for the development of new DPP-IV inhibitors. Future Med Chem 9:2129–2146
23. Dhiraviam KN, Balasubramanian S, Jayavel S (2015) Indole alkaloids as new leads for the design and development of novel DPP-IV inhibitors for the treatment of diabetes. Curr Bioinforma 13:157–169
24. Al-Masri IM, Mohammad MK, Tahaa MO (2009) Inhibition of dipeptidyl peptidase IV (DPP IV) is one of the mechanisms explaining the hypoglycemic effect of berberine. J Enzyme Inhib Med Chem 24:1061–1066
25. Mohanty I, Kumar S, Rajesh S (2017) Dipeptidyl peptidase IV inhibitory activity of Berberine and Mangiferin: an in Silico approach. Int J Clin Endocrinol Metab 3:018–022
26. Verma M, Gupta SJ, Chaudhary A, Garg VK (2017) Protein tyrosine phosphatase 1B inhibitors as antidiabetic agents – a brief review. Bioorg Chem 70:267–283
27. Liu L, Zhang J, Chen C, Teng J, Wang C, Luo D (2015) Structure and biosynthesis of fumosorinone, a new protein tyrosine phosphatase 1B inhibitor firstly isolated from the entomogenous fungus *Isaria fumosorosea*. Fungal Genet Biol 81:191–200

28. Choi JS, Ali MY, Jung HA, Oh SH, Choi RJ, Kim EJ (2015) Protein tyrosine phosphatase 1B inhibitory activity of alkaloids from Rhizoma Coptidis and their molecular docking studies. J Ethnopharmacol 171:28–36
29. Sasaki T, Li W, Higai K, Koike K (2015) Canthinone alkaloids are novel protein tyrosine phosphatase 1B inhibitors. Bioorganic Med Chem Lett 25:1979–1981
30. Bin CQ, Xin XL, Yang Y, Lee SS, Aisa HA (2014) Highly conjugated norditerpenoid and pyrroloquinoline alkaloids with potent ptp1b inhibitory activity from nigella glandulifera. J Nat Prod 77:807–812
31. Watson RT, Kanzaki M, Pessin JE (2004) Regulated membrane trafficking of the insulin-responsive glucose transporter 4 in adipocytes. Endocr Rev 25:177–204
32. Patel OPS, Mishra A, Maurya R et al (2016) Naturally occurring Carbazole alkaloids from *Murraya koenigii* as potential antidiabetic agents. J Nat Prod 79:1276–1284
33. Nooron N, Athipornchai A, Suksamrarn A, Chiabchalard A (2017) Mahanine enhances the glucose-lowering mechanisms in skeletal muscle and adipocyte cells. Biochem Biophys Res Commun 494:101–106
34. Rojas J, Bermudez V, Palmar J et al (2018) Pancreatic beta cell death: novel potential mechanisms in diabetes therapy. J Diabetes Res. https://doi.org/10.1155/2018/9601801
35. Cnop M, Welsh N, Jonas JC, Jörns A, Lenzen S, Eizirik DL (2005) Mechanisms of pancreatic β-cell death in type 1 and type 2 diabetes: many differences, few similarities. Diabetes 54:S97–S107
36. Shah SN, Bodhankar SL, Badole SL, Kamble HV, Mohan V (2006) Effect of trigonelline: an active compound from *Trigonella foenum graecum* Linn. In alloxan induced diabetes in mice. J Cell Tissue Res 6:585–590
37. Zhou JY, Du XH, Zhang Z, Qian GS (2017) Trigonelline inhibits inflammation and protects β cells to prevent fetal growth restriction during pregnancy in a mouse model of diabetes. Pharmacology 100:209–217
38. Zhou J, Zhou S, Tang J, Zhang K, Guang L, Huang Y, Xu Y, Ying Y, Zhang L, Li D (2009) Protective effect of berberine on beta cells in streptozotocin- and high-carbohydrate/high-fat diet-induced diabetic rats. Eur J Pharmacol 606:262–268
39. Chueh WH, Lin JY (2011) Berberine, an isoquinoline alkaloid in herbal plants, protects pancreatic islets and serum lipids in nonobese diabetic mice. J Agric Food Chem:8021–8027
40. Kojima I, Umezawa K (2006) Conophylline: a novel differentiation inducer for pancreatic β cells. Int J Biochem Cell Biol 38:923–930
41. Xiu LM, Miura AB, Yamamoto K, Kobayashi T, Song QH, Kitamura H, Cyong JC (2001) Pancreatic islet regeneration by ephedrine in mice with streptozotocin-induced diabetes. Am J Chin Med 29:493–500
42. Chandirasegaran G, Elanchezhiyan C, Ghosh K, Sethupathy S (2016) Determination of antidiabetic compounds from *Helicteres isora* fruits by oral glucose tolerance test. J Appl Pharm Sci 6:172–174
43. Al-Khateeb E, Hamadi SA, Al-Hakeemi AAN, Abu-Taha M, Al-Rawi N (2012) Hypoglycemic effect of trigonelline isolated from iraqi fenugreek seeds in normal and alloxan-diabetic rabbits. Eur Sci J December Ed 8:1857–7881
44. Dineshkumar B, Mitra A, Mahadevappa M (2010) Antidiabetic and hypolipidemic effects of mahanimbine (carbazole alkaloid) from *Murraya koenigii* (rutaceae) leaves. Int J Phytomedicine 2:22–30
45. Bobkiewicz-Kozłowska T, Dworacka M, Kuczyński S, Abramczyk M, Kolanoś R, Wysocka W, Garcia Lopez PM, Winiarska H (2007) Hypoglycaemic effect of quinolizidine alkaloids – lupanine and 2-thionosparteine on non-diabetic and streptozotocin-induced diabetic rats. Eur J Pharmacol 565:240–244
46. Tsutsumi T, Kobayashi S, Liu YY, Kontani H (2003) Anti-hyperglycemic effect of fangchinoline isolated from *Stephania tetrandra* Radix in streptozotocin-diabetic mice. Biol Pharm Bull 26:313–317

47. Hereu M, Ramos-Romero S, Busquets C, Atienza L, Amézqueta S, Miralles-Pérez B, Nogués MR, Méndez L, Medina I, Torres JL (2019) Effects of combined d-fagomine and omega-3 PUFAs on gut microbiota subpopulations and diabetes risk factors in rats fed a high-fat diet. Sci Rep 9:1–12
48. Ameyaw EO, Koffuor GA, Asare KK, Konja D, Du-Bois A, Kyei S, Forkuo AD, Mensah RNAO (2016) Cryptolepine, an indoloquinoline alkaloid, in the management of diabetes mellitus and its associated complications. J Intercult Ethnopharmacol 5:263–273
49. Zhai B, Zhang C, Sheng Y, Zhao C, He X, Xu W, Huang K, Luo Y (2018) Hypoglycemic and hypolipidemic effect of S-allyl-cysteine sulfoxide (alliin) in DIO mice. Sci Rep. https://doi.org/10.1038/s41598-018-21421-x
50. Sheela CG, Augusti KT (1992) Antidiabetic effects of S-allyl cysteine sulphoxide isolated from garlic *Allium sativum* Linn. Indian J Exp Biol 30:523–526
51. Guo K, Yu YH, Hou J, Zhang Y (2010) Chronic leucine supplementation improves glycemic control in etiologically distinct mouse models of obesity and diabetes mellitus. Nutr Metab 7:57
52. Doi M, Yamaoka I, Nakayama M, Mochizuki S, Sugahara K, Yoshizawa F (2005) Isoleucine, a blood glucose-lowering amino acid, increases glucose uptake in rat skeletal muscle in the absence of increases in AMP-activated protein kinase activity. J Nutr 135:2103–2108
53. Grizales AM, Patti ME, Lin AP et al (2018) Metabolic effects of betaine: a randomized clinical trial of betaine supplementation in prediabetes. J Clin Endocrinol Metab 103:3038–3049

Chapter 14
Plant Alkaloids with Antidiabetic Potential

Erick P. Gutiérrez-Grijalva, Laura A. Contreras-Angulo, Alexis Emus-Medina, and J. Basilio Heredia

Abstract Type-2 diabetes mellitus (DM) is a metabolic disorder characterized by hyperglycemia that arises by insufficient insulin secretion or action. DM is a non-communicable disease that has reached epidemic proportions; up to 2017, around 425 million people were living with diagnosed DM, and this number is expected to increase by 48% in 2045. The first line of treatment against DM consists of hypoglycemic drugs, such as acarbose and metformin. However, people from low- and middle-income countries often struggle to access these medications. This turns them to use either only medicinal plants or in combination with conventional treatments. Thus, this chapter aims to comprehensively review the current literature (2010–2020) regarding the potential pharmaceutical properties of plant alkaloids in the treatment/prevention of diabetes mellitus and its comorbidities.

Keywords Type-2 diabetes mellitus · Medicinal plants · Alkaloids · Treatment/prevention

1 Introduction

According to the International Diabetes Federation, it estimates that there are 430 million people with diabetes in 2019, and a growth of 51% (700 million) is expected by 2045. Worldwide, the regions with the highest prevalence are the Middle East and North Africa, with 55 million people and the countries with the

E. P. Gutiérrez-Grijalva
Cátedras CONACYT-Centro de Investigación en Alimentación y Desarrollo, Culiacán, Sinaloa, Mexico

L. A. Contreras-Angulo · J. B. Heredia (✉)
Centro de Investigación en Alimentación y Desarrollo, Culiacán, Sinaloa, Mexico
e-mail: jbheredia@ciad.mx

A. Emus-Medina
Universidad Autónoma de Sinaloa, Facultad de Medicina Veterinaria y Zootecnia, Culiacán, Sinaloa, Mexico

H. Chen, M. Zhang (eds.), *Structure and Health Effects of Natural Products on Diabetes Mellitus*, https://doi.org/10.1007/978-981-15-8791-7_14

highest incidence of adults between the ages of 20 and 79 are China, India, the USA, Pakistan, Brazil, and Mexico [1]. The World Health Organization reports 2.2 million deaths related to hyperglycemia in 2012 and 1.6 million deaths as a direct cause of diabetes in 2015. Diabetes is emerging to be the seventh leading cause of mortality by 2030 [2]. Annual costs derived from diabetes care worldwide amount to USD 760,000 billion, where the costs of treating diabetes complications represent more than 50% [1].

There are two different main types of diabetes, type 1 (damage or failure in insulin production) and 2 (decreased sensitivity of tissues to insulin) diabetes [3]. Type 1 diabetes or insulin-dependent accounts for between 5 and 10% of cases of diabetes in the world, usually occurs in people under 30, so it is also known as juvenile diabetes; however, it can be diagnosed at any age [4]. This type of diabetes is characterized by the destruction of β-cells, responsible for the production of insulin [5]. In most cases, the destruction occurs autoimmune, but there are several cases that the destruction or failure of β-cells is idiopathic [6, 7]. The destruction of β-cells is due to the presence of a chronic inflammatory infiltrate that affects the islets of Langerhans in the pancreas [8–10]. When the β-cells are damaged, control of blood glucose regulation is lost, through the progressive decrease in insulin production, which could generate various conditions such as ketoacidosis, blindness, heart and kidney diseases, among others [4, 9]. Several risk factors could influence the development of type 1 diabetes, such as age, gender, race, genotype, and geographic location [6]. The main symptoms associated with type 1 diabetes are polydipsia, polyphagia, and polyuria [10, 11]. People diagnosed with type 1 diabetes should have a series of constant care and precautions such as the administration of insulin, testing blood glucose levels, adequate food, and physical activity monitoring [12].

On the other hand, type 2 diabetes or non-insulin dependent is the most frequent type of diabetes, which represents between 90 and 95% of diabetes cases in the world [13]. It is characterized by a decrease in glucose absorption due to insulin deficiency caused by pancreatic β-cell dysfunction and insulin resistance of the organs [14]. Insulin resistance refers to the fact that it does not exert an action corresponding to its concentration in blood [15], which reduces the transport of glucose to the liver, muscle cells, and fat cells [16]. Insulin resistance has been attributed to B cell dysfunction due to oxidative stress, amyloid deposits in the pancreas, deposits of ectopic lipids, lipotoxicity, and glucotoxicity [17]. This disease can cause various complications such as kidney failure, blindness, heart failure, among others [18, 19]. This type of diabetes can be caused by the combination of different factors such as obesity, stress, sedentary lifestyle, and genetic factors related to insulin secretion or resistance [15, 20].

Despite the great scientific advances, a cure for diabetes has not been achieved. However, there are several treatments for its control from lifestyle changes to surgeries; nevertheless, the most commonly used treatments are oral antidiabetic drugs. Some of them are sulfonylureas, biguanides, thiazolidinediones, meglitinides, α-glucosidase inhibitors, dipeptidyl peptidase 4 (DPP4) inhibitors, incretin mimetics, and sodium-glucose cotransporter-2 (SGLT2) inhibitors [21, 22].

- Sulfonylureas stimulate the production of insulin by binding to specific receptors in β-cells, avoiding the flow of potassium (K^+) from the ATP-dependent channel [23]. Then, the membrane depolarizes, opens the channels of Ca^{++} in the β-cells, regulated by voltage, and causes the contraction of the actomyosin filaments responsible for the exocytosis of insulin [24, 25]. Hypoglycemia is an adverse effect of sulfonylureas mainly observed in people with renal, liver, or elderly dysfunction [26, 27].
- Meanwhile, the biguanides reduce hyperglycemia by an increment of sensitivity to insulin, decrease absorption of glucose, and inhibit hepatic gluconeogenesis. Metformin is the most widely prescribed biguanide [28]. Metformin activates AMPK (AMP-activated protein kinase) [29] that partially inhibits mitochondrial respiratory-complex 1, reducing ATP levels and increasing AMP levels, which limits gluconeogenesis [30]. One of the adverse effects of biguanides is lactic acidosis, which is avoided in people with kidney failure [31].
- Thiazolidinediones (TZDs) are drugs used whose function is to increase insulin sensitivity. This effect is exerted directly on adipose tissue and indirectly by the reduction of ectopic lipotoxicity in the liver and muscles [32]. The mechanism by which thiazolidinediones act is due to the union of nuclear receptor transcription factor PPARγ [33], this receptor controls genetic expressions related to lipids, cell differentiation, and inflammation [34]. Troglitazone, rosiglitazone, pioglitazone are some examples [35]. Weight gain and fluid retention are some possible adverse effects. Also, cardiovascular problems have been associated with rosiglitazone [35, 36].
- Meglitinides (repaglinide and nateglinide) are used against hyperglycemia, have a rapid and short-lasting effect, are secretors of insulin [37]. They are structurally different from sulfonylureas but have a similar mechanism, since they act on pancreatic β-cell receptors, regulating ATP-sensitive potassium channels in pancreatic β-cells, increasing insulin secretion [38]. They are administered after meals, which helps the physiological release of insulin; some disadvantages are high costs and doses [39].
- α-glucosidase inhibitors such as voglibose, acarbose, and miglitol are useful for postprandial hyperglycemia and act competitively inhibiting the active sites α-glucosidase in the small intestine. This enzyme is responsible for the hydrolysis of disaccharides and oligosaccharides to release of glucose. By inhibiting this enzyme, blood glucose is reduced, and there is a reduction in postprandial insulin levels. The main adverse effects are flatulencies and diarrhea [21, 40, 41].
- Another target enzyme for diabetes control is dipeptidyl peptidase 4 (DPP-4). This enzyme is responsible for inhibiting incretins such as glucagon-like peptide-1 (GLP-1) and glucose-dependent insulinotropic peptide (GIP) [42]. These incretins stimulate insulin secretion and prevent postprandial hyperglycemia. Besides, GLP-1 reduces glucagon secretion and slows gastric emptying [43]. Therefore, by inhibiting DPP-4, an increase in the levels of GLP-1 and GIP incretins is achieved, resulting in an adequate glycemic control [44]. Some DPP-4 inhibitors are sitagliptin, saxagliptin, and vildagliptin, which can be used as monotherapies or combined [45]. Other medications used are incretin

mimetics; this is based on the importance of incretins such as GIP and GLP-1, responsible for approximately 50% of postprandial insulin secretion [46]. Exenatide and liraglutide are some examples, which have a sequence of 53 and 97% homologous to GLP-1 (30 amino acids polypeptide), respectively. These incretin mimetics bind and activate the GLP-1 receptors of pancreatic β-cells, which lead to stimulation of insulin secretion and synthesis. Some adverse effects they present are nausea, vomiting and to lesser extent cases of hypoglycemia when used in combination with other medications [47].
- The sodium-glucose cotransporter-2 (SGLT2) inhibitors are another class of medications used to lower blood glucose levels. The SGLT2 and SGLT1 proteins are responsible for the reabsorption of renal glucose by 90 and 10%, respectively [48]. Therefore, inhibition of SGLT2 decreases glucose reabsorption and increases renal glucose excretion [22]. Some SGLT2 inhibitors are canagliflozin, dapagliflozin, and empagliflozin [21]. These inhibitors act independently of insulin secretion by pancreatic β-cells, so hypoglycemia is not expected [49]. On the other hand, some adverse effects are urinary infections [21, 50].

Despite the variety of drugs used for the treatment of diabetes, their adverse effects, and high costs, currently, the use of natural alternatives is on the rise, which is related to the growing interest in the search for new alternatives based on herbal medicine [51, 52]. These therapeutic properties are related to the different compounds of plants, such as polyphenols, flavonoids, terpenoids, polysaccharides, glycosides, and alkaloids [53–55].

In recent years, alkaloids have been related to some antidiabetic properties through various mechanisms, which will be addressed in the sections below [56–60].

2 Alkaloids

Wilhelm Meissner apothecary was the first person who coined the term alkaloid, a term derived from "alkali" and Greek "oiedes" (like), in 1819 [61]. Alkaloids are secondary metabolites widely distributed in nature, including bacteria, fungi, animals, and mainly in plants (root, leaf, stem, fruits, and seed), are present in almost any organ of cell, in mitochondria, vesicles, chloroplast, and vacuoles [62–64]. Alkaloids are present in diverse plant families, and it is estimated that around 40% have at least one species producer of this metabolite. Besides, due to their potential biological activity, about 20,000 compounds have been explored, some are known for their effect on pharmaceutical, stimulant, narcotic and poisonous activity; and these represent only half of the reported by the dictionary of alkaloids [65, 66]. There is not an exact definition of alkaloids. However, the following is suggested: "An alkaloid is a cyclic compound containing nitrogen in a negative state which is of limited distribution in living organisms." In plants, alkaloids can exist in a free state, as salt or as N-oxides. Chemically, alkaloids are low-molecular weight compounds, alkaline, insoluble in water, optically active (a few are dextrorotatory), bitter in taste, colorless, crystalline, nonvolatile solids, and just a few are liquids and colored.

The structure of alkaloids consists of a heterocyclic ring with a nitrogen atom (some are aliphatic nitrogen compounds—noncyclic), which is their main characteristic [65, 67–70]. Almost all alkaloids have a nitrogen atom, and some have more than 5, which can be in the form of primary amine, secondary amine, or tertiary amine; they also can be found as monomers, dimers, trimers, or tetramers [71]. Alkaloids are synthesized from the amino acids lysine, tyrosine, tryptophan, phenylalanine, and anthranilic acid. Some are derived from ornithine by the metabolic pathways: Glycolysis, Krebs cycle, Shikimate pathway [68, 72, 73].

2.1 Classification of Alkaloids

There are different ways to classify alkaloids, whether by their natural source or chemical nature, however the most accepted and used is based on their main structure (carbon-nitrogen skeleton) [71, 74]. Alkaloids are classified into three main types, true alkaloids (heterocyclics), protoalkaloids (non-heterocyclics), and pseudo alkaloids (Table 14.1). This classification depends on the heterocyclic ring structure and the amino acid from which they are originated (Table 14.1) [66, 69, 75].

2.1.1 True Alkaloids (Heterocyclics)

True alkaloids, or heterocyclic alkaloids, constitute the main type of alkaloids. True alkaloids can be found in some species and families. They are basic in nature with a complex molecular structure, very reactive even at low doses, and have an intracyclic nitrogen in their structure. They are derived from the amino acids L-ornithine, L-tyrosine, L-phenylalanine, L-lysine, L-histidine, L-tryptophan, L-arginine, and glycine/aspartic acid [69, 73, 76, 77]. Some of the most common heterocyclic alkaloids are pyrrolidine, pyrrolizidine, pyridine, piperidine, tropane, quinolone, isoquinoline, aporphine, quinolizidine, indole, indolizidine, and imidazole.

The main class is *Pyridine* alkaloids, which are derived from amino acid L-ornithine and have a structure of five carbon atoms and one nitrogen atom. Their structure is similar to benzene. Some compounds have a pyridine nucleus like vitamin B3 and vitamin B6. Pyridine alkaloids are sub-classified as piperine, coniine, trigonelline, arecoline, arecaidine, nicotine, and lobeline [78, 79]. *Piperidine* structure is similar to cyclohexane, but with a secondary amino group (NH) replacing one methylene group, it is abundant in black pepper. In this class, we can find the alkaloids quinine, pergolide, and solenopsin [80, 81].

Quinoline is an aromatic compound obtained by ortho-condensation of the benzene ring of pyridine, and is an alkaloid derived from tryptophan. The primary

Table 14.1 Classification of alkaloids according to the type, precursor and class, and their structure) [66, 69, 75]

Alkaloid type	Precursor compound	Alkaloid class	Examples
True alkaloids (heterocyclic)	Tryptophan 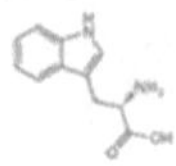	Quinoline 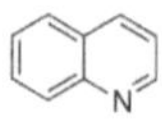 Indole 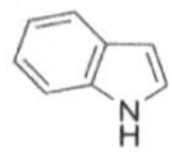	Pamaquine, quinine, chloroquine, bulaquine, serpentinine
	Tyrosine 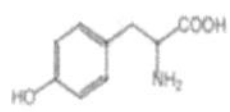	Isoquinoline 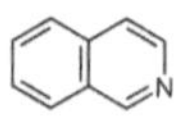	Dopamine, aporphine, morphine, berberine
	Lysine 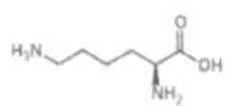	Quinolizidine 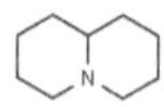 Piperidine 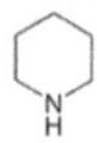	Solenopsin, cynapine, lupanine, sparteine
	Aspartate 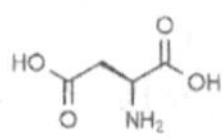	Pyridine 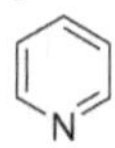	Nicotine, lobeline, piperine, cytisine
	Ornithine 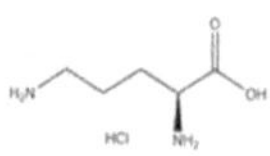	Tropane 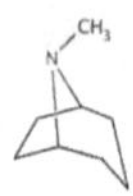 Pyrrolizidine 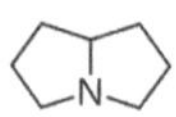 Pyrrolidine 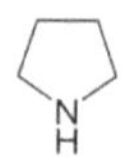	Atropine, hygrine, cocaine, necine, stachydrine, symphitine
	Anthranilic acid 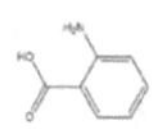	Quinazoline 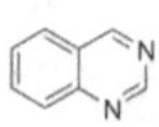 Acridone 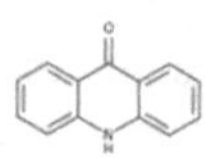	Acronycine
	Hstidine 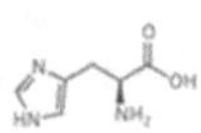	Imidazole 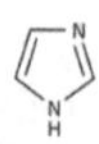	Pilocarpine, histamine, pilose histidine
Protoalkaloids (non-heterocyclics)	Tryptophan 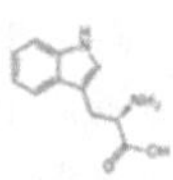	Terpenoid índole (Strictosidine) 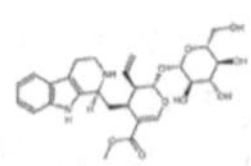	Vinblastine, vincristine
	Phenylalanine 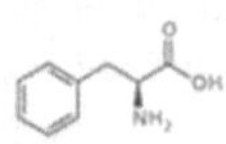	Phenylethylamine 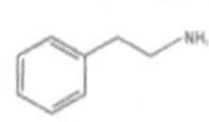	Mescaline, dopamine, ephedrine, cathinone

(continued)

Table 14.1 (continued)

Alkaloid type	Precursor compound	Alkaloid class	Examples
	Tyrosine 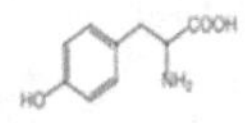	Phenethyisoquinoline 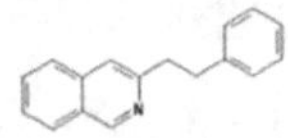	Colchicine
Pseudoalkaloids	Saponins 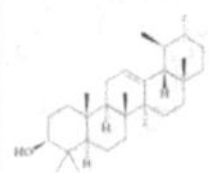	Steroidal (cyclopentanophenanthrene) 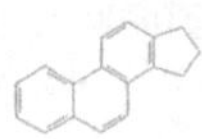	Solanidine, chaconine, solasodine, tomatine
	Acetic acid 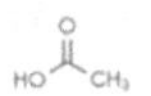	Terpene alkaloid 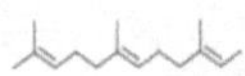	Aconitine, atisine, lyctonine, nenufares, veratramine

classes of this group are quinine and quinidine, and also exist pamaquine, chloroquine in nature [79, 81]. *Isoquinoline*, an isomer of quinolone, is derived from tyrosine and phenylalanine. Its nucleus is mainly present in berberine and pavarine. It is classified into benzoquinone, aporphine, morphine, and the last one was obtained from opium [81–83].

Quinazoline contains nitrogen in its heterocyclic structure, which is based on the electrophilicity of the C-4 position and is a fusion of a benzene ring with a pyrimidine. The first alkaloid isolated of this class was vasicine [81, 84]. *Pyrrole* is a heterocyclic aromatic alkaloid that has an unsaturated five-membered ring. Pyrrole alkaloids are present in nature and are part of the structure of some proteins. Pyrroles are very weak bases, and nicotine is the most-known pyrrole alkaloid [78, 85–87]. *Pyrrolidine* is a derivative of the amino acid ornithine, or arginine and lysine, and is a saturated five-membered cyclic compound with a nitrogen atom. Some alkaloids of this type are putrescine, hygrine, and cuscohygrine [78, 79]. *Pyrrolizidine* structurally has two conjoined five-membered rings with a bridgehead nitrogen atom, consists of a necine base esterified with a necic acid. This group is divided into retronecine, heliotridine, and otonecine [83, 88, 89].

Tropane is derived from the amino acid ornithine, has a complex structure, which is characterized by the N-methyl-8-azabicyclo[3,2,1]octane skeleton. These alkaloids are either esters of tropine or pseudotropine (lesser extent). They can be distinguished into three groups hyoscyamine, scopolamine, and cocaine [83, 90]. *Indole* alkaloids are a group of compounds not homogenous with a bicyclic structure, consisting of a six-membered benzene ring fused to a five-membered nitrogen-containing pyrrole ring, are derived from the amino acid tryptophan. Different subclasses were distinguished as strychnine, brucine, reserpine, vinblastine, vincristine, harmine, serotonin [91]. Imidazols are alkaloids with a planer five-member heterocyclic with 3 carbons and 2 nitrogen atoms, and are derived from

L-histidine. They are classified as aromatic, and some alkaloids of this group are pilocarpine and pilose [73, 79, 92].

2.1.2 Protoalkaloids (Non-heterocyclics)

The protoalkaloids are formed from the amino acids tyrosine, tryptophan, and phenylalanine; however, they differ from true alkaloids by a nitrogen atom that has a side chain of the molecule (extra cyclic). They belong to a small group of alkaloids that could be found in some families of plants like Capparidaceae, Cactaceae, Colchicaceae, and Apocynaceae [73, 76, 93, 94]. Some known protoalkaloids are vinblastine, and vincristine, which are considered as anticancer dimeric alkaloids [94].

2.1.3 Pseudoalkaloids

Pseudoalkaloids are physiologically active compounds that have a nitrogen atom in their structure but are not derived from amino acids. The precursors of pseudoalkaloids are acetate, terpenoids, and aliphatic polyketoacids. Nonetheless, pseudoalkaloids can be derived from amino acids through different reactions of amination and transamination [73, 77, 95]. Pseudoalkaloids are distributed in diverse botanical families like Nymphaeaceae, Solanaceae, Pinaceae, Ranunculaceae, Liliaceae [73, 76, 96]. The terpenoid alkaloids are part of this group, the sources of nitrogen in their structure are methylamine, ethylamine, and beta-aminoethanol, and its biosynthesis begins with prenyl units, this class have been used traditionally in China, Japan, Russia, Mongolia, and India [97].

3 Alkaloids and Diabetes

3.1 In Vitro Studies

Isolated alkaloids from *Catharanthus roseus* (L.) leaves extracts, namely vindoline I, vindolidine II, vindolicine III, and vindolinine IV were tested by in vitro cell-based methods for their antidiabetic potential. A cytotoxicity assay showed that compounds vindolidine, vindolicine, and vindolinine were not cytotoxic at 25 μg/mL in pancreatic β-TC6 cells. Moreover, the compounds induced glucose uptake in pancreatic cells, and vindolicine showed the highest potential. The alkaloid molecules might be target-specific because, on the other hand, compounds vindolidine, vindolicine, and vindolinine showed protein tyrosine phosphatase-1B inhibitory activity. The authors, nonetheless, did not determine the IC_{50} values [98].

The canthinone alkaloids picrasidine L, 3,4-dimethyl-canthin-5,6-dione, 4-ethyl-3-methyl-canthin-5,6-dione, eurycomine E, 5-methoxy-canthin-6-one, and 5-acethoxy-canthin-6-one are inhibitors of the protein tyrosine phosphatase 1B

(PTP1B). Thus the authors claim they are molecules with potential antidiabetic effect [99]. The authors used an in vitro approach to evaluate the inhibitory mode of action of these alkaloids and found that picrasidine L has a competitive and selective inhibitory action with an IC_{50} value of 19.80 μM, while the rest were non-competitive PTP1B inhibitors. Moreover, a cell-based assay with HepG2 cells showed that picrasidine L promotes activity on the insulin signaling pathway.

A standard study performed to evaluate the potential hypoglycemic properties of a molecule is to measure its inhibitory capacity on carbohydrate metabolic enzymes. Moreover, in silico studies can help minimize time and costs in the evaluation of structure interaction of potential inhibitors of α-glucosidase.

In this sense, Rahman et al. [100] recently reported an in silico study, using the MODELLER 9.20. They showed that the alkaloids nummularine-R and vindoline interact with the active sites of α-glucosidase through hydrogen bonds with the N_2 atom of the pentene ring of the ligand. This interaction happens in the regions where miglitol and acarbose also bind. A similar study by Liu et al. [101] reported that mulberry alkaloids, which have shown hypoglycemic properties in phase II/III clinical trials, have selective inhibitory activity towards α-maltase and sucrase with IC_{50} values of 0.06 and 0.03 μg/mL, respectively. The in silico assay showed that *Ramulus mori* alkaloids have a high affinity to the active sites of the enzymes binding through hydrogen bonds, with a lower affinity towards α-amylase. On this subject, the alkaloids berberine, epiberberine, magnoflorine, and coptisine from *Rhizoma coptidis* have shown inhibitory capacity against protein tyrosine phosphatase 1B, with IC_{50} values of 16.43, 24.19, 28.14, and 51.04 μM, respectively. The in silico study with Autodock 4.0 and Fred 2.0 showed that these molecules bind to the amino acid Cys215 in the active sites with high affinity [102].

Also, *Coptis chinensis* alkaloids like berberine, epiberberine, coptisine, palmatine, and magnoflorine showed the ability to inhibit lipid accumulation in 3T3-L1 cells. This was partially attributed to the decreased expression of the adipocyte markers like the proliferator-activated receptor γ and CCAAT/enhancer-binding protein-α [103]. Cell-based evidence from HepG2 cells has shown that the indoloquinoline alkaloid isocryptolepine isolated from *Cryptolepis sanguinolenta* promotes low lipid lipoprotein (LDL) uptake by 1.85 fold [104]. This research group also showed that the indole alkaloid 3R-3,14-dihydroangustoline has activity promoting LDL uptake by 1.26-fold [105]. Furthermore, the cyclopeptide alkaloids nummularine-R, nummularin-C, and hemsine-A isolated from Ziziphus oxyphylla have potent in vitro α-glucosidase inhibitory rate with IC_{50} values of 212.1, 215.1, and 394.0 μM. The authors also showed that these compounds are not cytotoxic in PC-3 cell lines at the tested concentrations [51].

3.2 *In Vivo Studies*

Alkaloid-rich extracts from barks of *Litsea glutinosa* at concentrations of 50, 100, and 200 mg/kg were orally administrated to ob/ob mice for 4 weeks. The treatment was significant at concentrations of 100 and 200 mg/kg, where the reduction of the parameters of serum levels of fasting glucose, glycosylated hemoglobin, and glycosylated serum protein was observed. Also, the treatment with alkaloid-rich extracts increased insulin sensitivity and improved oral glucose tolerance. In this study, eight alkaloids were identified, namely laurelliptine, 6-isoquinolinol, laurolitsine, isoboldine, N-methyl laurolitsine, laurolitsine, boldine, and litseglutine A [106].

The alkaloid-rich fraction from *Capparis decidua* was evaluated against diabetic mice at a concentration of 50 mg/kg for 28 d showed a potential antidiabetic and anti-hyperlipidemic effect through various mechanisms. This particular study reports that there was no sign of hematological toxicity caused by the administered doses of the alkaloid-rich fraction. Alkaloid treatment decreased the total cholesterol and triglyceride content, the activity of glucose-6-phosphatase and the expression of different genes was downregulated (G6Pase, PECK, aldose reductase, and TNF-α) [107]. An in vivo study reported by Li et al. [108] in type-2 diabetic rats with *Coptis chinensis* alkaloids showed that these molecules have a protective effect on the cognitive deficits of diabetic rats. The authors attribute to the ability of these alkaloids to reverse the loss of neurons and Aβ deposition and might regulate the IRS1/PI3K/Akt insulin signaling and GSK3β activity.

Furthermore, doses of 5 mg/kg of steroidal alkaloids from *Sarcococca saligna,* namely sarcovagine-D, saracodine, and holaphylline, decreased glucose levels in blood in streptozotocin-induced diabetic rats. Rats treated with holaphylline and sarcovagine-D additionally showed improvement in blood lipids but no significant effect in weight loss. The reduction of the blood lipid content as a result of *S. saligna* alkaloid treatment might indicate a potential effect on diabetes and some of its comorbidities [109].

4 Conclusions

Alkaloids have been of interest to research since ancient times due to their ethnopharmacological properties. Current research has shown that these molecules have the potential to be used as biopharmaceuticals against many non-communicable diseases like diabetes. However, during the data research, the lack of in vivo and clinical studies was evident. Studies mainly focused on alkaloids from different sources that have shown potential in vitro biopharmaceutical effect on diabetes and some of its complications. Up to date, most studies have focused on in vitro, in silico, and in vivo approaches. However, preclinical and clinical studies with their pharmacokinetic and toxicology reports are still needed to consider alkaloids as safe and efficient biopharmaceutical agents.

References

1. IDF (2019) Diabetes atlas, 9th edn. International Diabetes Federation
2. OMS (2018) Diabetes fact-sheets. World Health Organization. https://www.who.int/es/news-room/fact-sheets/detail/diabetes. Accessed 18 Feb 2020
3. Ozougwu J, Obimba K, Belonwu C, Unakalamba C (2013) The pathogenesis and pathophysiology of type 1 and type 2 diabetes mellitus. J Physiol Pathophysiol 4(4):46–57. https://doi.org/10.5897/JPAP2013.0001
4. Van Belle TL, Coppieters KT, Von Herrath MG (2011) Type 1 diabetes: etiology, immunology, and therapeutic strategies. Physiol Rev 91(1):79–118. https://doi.org/10.1152/physrev.00003.2010
5. Lehuen A, Diana J, Zaccone P, Cooke A (2010) Immune cell crosstalk in type 1 diabetes. Nat Rev Immunol 10(7):501–513. https://doi.org/10.1038/nri2787
6. Maahs DM, West NA, Lawrence JM, Mayer-Davis E (2010) Epidemiology of type 1 diabetes. Endocrinol Metabol Clin North America 39(3):481–497. https://doi.org/10.1016/j.ecl.2010.05.011
7. AMD ADA (2017) 2. Classification and diagnosis of diabetes. J Diab Care 40:S11–S24. https://doi.org/10.2337/dc17-S005
8. Pixley JS (2018) Mesenchymal stem cells to treat type 1 diabetes. Biochim Biophys Acta Mol basis Dis:165315. https://doi.org/10.1016/j.bbadis.2018.10.033
9. Bluestone JA, Herold K, Eisenbarth G (2010) Genetics, pathogenesis and clinical interventions in type 1 diabetes. Nature 464(7293):1293–1300. https://doi.org/10.1038/nature08933
10. Atkinson MA, Eisenbarth GS, Michels AW (2014) Type 1 diabetes. J Lancet 383 (9911):69–82. https://doi.org/10.1016/S0140-6736(13)60591-7
11. Sonia TA, Sharma CP (2014) Oral delivery of insulin. Elsevier, Amsterdam. https://doi.org/10.1533/9781908818683.1
12. Pals RAS, Skinner T, Velasco ER, Grabowski D (2019) The role of theories in interventions targeting preteens with type 1 diabetes: a critical literature review. Child Care Health Dev 46 (2):155–174. https://doi.org/10.1111/cch.12730
13. Faselis C, Katsimardou A, Imprialos K, Deligkaris P, Kallistratos M, Dimitriadis K (2020) Microvascular complications of type 2 diabetes mellitus. Curr Vasc Pharmacol 18 (2):117–124. https://doi.org/10.2174/1570161117666190502103733
14. Chatterjee S, Khunti K, Davies MJ (2017) Type 2 diabetes. Lancet 389(10085):2239–2251. https://doi.org/10.1016/S0140-6736(17)30058-2
15. Kohei K (2010) Pathophysiology of type 2 diabetes and its treatment policy. Jpn Med Assoc J 53(1):41–46
16. Olokoba AB, Obateru OA, Olokoba LB (2012) Type 2 diabetes mellitus: a review of current trends. Oman Med J 27(4):269. https://doi.org/10.5001/omj.2012.68
17. Donath MY, Shoelson SE (2011) Type 2 diabetes as an inflammatory disease. Nat Rev Immunol 11(2):98–107. https://doi.org/10.1038/nri2925
18. Lin Y, Sun Z (2010) Current views on type 2 diabetes. J Endocrinol 204(1):1. https://doi.org/10.1677/JOE-09-0260
19. Wilkinson MJ, Zadourian A, Taub PR (2019) Heart failure and diabetes mellitus: defining the problem and exploring the interrelationship. Am J Cardiol 124:S3–S11. https://doi.org/10.1016/j.amjcard.2019.10.024
20. McCarthy MI (2010) Genomics, type 2 diabetes, and obesity. N Engl J Med 363 (24):2339–2350. https://doi.org/10.1056/NEJMra0906948
21. Habtemariam S (2019) Medicinal foods as potential therapies for Type-2 diabetes and associated diseases: the chemical and pharmacological basis of their action. Academic Press, pp 89–106. https://doi.org/10.1016/B978-0-08-102922-0.00005-5
22. Frandsen CS, Dejgaard TF, Madsbad S (2016) Non-insulin drugs to treat hyperglycaemia in type 1 diabetes mellitus. Lancet Diab Endocrinol 4(9):766–780. https://doi.org/10.1016/S2213-8587(16)00039-5

23. Monami M, Genovese S, Mannucci E (2013) Cardiovascular safety of sulfonylureas: a meta-analysis of randomized clinical trials. Diabetes Obes Metab 15(10):938–953. https://doi.org/10.1111/dom.12116
24. Sola D, Rossi L, Schianca GPC, Maffioli P, Bigliocca M, Mella R, Corlianò F, Fra GP, Bartoli E, Derosa G (2015) Sulfonylureas and their use in clinical practice. Arch Med Sci 11 (4):840. https://doi.org/10.5114/aoms.2015.53304
25. Thulé PM, Umpierrez G (2014) Sulfonylureas: a new look at old therapy. Curr Diab Rep 14 (4):473. https://doi.org/10.1007/s11892-014-0473-5
26. Deusenberry CM, Coley KC, Korytkowski MT, Donihi AC (2012) Hypoglycemia in hospitalized patients treated with sulfonylureas. Pharmacotherapy 32(7):613–617. https://doi.org/10.1002/j.1875-9114.2011.01088.x
27. Douros A, Yin H, Yu OHY, Filion KB, Azoulay L, Suissa S (2017) Pharmacologic differences of sulfonylureas and the risk of adverse cardiovascular and hypoglycemic events. Diabetes Care 40(11):1506–1513. https://doi.org/10.2337/dc17-0595
28. Perla V, Jayanty SS (2013) Biguanide related compounds in traditional antidiabetic functional foods. Food Chem 138(2–3):1574–1580. https://doi.org/10.1016/j.foodchem.2012.09.125
29. Bridges HR, Jones AJ, Pollak MN, Hirst J (2014) Effects of metformin and other biguanides on oxidative phosphorylation in mitochondria. Biochem J 462(3):475–487. https://doi.org/10.1042/BJ20140620
30. Foretz M, Guigas B, Bertrand L, Pollak M, Viollet B (2014) Metformin: from mechanisms of action to therapies. Cell Metab 20(6):953–966. https://doi.org/10.1016/j.cmet.2014.09.018
31. Graham GG, Punt J, Arora M, Day RO, Doogue MP, Duong J, Furlong TJ, Greenfield JR, Greenup LC, Kirkpatrick CM (2011) Clinical pharmacokinetics of metformin. Clin Pharmacokinet 50(2):81–98. https://doi.org/10.2165/11534750-000000000-00000
32. Tolman KG (2011) The safety of thiazolidinediones. Expert Opin Drug Saf 10(3):419–428. https://doi.org/10.1517/14740338.2011.534982
33. Soccio RE, Chen ER, Lazar MA (2014) Thiazolidinediones and the promise of insulin sensitization in type 2 diabetes. Cell Metab 20(4):573–591. https://doi.org/10.1016/j.cmet.2014.08.005
34. Divakaruni AS, Wiley SE, Rogers GW, Andreyev AY, Petrosyan S, Loviscach M, Wall EA, Yadava N, Heuck AP, Ferrick DA (2013) Thiazolidinediones are acute, specific inhibitors of the mitochondrial pyruvate carrier. Proc Natl Acad Sci 110(14):5422–5427. https://doi.org/10.1073/pnas.1303360110
35. Krishnaswami A, Ravi-Kumar S, Lewis JM (2010) Thiazolidinediones: a 2010 perspective. Perm J 14(3):64
36. Elte J, Blickle J (2011) Reprint of: thiazolidinediones for the treatment of type 2 diabetes. Eur J Intern Med. https://doi.org/10.1016/j.ejim.2011.08.002
37. Li X, Liu Z (2017) Pharmacogenetic factors that affect drug metabolism and efficacy in type 2 diabetes mellitus. In: Xie W (ed) Drug metabolism in diseases. Elsevier, pp 157–179. https://doi.org/10.1016/B978-0-12-802949-7.00007-9
38. Srinivasan S, Yee SW, Giacomini KM (2018) Pharmacogenetics of antidiabetic drugs. In: Advances in pharmacology, vol 83. Elsevier, Amsterdam, pp 361–389. https://doi.org/10.1016/bs.apha.2018.04.005
39. Varghese S (2017) Noninsulin diabetes medications. Nurs Clin N Am 52(4):523–537. https://doi.org/10.1016/j.cnur.2017.07.003
40. Derosa G, Maffioli P (2012) α-Glucosidase inhibitors and their use in clinical practice. Arch Med Sci 8(5):899. https://doi.org/10.5114/aoms.2012.31621
41. Joshi SR, Standl E, Tong N, Shah P, Kalra S, Rathod R (2015) Therapeutic potential of α-glucosidase inhibitors in type 2 diabetes mellitus: an evidence-based review. Expert Opin Pharmacother 16(13):1959–1981. https://doi.org/10.1517/14656566.2015.1070827
42. Shirakawa J, Fujii H, Ohnuma K, Sato K, Ito Y, Kaji M, Sakamoto E, Koganei M, Sasaki H, Nagashima Y (2011) Diet-induced adipose tissue inflammation and liver steatosis are

prevented by DPP-4 inhibition in diabetic mice. Diabetes 60(4):1246–1257. https://doi.org/10.2337/db10-1338
43. Gerich J (2010) DPP-4 inhibitors: what may be the clinical differentiators? Diab Res Clin Pract 90(2):131–140. https://doi.org/10.1016/j.diabres.2010.07.006
44. Shah Z, Pineda C, Kampfrath T, Maiseyeu A, Ying Z, Racoma I, Deiuliis J, Xu X, Sun Q, Moffatt-Bruce S (2011) Acute DPP-4 inhibition modulates vascular tone through GLP-1 independent pathways. Vasc Pharmacol 55(1–3):2–9. https://doi.org/10.1016/j.vph.2011.03.001
45. Dicker D (2011) DPP-4 inhibitors: impact on glycemic control and cardiovascular risk factors. Diabetes Care 34(Supplement 2):S276–S278. https://doi.org/10.2337/dc11-s229
46. Zettl H, Schubert-Zsilavecz M, Steinhilber D (2010) Medicinal chemistry of incretin mimetics and DPP-4 inhibitors. ChemMedChem Chem Enabl Drug Disc 5(2):179–185. https://doi.org/10.1002/cmdc.200900448
47. Hansen KB, Vilsbøll T, Knop FK (2010) Incretin mimetics: a novel therapeutic option for patients with type 2 diabetes–a review. Diab Metabol Synd Obes Targets Ther 3:155. https://doi.org/10.2147/DMSOTT.S7004
48. Dowarah J, Singh VP (2020) Anti-diabetic drugs recent approaches and advancements. Bioorg Med Chem:115263. https://doi.org/10.1016/j.bmc.2019.115263
49. Chao EC, Henry RR (2010) SGLT2 inhibition—a novel strategy for diabetes treatment. Nat Rev Drug Discov 9(7):551–559. https://doi.org/10.1038/nrd3180
50. Kalra S (2014) Sodium glucose co-transporter-2 (SGLT2) inhibitors: a review of their basic and clinical pharmacology. Diab Ther 5(2):355–366. https://doi.org/10.1007/s13300-014-0089-4
51. Choudhury H, Pandey M, Hua CK, Mun CS, Jing JK, Kong L, Ern LY, Ashraf NA, Kit SW, Yee TS (2018) An update on natural compounds in the remedy of diabetes mellitus: a systematic review. J Tradit Complement Med 8(3):361–376. https://doi.org/10.1016/j.jtcme.2017.08.012
52. Borokini TI, Omotayo FO (2012) Phytochemical and ethnobotanical study of some selected medicinal plants from Nigeria. J Med Plants Res 6(7):1106–1118. https://doi.org/10.5897/JMPR09.430
53. Gutierrez-Grijalva EP, Criollo-Mendoza MS, Cabanillas-Bojorquez LA, Heredia JB (2020) Mexican medicinal plants with anti-cancer properties. In: Heredia JB, Gutiérrez-Grijalva EP, López-Martínez LX (eds) Phytochemicals from Mexican medicinal plants: potential biopharmaceuticals against noncommunicable diseases. Nova Science Publishers, New York, pp 229–262
54. Iid II, Kumar S, Shukla S, Kumar V, Sharma R (2020) Putative antidiabetic herbal food ingredients: Nutra/functional properties, bioavailability and effect on metabolic pathways. Trends Food Sci Technol. https://doi.org/10.1016/j.tifs.2020.01.017
55. Hung H-Y, Qian K, Morris-Natschke SL, Hsu C-S, Lee K-H (2012) Recent discovery of plant-derived anti-diabetic natural products. Nat Prod Rep 29(5):580–606. https://doi.org/10.1039/c2np00074a
56. Li W, Zheng H, Bukuru J, De Kimpe N (2004) Natural medicines used in the traditional Chinese medical system for therapy of diabetes mellitus. J Ethnopharmacol 92(1):1–21. https://doi.org/10.1016/j.jep.2003.12.031
57. Agrawal R, Sethiya NK, Mishra S (2013) Antidiabetic activity of alkaloids of Aerva lanata roots on streptozotocin-nicotinamide induced type-II diabetes in rats. J Pharm Biol 51(5):635–642. https://doi.org/10.3109/13880209.2012.761244
58. Baldeón ME, Castro J, Villacrés E, Narváez L, Fornasini M (2012) Hypoglycemic effect of cooked Lupinus mutabilis and its purified alkaloids in subjects with type-2 diabetes. Nutr Hosp 27(4):1261–1266. https://doi.org/10.3305/nh.2012.27.4.5761
59. Kumar S, Narwal S, Kumar V, Prakash O (2011) α-Glucosidase inhibitors from plants: a natural approach to treat diabetes. Pharmacogn Rev 5(9):19. https://doi.org/10.4103/0973-7847.79096

60. Zhang T-T, Jiang J-G (2012) Active ingredients of traditional Chinese medicine in the treatment of diabetes and diabetic complications. Expert Opin Investig Drugs 21 (11):1625–1642. https://doi.org/10.1517/13543784.2012.713937
61. Talapatra SK, Talapatra B (2015) Alkaloids. General introduction. In: Chemistry of plant natural products: stereochemistry, conformation, synthesis, biology, and medicine. Springer, Berlin, pp 717–724. https://doi.org/10.1007/978-3-642-45410-3_15
62. Roberts MF (2013) Alkaloids: biochemistry, ecology, and medicinal applications. Springer, Berlin
63. Goyal S (2013) Ecological role of alkaloids. In: Natural products: phytochemistry, botany and metabolism of alkaloids, phenolics and terpenes, pp 149–171
64. Perviz S, Khan H, Pervaiz A (2016) Plant alkaloids as an emerging therapeutic alternative for the treatment of depression. Front Pharmacol 7:–28. https://doi.org/10.3389/fphar.2016.00028
65. Yang L, Stöckigt J (2010) Trends for diverse production strategies of plant medicinal alkaloids. Nat Prod Rep 27(10):1469–1479
66. Buckingham J, Baggaley KH, Roberts AD, Szabo LF (2010) Dictionary of alkaloids, 2nd edn. CRC Press
67. Matsuura H, Fett-Neto A (2015) Plant alkaloids: main features, toxicity, and mechanisms of action:1–15. https://doi.org/10.1007/978-94-007-6728-7_2-1
68. García AÁ, Carril EP-U (2011) Metabolismo secundario de plantas. Reduca (biología) 2(3)
69. Srivastava S, Srivastava A (2013) Biotechnology and genetic engineering for alkaloid production. In: Ramawat GK, Mérillon JM (eds) Natural products: phytochemistry, botany and metabolism of alkaloids, phenolics and terpenes. Springer, Berlin, pp 213–250
70. Pelletier SW (1983) The nature and definition of an alkaloid. In: Pelletier SW (ed) Alkaloids: chemical biological perspectives, vol 7. Springer, New York, pp 1–31
71. Cushnie TT, Cushnie B, Lamb AJ (2014) Alkaloids: an overview of their antibacterial, antibiotic-enhancing and antivirulence activities. Int J Antimicrob Agents 44(5):377–386
72. Wink M (2010) Annual plant reviews, functions and biotechnology of plant secondary metabolites, vol 39. Wiley
73. Aniszewski T (2015) Alkaloids: chemistry, biology, ecology, and applications. Elsevier, Amsterdam
74. Bribi N (2018) Pharmacological activity of alkaloids: a review. Asian J Bot 1
75. Cooper R, Nicola G (2014) Natural products chemistry: sources, separations and structures. CRC Press
76. Henning CP (2013) Compuestos secundarios nitrogenados: alcaloides. Productos Naturales Vegetales, p 18
77. Ranjitha D, Sudha K (2015) Alkaloids in foods. Int J Pharm Chem Biol Sci 5(4)
78. Ji Ram V, Sethi A, Nath M, Pratap R (2019) Chapter 5 - Five-membered heterocycles. In: Ji Ram V, Sethi A, Nath M, Pratap R (eds) The chemistry of heterocycles. Elsevier, Amsterdam, pp 149–478. https://doi.org/10.1016/B978-0-08-101033-4.00005-X
79. Kaur R, Arora S (2015) Alkaloids-important therapeutic secondary metabolites of plant origin. J Crit Rev 2(3):1–8
80. Böttger A, Vothknecht U, Bolle C, Wolf A (2018) Alkaloids. In: Lessons on caffeine, Cannabis & Co. Springer, Berlin, pp 179–203
81. Ji Ram V, Sethi A, Nath M, Pratap R (2019) Chapter 2 - Six-membered heterocycles. In: Ji Ram V, Sethi A, Nath M, Pratap R (eds) The chemistry of heterocycles. Elsevier, Amsterdam, pp 3–391. https://doi.org/10.1016/B978-0-12-819210-8.00002-3
82. Baros S, Karsayová M, Jomová K, Gáspár A, Valko M (2012) Free radical scavenging capacity of Papaver somniferum L. and determination of pharmacologically active alkaloids using capillary electrophoresis. J Microbiol Biotechnol Food Sci 1:725
83. Koleva II, van Beek TA, Soffers AE, Dusemund B, Rietjens IM (2012) Alkaloids in the human food chain–natural occurrence and possible adverse effects. Mol Nutr Food Res 56(1):30–52

84. Hameed A, Al-Rashida M, Uroos M, Ali SA, Arshia IM, KMJEootp K (2018) Quinazoline and quinazolinone as important medicinal scaffolds: a comparative patent review (2011–2016). Exp Opin Ther Patents 28(4):281–297
85. Bauer I, Knölker H-J (2012) Synthesis of pyrrole and carbazole alkaloids. In: Knölker H-J (ed) Alkaloid synthesis. Springer, Berlin, pp 203–253. https://doi.org/10.1007/128_2011_192
86. Bhardwaj V, Gumber D, Abbot V, Dhiman S, Sharma PJRA (2015) Pyrrole: a resourceful small molecule in key medicinal hetero-aromatics. RSC Adv 5(20):15233–15266
87. Estévez V, Villacampa M, Menéndez JC (2014) Recent advances in the synthesis of pyrroles by multicomponent reactions. Chem Soc Rev 43(13):4633–4657. https://doi.org/10.1039/C3CS60015G
88. Schramm S, Köhler N, Rozhon W (2019) Pyrrolizidine alkaloids: biosynthesis, biological activities and occurrence in crop plants. Molecules 24(3):498
89. Kukula-Koch WA, Widelski J (2017) Chapter 9 - Alkaloids. In: Badal S, Delgoda R (eds) Pharmacognosy. Academic Press, Boston, pp 163–198. https://doi.org/10.1016/B978-0-12-802104-0.00009-3
90. Kohnen-Johannsen KL, Kayser O (2019) Tropane alkaloids: chemistry, pharmacology, biosynthesis and production. Molecules 24(4):796. https://doi.org/10.3390/molecules24040796
91. Hamid HA, Ramli ANM, Yusoff MM (2017) Indole alkaloids from plants as potential leads for antidepressant drugs: a mini review. Front Pharmacol 8:96. https://doi.org/10.3389/fphar.2017.00096
92. Shalini K, Sharma PK, Kumar N (2010) Imidazole and its biological activities: a review. J Der Chem Sinica 1(3):36–47
93. Larsson S, Ronsted N (2014) Reviewing colchicaceae alkaloids-perspectives of evolution on medicinal chemistry. Curr Top Med Chem 14(2):274–289
94. Barrales-Cureño HJ (2015) Pharmacological applications and in vitro biotechnological production of anticancer alkaloids of *Catharanthus roseus*. Biotecnol Apl 32:1101–1110
95. Sabapathy PC, Devaraj S (2016) Alkaloids. In: Preethi K (ed) Secondary metabolites. Darshan
96. Debnath B, Singh WS, Das M, Goswami S, Singh MK, Maiti D, Manna K (2018) Role of plant alkaloids on human health: a review of biological activities. Mater Today Chem 9:56–72. https://doi.org/10.1016/j.mtchem.2018.05.001
97. Cherney EC, Baran PS (2011) Terpenoid-alkaloids: their biosynthetic twist of fate and total synthesis. Isr J Chem 51(3–4):391–405. https://doi.org/10.1002/ijch.201100005
98. Tiong HS, Looi YC, Hazni H, Arya A, Paydar M, Wong FW, Cheah S-C, Mustafa RM, Awang K (2013) Antidiabetic and antioxidant properties of alkaloids from Catharanthus roseus (L.) G. Don. Molecules 18(8). https://doi.org/10.3390/molecules18089770
99. Sasaki T, Li W, Higai K, Koike K (2015) Canthinone alkaloids are novel protein tyrosine phosphatase 1B inhibitors. Bioorg Med Chem Lett 25(9):1979–1981. https://doi.org/10.1016/j.bmcl.2015.03.014
100. Rahman N, Muhammad I, Gul EN, Khan H, Aschner M, Filosa R, Daglia M (2019) Molecular docking of isolated alkaloids for possible α-glucosidase inhibition. Biomol Ther 9(10). https://doi.org/10.3390/biom9100544
101. Liu Z, Yang Y, Dong W, Liu Q, Wang R, Pang J, Xia X, Zhu X, Liu S, Shen Z, Xiao Z, Liu Y (2019) Investigation on the enzymatic profile of mulberry alkaloids by enzymatic study and molecular docking. Molecules 24(9). https://doi.org/10.3390/molecules24091776
102. Choi JS, Ali MY, Jung HA, Oh SH, Choi RJ, Kim EJ (2015) Protein tyrosine phosphatase 1B inhibitory activity of alkaloids from Rhizoma Coptidis and their molecular docking studies. J Ethnopharmacol 171:28–36. https://doi.org/10.1016/j.jep.2015.05.020
103. Choi JS, Kim J-H, Ali MY, Min B-S, Kim G-D, Jung HA (2014) Coptis chinensis alkaloids exert anti-adipogenic activity on 3T3-L1 adipocytes by downregulating C/EBP-α and PPAR-γ. Fitoterapia 98:199–208. https://doi.org/10.1016/j.fitote.2014.08.006
104. Aggrey MO, Li H-H, Wang W-Q, Song W, Wang Y, Xuan L-J (2018) Isocryptolepine, an indoloquinoline alkaloid from cryptolepis sanguinolenta promotes LDL uptake in HepG2 cells. Rev Bras 28(6):654–657. https://doi.org/10.1016/j.bjp.2018.08.008

105. Aggrey MO, Li H-H, Wang W-Q, Wang Y, Xuan L-J (2019) Indole alkaloid from Nauclea latifolia promotes LDL uptake in HepG2 cells by inhibiting PCSK9. Phytomedicine 55:264–268. https://doi.org/10.1016/j.phymed.2018.06.041
106. Zhang X, Jin Y, Wu Y, Zhang C, Jin D, Zheng Q, Li Y (2018) Anti-hyperglycemic and anti-hyperlipidemia effects of the alkaloid-rich extract from barks of Litsea glutinosa in Ob/Ob mice. Sci Rep 8(1):12646. https://doi.org/10.1038/s41598-018-30823-w
107. Sharma B, Salunke R, Balomajumder C, Daniel S, Roy P (2010) Anti-diabetic potential of alkaloid rich fraction from Capparis decidua on diabetic mice. J Ethnopharmacol 127 (2):457–462. https://doi.org/10.1016/j.jep.2009.10.013
108. Li JC, Shen XF, Shao JA, Tao MM, Gu J, Li J, Huang N (2018) The total alkaloids from *Coptis chinensis* Franch improve cognitive deficits in type 2 diabetic rats. Drug Des Dev Ther 12:2695–2706. https://doi.org/10.2147/dddt.S171025
109. Ullah Jan N, Ali A, Ahmad B, Iqbal N, Adhikari A, Inayat ur R, Ali A, Ali S, Jahan A, Ali H, Ali I, Ullah A, Musharraf SG (2018) Evaluation of antidiabetic potential of steroidal alkaloid of Sarcococca saligna. Biomed Pharmacother 100:461–466. https://doi.org/10.1016/j.biopha.2018.01.008

Chapter 15
The Role of Alkaloids in the Management of Diabetes Mellitus

Sinmisola Aloko and M. Oluwasesan Bello

Abstract Diabetes mellitus (DM) is a metabolic disorder with a rising prevalence globally. Currently about 400 million people are affected. In 2016, it was the seventh leading cause of death worldwide. Diabetes mellitus is prevalent in most countries. Its major feature is the high level of blood glucose which can be caused by insulin resistance or reduced insulin production. As the disease progresses, complications like diabetic neuropathy, nephropathy, and retinopathy occur. The use of herbs and natural products in the management of diabetes mellitus dates to the prehistoric era. Eighteen studies from about nine countries showed that between 18% and 72% use traditional and complementary medicine in the management of diabetes. In developing countries where the disease is highly prevalent natural products are very common because of ready availability, cheapness, and minimal side effects. This review highlights alkaloids, one of the most important secondary metabolites isolated from plants which have been studied to have anti-diabetes effects. Alkaloids are a group of highly diverse natural products that contain one or more basic nitrogen atoms in a heterocyclic ring. The various plants that contain them, their pharmacological actions studied thus far, mechanism of action will be extensively discussed. The aim of this review is to provide rich knowledge on research carried out thus far on plant alkaloids in the management of diabetes: their pharmacology, chemistry, and effectiveness which will serve as a platform for further research and development of novel drug molecules that can further help in the management of this disease.

Keywords Alkaloids · Diabetes mellitus · Hyperglycemia · Metabolic disorders · Antidiabetic agents · Blood sugar

---

S. Aloko (✉)
Department of Pharmaceutics and Pharmaceutical Technology, Faculty of Pharmacy, University of Lagos, Lagos, Nigeria
e-mail: saloko@unilag.edu.ng

M. O. Bello
Department of Applied Chemistry, Federal University, Dutsin-Ma, Katsina, Nigeria

H. Chen, M. Zhang (eds.), *Structure and Health Effects of Natural Products on Diabetes Mellitus*, https://doi.org/10.1007/978-981-15-8791-7_15

1 Introduction

Alkaloids denote a large group of chemical entities that were defined by Pelletier as thus: cyclic organic compounds that have nitrogen in a negative oxidation state. Chemically they are compounds that have basicity arising from a heterocyclic tertiary nitrogen atom (there are few exceptions like colchicine, xanthines). Alkaloids can be naturally occurring, biomimic, or synthetic based on their origin [1]. They are secondary metabolites that play a role in protection. They have significant biological action. There are about 16,000 known alkaloids and most originate from plants, others originate from algae, microbes, etc. we have the indole alkaloids, the isoquinoline alkaloids, tropane alkaloids, and steroidal alkaloids depending on their backbone. The indole and isoquinoline alkaloids constitute the largest numbers; about 8100 alkaloids belong to these groups [2].

Diabetes mellitus is a metabolic disorder that results in elevated blood sugar levels. It can result from resistance to insulin's action, insufficient production of insulin, glucagon secretion dysfunction. It usually results from a combination of more than one of these conditions. Complications like retinopathy, renal failure, neuropathy may result if poorly managed. According to the World Health Organization, in 2016, an estimated 1.6 million deaths were directly caused by diabetes. In Africa, there is an estimated prevalence of 3.8%, 7.3% in Europe, 10.7% in the Middle East and North America, 11.5% in North America and Caribbean, 9.6% and 9.1% in South America and Southeast Asia, respectively [3].

2 Current Management

Diabetes has been managed with conventional drugs and lifestyle modifications. Biguanides, sulfonylureas, meglitinide derivatives, alpha-glucosidase inhibitors, thiazolidinediones, glucagonlike peptide-1 agonists, insulins are some of the classes of drugs that are currently in use in the management of diabetes mellitus. The medications have been of great help in the management of diabetes mellitus but there are several drawbacks associated with their use. A good example is lactic acidosis risk associated with the use of metformin. Another example is the risk of cardiovascular effects in patients taking sulfonylureas and other insulin secretagogues [4]. Sulfonylureas have been said to lose their effectiveness in 44% of patients who have been taking the drug for more than 6 years. Most of these medications do not also have lipid lowering effects in diabetic patients [5]. Overall, diabetes mellitus is a public health issue that the desired control levels and efficacy have not been achieved.

3 Signal Pathways

Signal pathways involved in diabetes mellitus include protein tyrosine phosphatase 1B (PTP1B), AMPK (5'AMP-activated protein kinase), mitogen-activated protein kinase (MAPK), rapamycin (mTOR), and phosphoinositol 3-kinase (PI3K). These molecular mediators and enzymes are explored as potential targets in the management of diabetes mellitus [6].

Peroxisome proliferator activated receptor agonists, e.g. thiazolidinediones, sulfonylurea derivatives help reduce blood glucose level. Protein tyrosine phosphatase 1B (PTP1B) enzyme is a negative regulator of insulin. PTP1B inhibitors, e.g. sulfonamide derivatives, amide hybrids, phenol derivatives are used as antidiabetics. Aldol reductase is a key enzyme of the polyol pathway. Aldol reductase inhibitors are crucial in the treatment of diabetic complications such as cataract and glaucoma. A good example is epalrestat, some sulfonylurea derivatives, pyrrole and imidazole-based hybrids.

α-glucosidase is a hydrolase enzymc that breaks down carbohydrates. α glucosidase inhibitors like acarbose, flavonoid analogs, xanthone analogs have an antidiabetic effect. Dipeptidyl peptidase IV enzyme is also indicated in the metabolic pathway. Many antidiabetic drugs inhibit this enzyme. Sodium glucose co-transporter inhibitors, G protein coupled receptor agonists all serve as antidiabetic agents [7].

4 Alkaloids and Diabetes

Alkaloids have shown promising antidiabetic effects and with proper development may serve as alternatives or augmentation for current diabetic medications. There are about 50 plant alkaloids available as drugs currently. Examples are pilocarpine (**1**), atropine (**2**), berberine (**3**). Berberine (**3**) is already being marketed as an antidiabetic agent [8]. Naturally occurring compounds like alkaloids may produce less side or untoward effect in the body. They also have the additional benefit of providing lipid lowering effect. Numerous antidiabetic agents can still be produced from the numerous naturally occurring alkaloids with proper research. Many alkaloids were isolated from several medicinal plants and examined with different animal models for their potential antidiabetic function. Several alkaloids with antidiabetic potential are discussed below. All the chemical structures are represented in Fig. 15.1 with numbers in brackets ().

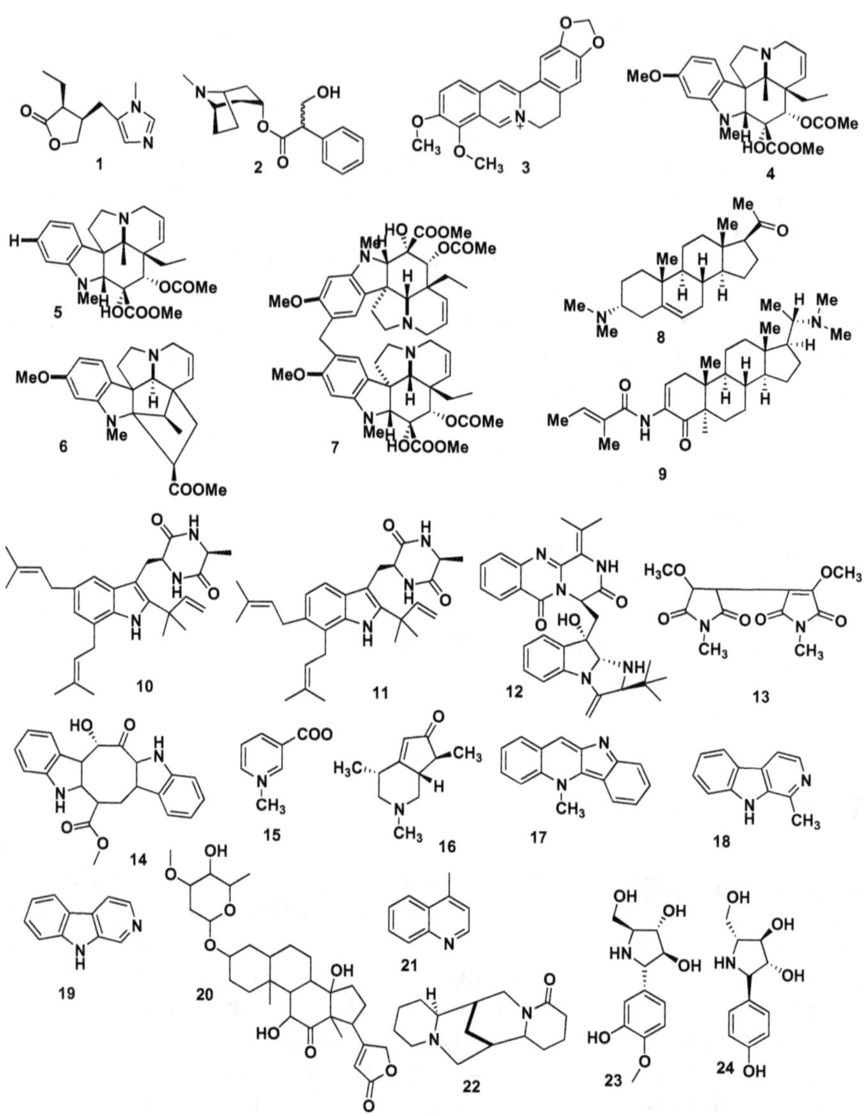

Fig. 15.1 Isolated alkaloids from natural sources

4.1 Alkaloids Part A

Catharanthus roseus: A study by [9] was carried out on its pure isolates; vindoline (**4**), vindolidine (**5**), vindolinine (**6**), and vindolicine (**7**) showed that all four alkaloids induced relatively high glucose uptake in B-TC6 cells and C2C12 cells. Vindolicine (**7**) showed the highest antioxidant and hypoglycemic property through

ORAC, DPPH, ROS production, glucose uptake, and PTP-1B inhibition. All four alkaloids showed far less toxicity than doxorubicin [9]. All four alkaloids are indole alkaloids. The plant is commonly known as cape periwinkle and native to Madagascar. It is the source of vincristine and vinblastine which are anti-neoplastic agents.

Sarcococca saligna: Holaphylline (**8**) and sarcovagine-D (**9**) are steroidal alkaloids isolated from the chloroform extract of the plant. Both alkaloids showed hypoglycemic properties and showed potential against complications associated with diabetes. Glibenclamide was used as a standard in a study carried out using rats. They elevated blood HDL cholesterol levels and improved blood lipids. Previous studies confirmed the same effects with ethyl acetate and petroleum ether extracts of the plant [10]. Antidiabetic effects compared well with that of glibenclamide but were not superior. It has been hypothesized that they cause these effects by preventing diffusion of glucose through the enteric epithelium. This results in a decrease in blood glucose levels [11]. The plant is commonly known as the sweet box and it is native to Pakistan.

Combretum dolichopetalum: Echinulin (**10**) and arestictin B (**11**) are two alkaloids isolated from the roots of the plant and their antidiabetic properties tested using rats. Alloxan monohydrate was used to induce diabetes and glibenclamide was used as a standard antidiabetic agent. The alkaloids were obtained as an isomeric mixture indole-containing diketopiperazine alkaloids in the following ratio: 56:54. Compared to other isolated parts of the plants, the alkaloids mixture produced the most marked and sustained hypoglycemic effect. It produces a 53.7% reduction in blood sugar after just three hours and compared well with the standard drug, glibenclamide which produced a 48.72% reduction in 9 h [12]. The plant genus is commonly called the bushwillows.

Scedosporium apiospermum: This marine fungus contains several fumiquinazoline alkaloids. An alkaloid, scequinadoline D (**12**) showed the highest antidiabetic potential with an EC_{50} value of 0.27 ± 0.03 μM in 3T3-L1 adipocytes cells where its triglyceride promoting activity was tested. The study showed it has potential as an insulin sensitizer [13].

Chrozophora plicata: A pyrrole alkaloid, plicatanin C (**13**) was isolated from the whole plant and it showed the best antidiabetic potential among the isolated compounds. It had an IC_{50} value of 27.8 μM when its α-glucosidase inhibitory effect was tested [14].

Caulerpa racemosa: A green alga from which racemosin C (**14**) was isolated. It is a bisindole alkaloid. The alkaloid showed marked PTP1B inhibitory activity with an IC_{50} value of 5.86 ± 0.57 μM compared to the positive control of oleanolic acid ($IC_{50} = 3.03 \pm 0.20$ μM) [15]. The alga is edible and commonly known as sea grape.

Rhizoma coptidis: Berberine (**3**), an isoquinoline alkaloid isolated from this plant has been established to have antidiabetic properties. It is a benzylisoquinoline alkaloid. It also shows lipid lowering effects. The alkaloid is presently being marketed as an antidiabetic drug after undergoing a series of clinical trials. It is about 5.2–7.7% of the plant. It has been hypothesized to enhance insulin secretion, lower lipid levels, and improve resistance to insulin. It is also said to have an anti-inflammatory mechanism of reducing glycemic levels [16]. Studies have also shown that it has the ability to alleviate platelet aggregation that is potentiated by high glucose levels. It does this by inhibition of the enzymes aldose reductase and NADPH oxidase [17]. The plant is a traditional Chinese herb.

Trigonella foenum-graecum L: This plant is popularly known as fenugreek. The alkaloid trigonelline (**15**) isolated from it has shown remarkable antidiabetic properties. It showed a marked ability to protect against β cell apoptosis which results in peripheral neuropathy in patients with diabetes. In a study where insulin was used as standard and streptozocin was used to induce diabetes in rats, trigonelline decreased blood glucose, increased insulin expression in β cells. It suppressed caspase 3 expression and raised the activities of antioxidant enzymes [18].

4.2 Alkaloids Part B

Tecoma stans L: Three alkaloids were isolated from this plant: boschniakine, 5-hydroxyskitanthine, and tecomine but only tecomine (**16**) had an antidiabetic effect. This plant is from the family Bignoniaceae, it is used traditionally in the Ayurvedic system of medicine as an antihyperglycaemic agent. Tecomine (**16**) showed in vitro effects on glucose uptake in white adipocytes but did not show any effect in animal models. However, it showed a decrease in plasma cholesterol levels in vivo [19]. *Cryptolepis sanguinolenta* (Lindl.) Schltr.: An indoloquinolone alkaloid. Cryptolepine (**17**) was isolated from *C. sanquinolenta,* this compound was assessed for antihyperglycemic activity in vitro and in a non-insulin dependent diabetes mellitus (NIDDM) mouse model. When given orally to diabetic mice, cryptolepine (**17**) reduced the glucose level and decreased the insulin plasma concentration, with evidence of a rise in insulin mediated glucose disposal. Cryptolepine (**17**) improved glucose absorption by employing 3T3-L1 cells [20, 21]. The plant belongs to the family apocynaceae (Fig. 15.2).

Tribulus terrestris L: Two alkaloids were isolated from *Tribulus terrestris* L. (Zygophyllaceae). Harmane (**18**) and norharmane (**19**) which are imidazoline alkaloids. They have a stimulating effect on the secretion of insulin in the pancreatic cell by activating imidazoline binding sites. Harmane (**18**) and norharmane (**19**) increased insulin secretion in two to threefolds in isolated human islets of Langerhans. Harmane invigorates and speeds up the secretion from insulin in a glucose-dependent way [22, 23].

Syzygium cumini L: It is synonymous to *Eugenia jambolana* Lam. belonging to the family Myrtaceae, it is an imperative plant to India's traditional medicine for the management of diabetes mellitus. An alkaloid called Jambosine (**20**) has been reported isolated from the barks, fruits, and seeds of *S. cumin.* It was claimed to halt the conversion of starch to glucose; however, there is the need to ascertain this claim [24].

Lepidium sativum: A member of Cruciferae called *Lepidium sativum* Linn. But is commonly referred to as Garden cress. It has been employed in traditional and folklore medicine for the management of diabetes for so many years. Lepidine

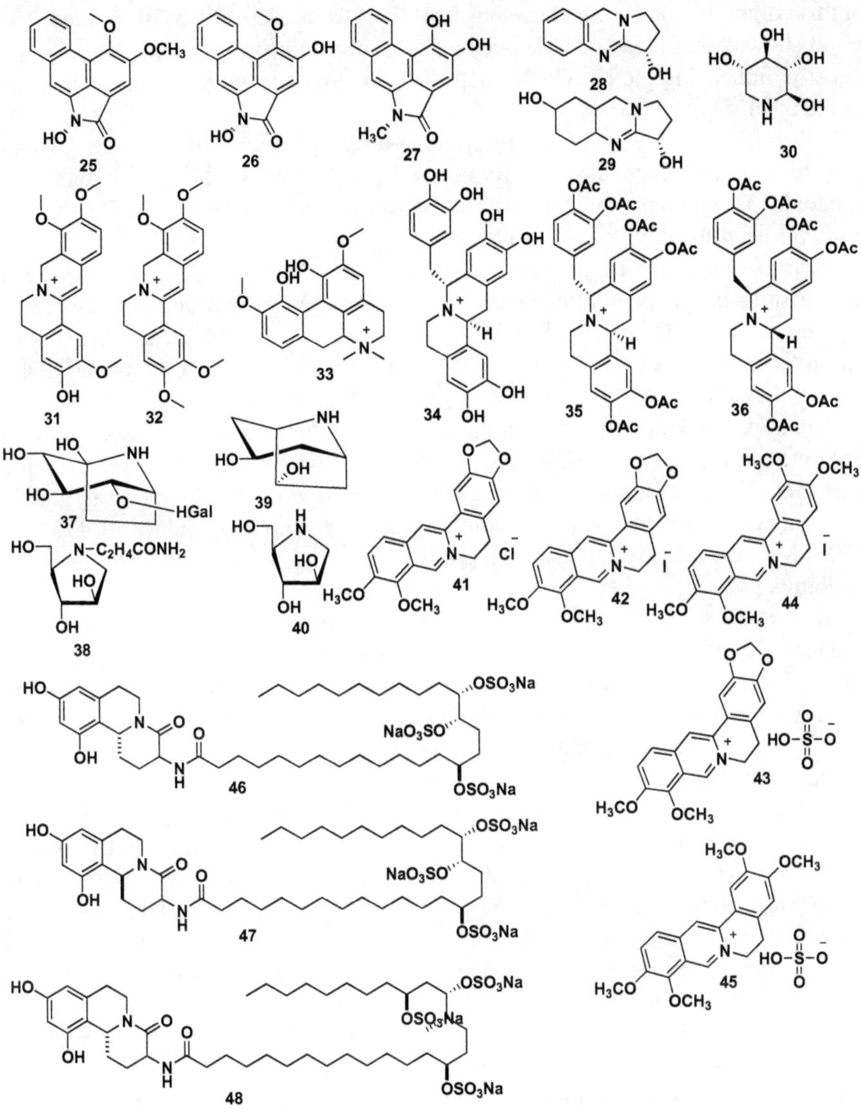

Fig. 15.2 Continuation of structures of isolated structures from natural sources

(**21**) belongs to the usual group of imidazole alkaloids isolated from *Lepidium sativum*. The antidiabetic effect of the seed total alkaloid on alloxan-induced diabetic mice was evaluated for 21 days. The 250 mg/kg dose showed an antidiabetic effect. It may be attributed to the elimination of oxidative stress and the regulation of antioxidant enzymes [25].

Lupanine is quinolizidine alkaloid isolated from *Sophora flavescens* Aiton. Wiedemann et al. [26] investigated the in vitro and in vivo antihyperglycemic effect of this compound on insulin-secreting INS-1E cells or islets of C57BL/6 mice. The result showed that lupanine (**22**) potentiates insulin (glucose-stimulated) release by directly influencing the KATP channels. The alkaloid also increases the expression of the insulin gene [26].

Lobelia chinensis Lour: Two pyrrolidine alkaloids were isolated from *Lobelia chinensis* Lour (Campanulaceae). Radicamines A (**23**) and B (**24**) are polyhydroxyl alkaloids. These compounds are said to be α-glucosidase inhibitors [27]. There is however need for in vivo tests to ascertain the claim.

Piper umbellatum: Three alkaloids isolated from the parts of *Piper umbellatum* are aristolactams: piperumbellactam A (**25**), piperumbellactam B (**26**), and piperumbellactam C (**27**). They displayed a significant α-glucosidase enzyme inhibition with IC_{50} values 98.07 ± 0.44, 43.80 ± 0.56, and 29.64 ± 0.46, respectively [28].

Adhatoda vasica Nees: Methanolic extract of *Adhatoda vasica* Nees has been examined in rat intestinal α-glucosidase screening experiments. Vasicine (**28**) and vasicinol (**29**) were isolated from the extract of *Adhatoda vasica* by bioassay-guided fractionation. They had glucose inhibitory activity with IC_{50} values of 125 and 250 μM, respectively. These compounds are said to be reversible sucrase inhibitors [29].

Morus alba: An alkaloid was isolated among other compounds from the barks and leaves of *Morus alba*. 1-deoxynojirimycin (**30**) is a polyhydroxylated piperidine alkaloid present in both leaves and bark of *M. alba*. This compound is said to be an effective α-glucosidase inhibitor with activity like radicamines A (**23**) and B (**24**) [30, 31]. Three new and previously known polyhydroxylated alkaloids were also isolated from *Morus alba* L. These alkaloids are 4-*O*-*R*-*D*-galactopyranosyl-calystegine B2 (**37**), (2R,3R,4R)-2-hydroxymethyl-3,4-dihydroxy-pyrrolidine-N-propionamide (**38**), 3,6-dihydroxynortropane (**39**), and 1,4-dideoxy-1,4-imino-D-arabinotol (the known compound) (**40**). They were isolated and structures elucidated using chromatographic and spectroscopic techniques are available. The study showed that compound **40** displayed a significant inhibitory activity towards yeast R-glucosidase and mammalian isomaltase. Compound **40** also showed an inhibitory effect towards mammalian Rhamnosidase and porcine kidney trehalase. The study further revealed that *N*-alkylation of compound **40** reduces or stops inhibition of glycosidase. R-glucosidase inhibitors decrease diet-induced hyperglycemia and endogenous secretion from insulin by hindering R-glucosidase in the intestines [32].

Tinospora cordifolia: The alkaloids jatrorrhizine (**31**), palmatine (**32**), magnoflorine (**33**) were isolated from the plant species called *Tinospora cordifolia*. These compounds were investigated in vitro for antihyperglycemic effects using insulin-secreting RINm5F cell line. The study concluded that these compounds were able to activate insulin secretion from RINm5F cell line [33].

Talinum paniculatum Gaertner: Family Portulacaceae. Three quinolizidine alkaloids were isolated from the roots of *T. paniculatum*, these compounds are javaberine A (**34**), javaberine A hexaacetate (**35**), and javaberine B hexaacetate (**36**). They are known as good inhibitors of TNF-a production by macrophages and fat cells [34, 35].

Coptis japonica: Five isoquinoline alkaloids were isolated from the root of *Coptis japonica* makino var. dissecta nakai from the family Ranunculaceae. The inhibitory effects of these compounds against aldose reductase were evaluated. The alkaloids are: berberine chloride (**41**), berberine iodide (**42**), berberine sulfate (**43**), palmatine iodide (**44**), palmatine sulfate (**45**). Compounds **41**, **42,** and **43** gave IC_{50} of 13.98, 13.45, and 32.84 nM, respectively, while compounds **44** and **45** gave IC_{50} of 51.78 and 68.0 nM, respectively [36].

Marine sponges: From a marine sponge named Penares schulzei, three isoquinoline alkaloids Schulzeines A (**46**), B (**47**), and C (**48**) [37] were isolated. The study claimed that these compounds gave a significant inhibitory activity on α-glucosidase with IC_{50} values of 48–170 nM [37].

5 Recent Advances

The alkaloids discussed in part A were tested for antidiabetic potential between 2010 and 2020. It was observed that the alkaloids were tested in in vivo or in vitro models. The researchers also took a step further in determining the signal pathway and mechanism of antidiabetic action. In some of the articles, the alkaloids were compared with well-known antidiabetic agents. A good example is the alkaloid holaphylline, sarcovagine-D is another alkaloid that was compared with glibenclamide, a renowned antidiabetic agent [11]. These make it easier to carry out further research on such alkaloids and extrapolate their use to humans in the management of diabetes mellitus.

Older publications (part B) also highlighted numerous alkaloids with good antidiabetic properties. These alkaloids can also translate to new antidiabetic agents; however, it is important to use animal models where it has not been used. This is bearing in mind that in vitro effects do not always translate to in vivo effect. There is also the need to state the effective dose, compare with standard antidiabetic agents, propose a signal pathway to make further research easier.

Overall, several alkaloids had the ability to inhibit α glucosidase and aldose reductase enzymes, could alkaloids be a lead for the production of new α glucosidase/aldose reductase inhibitors?

6 Conclusion

Diabetes is probably one of the fastest growing metabolic diseases globally and as awareness of this disorder's complexity grows, so does the need for more effective treatments. Medicinal plants and herbs products are employed worldwide for diabetes control, as they are conveniently and easily available. Scientific analysis of many plant species has shown the efficacy of the herbs in controlling diabetes through various mechanisms.

Alkaloids undoubtedly have significant antidiabetic properties from studies carried out so far. This review highlights diverse isolated natural alkaloids, their structures, signal pathways, doses, and how well the alkaloids compare with known antidiabetic agents where it was documented. α glucosidase and aldose reductase inhibition appeared to be quite a common feature of alkaloids. Some alkaloids also have lipid lowering effect, antioxidant properties which are beneficial.

It is intended to be expository and shows gray research areas that can be exploited in the management of diabetes. The end goal is to transform the alkaloids to drug candidates that can augment or produce better results. It also has fewer side effects on the growing diabetic population.

Conflicts of Interest No conflicts of interest among the authors.

Funding This study receives no funding or grant.

References

1. Aniszewski T (2015) Introduction. In: Alkaloids. Elsevier, Amsterdam, pp 1–20
2. Verpoorte R (2000) Alkaloids | liquid chromatography. In: Poole C, Cooke M (eds) Encyclopedia of separation science. Academic Press, Cambridge, MA, pp 1949–1956
3. Fan W (2017) Epidemiology in diabetes mellitus and cardiovascular disease. Cardiovasc Endocrinol 6:8–16
4. Mendez CE, Umpierrez GE (2014) Pharmacotherapy for hyperglycemia in noncritically ill hospitalized patients. Diabetes Spectr 27:180–188. https://doi.org/10.2337/diaspect.27.3.180
5. Asadi-Samani M, Kooti W, Farokhipour M et al (2016) The role of medicinal plants in the treatment of diabetes: a systematic review. Electron Physician 8(1):1832–1842. https://doi.org/10.19082/1832
6. Fernando IPS, Ryu BM, Ahn G et al (2020) Therapeutic potential of algal natural products against metabolic syndrome: a review of recent developments. Trends Food Sci Technol 97:286–299
7. Dowarah J, Singh VP (2020) Anti-diabetic drugs recent approaches and advancements. Bioorganic Med Chem 28:115263
8. Debnath B, Singh S, Das M et al (2018) Role of plant alkaloids on human health: a review of biological activities. Mater Today 9:56–72. https://doi.org/10.1016/j.mtchem.2018.05.001
9. Tiong SH, Looi CY, Hazni H et al (2013) Antidiabetic and antioxidant properties of alkaloids from Catharanthus roseus (L.) G. Don. Molecules 18:9770–9784. https://doi.org/10.3390/molecules18089770
10. Ullah Jan N, Ali A, Ahmad B et al (2018) Evaluation of antidiabetic potential of steroidal alkaloid of Sarcococca saligna. Biomed Pharmacother 100:461–466. https://doi.org/10.1016/j.biopha.2018.01.008
11. Sah SP, Sah ML, Juyal V, Pandey S (2010) Hypoglycemic activity of aqueous extract of *Urtica parviflora* roxb. in normoglycemic rats. Int J Phytomed 2:47–51
12. Uzor PF, Osadebe PO (2016) Antidiabetic activity of the chemical constituents of Combretum dolichopetalum root in mice. EXCLI J 15:290–296. https://doi.org/10.17179/excli2016-252
13. Li CJ, Chen PN, Li HJ et al (2020) Potential antidiabetic fumiquinazoline alkaloids from the marine-derived fungus scedosporium apiospermum F41-1. J Nat Prod 83(4):1082–1091. https://doi.org/10.1021/acs.jnatprod.9b01096

14. Tabussum A, Riaz N, Saleem M et al (2013) α-Glucosidase inhibitory constituents from Chrozophora plicata. Phytochem Lett 6:614–619. https://doi.org/10.1016/j.phytol.2013.08.005
15. Yang H, Liu DQ, Liang TJ et al (2014) Racemosin C, a novel minor bisindole alkaloid with protein tyrosine phosphatase-1B inhibitory activity from the green alga Caulerpa racemosa. J Asian Nat Prod Res 16:1158–1165. https://doi.org/10.1080/10286020.2014.965162
16. Dong H, Wang N, Zhao L, Lu F (2012) Berberine in the treatment of type 2 Diabetes mellitus: a systemic review and meta-analysis. Evid Based Complement Alternat Med 2012:591654. https://doi.org/10.1155/2012/591654
17. Paul M, Hemshekhar M, Kemparaju K, Girish KS (2019) Berberine mitigates high glucose-potentiated platelet aggregation and apoptosis by modulating aldose reductase and NADPH oxidase activity. Free Radic Biol Med 130:196–205. https://doi.org/10.1016/j.freeradbiomed.2018.10.453
18. Liu L, Du X, Zhang Z, Zhou J (2018) Trigonelline inhibits caspase 3 to protect β cells apoptosis in streptozotocin-induced type 1 diabetic mice. Eur J Pharmacol 836:115–121. https://doi.org/10.1016/j.ejphar.2018.08.025
19. Costantino L, Laura R, Renato P, Tiziana B, Pompeo P, Fabio G (2003) Isolation and pharmacological activities of the Tecoma stans alkaloids. Il Farmaco 9:781–785
20. Luo J, Fort DM, Carlson TJ, Noamesi BK, Nii-Amon-Kotei D, King SR, Tsai J, Quan J, Hobensack C, Lapresca P, Waldeck N, Mendez CD, Jolad SD, Bierer DE, Reaven GM (1998) Cryptolepis sanguinolenta: an ethnobotanical approach to drug discovery and the isolation of a potentially useful new antihyperglycaemic agent. Diabet Med 15(5):367–374
21. Nadkarni KM (1992) Dr. K. M. Nadkarni's Indian materia medica, with Ayurvedic, Unani-Tibbi, Siddha, Allopathic, Homeopathic, Naturopathic & home remedies, appendices & indexes. In: Nadkarni KM (ed) The Indian plants and drugs. Indian materia medica, vol 2. Popular Prakashan, Bombay
22. Cooper EJ, Hudson AL, Parker CA, Morgan NG (2003) Effects of the beta-carbolines, harmane and pinoline, on insulin secretion from isolated human islets of Langerhans. Eur J Pharmacol 482:189–196
23. Nandkarni AK (1992) Indian Materia Medica, vol 1. Popular Prakashan, Bombay
24. Ayyanar M, Subash-Babu P (2012) Syzygium cumini (L.) Skeels: a review of its phytochemical constituents and traditional uses. Asian Pac J Trop Biomed 2(3):240–246. https://doi.org/10.1016/S2221-1691(12)60050-1
25. Shukla AK, Bigoniya P, Srivastava B (2012) Hypoglycemic activity of *Lepidium sativum* Linn seed total alkaloid on alloxan induced diabetic rats. Res J Med Plant 6(8):587–596
26. Wiedemann M, Gurrola-Díaz CM, Vargas-Guerrero B, Wink M, García-López PM, Lupanine MD (2015) Improves glucose homeostasis by influencing KATP channels and insulin gene expression. Molecules 20(10):19085–19100. https://doi.org/10.3390/molecules201019085
27. Shibano M, Tsukamoto D, Masuda A, Tanaka Y, Kusano G (2001) Two new pyrrolidine alkaloids, radicamines A and B, as inhibitors of alpha-glucosidase from Lobelia chinensis Lour. Chem Pharm Bull 49(10):1362
28. Tabopda TK, Ngoupayo J, Liu J, Mitaine-Offer AC, Tanoli SA, Khan SN, Ali MS, Ngadjui BT, Tsamo E, Lacaille-Dubois MA, Luu B (2008) Bioactive aristolactams from *Piper umbellatum*. Phytochemistry 69(8):1726–1731
29. Gao H, Huang YN, Gao B, Li P, Inagaki C, Kawabata J (2008) Inhibitory effect on α-glucosidase by Adhatoda vasica Nees. Food Chem 108(3):965–972
30. Brahmachari G (2012) Bioactive natural products: opportunities and challenges in medicinal chemistry. World Scientific Publishing, Singapore
31. Oku T, Yamada M, Nakamura M, Sadamori N, Nakamura S (2006) Inhibitory effects of extractives from leaves of Morus alba on human and rat small intestinal disaccharidase activity. Br J Nutr 95:933–938
32. Asano N, Yamashita T, Yasuda K, Ikeda K, Kizu H, Kameda Y, kato A, Nash R-J, Lee HS, Ryu KS (2001) Polyhydroxylated alkaloids isolated from mulberry trees (Morusalba L.) and silk-worms (Bombyx mori L.). J Agric Food Chem 49:4208–4213

33. Patel MB, Mishra S (2011) Hypoglycemic activity of alkaloidal fraction of *Tinospora cordifolia*. Phytomedicine 18(12):1045–1052
34. Shimoda H, Nishida N, Ninomiya K, Matsuda H, Yoshikawa M (2001) Javaberine A, new TNF-alpha and nitric oxide production inhibitor, from the roots of *Talinum paniculatum*. Heterocycles 55(11):2043–2050
35. Catthareeya T, Papirom P, Chanlun S, Kupittayanant S (2013) Talinum paniculatum (jacq.) GERTN: a medicinal plant with potential estrogenic activity in ovariectomized rats. Int J Pharm Pharm Sci 5(2):478–485
36. Lee H-S (2002) Rat lens aldose reductase inhibitory activities of *Coptis japonica* root-derived isoquinoline alkaloids. J Agric Food Chem 50:7013–7016
37. Takada K, Uehara T, Nakao Y, Matsunaga S, van Soest WM, Fusetani N (2004) Schulzeines A-C, new alpha-glucosidase inhibitors from the marine sponge Penares schulzei. J Am Chem Soc 126:187–193

Chapter 16
Traditional Indian Herbs for the Management of Diabetes Mellitus and their Herb–Drug Interaction Potentials: An Evidence-Based Review

Myrene Roselyn D'souza

Abstract The growing trend of nutraceuticals in diabetes management makes it imperative to document traditional knowledge of medicines under a single heading as it could help researchers formulate new functional foods. According to the World Health Organization, about 90% of the population in developing countries use plant products in primary health care either alone or in conjunction with western medicine. About 800 plants worldwide have been reported to possess powerful antidiabetic potential directed mainly at enhancing insulin secretion, improving insulin resistance, α-glucosidase enzyme inhibitory activity, anti-inflammatory effects, regenerating pancreatic β-cells and mitigating diabetes-associated oxidative stress. Most natural remedies mediate enhancement of glucose uptake and suppression of hepatic glucose output by stimulating glycolysis, glucose oxidation and glycogenesis, along with reducing glycogen degradation and gluconeogenesis. Herbal treatment for managing diabetic symptomatologies has been a part of the Indian traditional system of medicine. This chapter aims at evaluating the efficacy and safety of the most commonly used ethnomedicinal plants in India by providing an insight into the molecular and cellular mechanisms of action of key bioactives. As herbal remedies are complex mixtures of various bioactive entities, there is high probability of interaction with prescription drugs via pharmacokinetic or pharmacodynamic mechanisms. In addition, diabetic patients present different comorbidities and hence may be treated with multiple medications, increasing the chances of herb–drug interaction. This chapter discusses the mechanism of interactions of herbal remedies with antidiabetic medicines, evidence and management strategies.

Keywords Diabetes · Drug interaction · Ethnomedicine · Herbal medicine · Hyperglycaemia · Hypoglycaemia

---

M. R. D'souza (✉)
Department of Biochemistry, Mount Carmel College (Autonomous), Bengaluru, India

H. Chen, M. Zhang (eds.), *Structure and Health Effects of Natural Products on Diabetes Mellitus*, https://doi.org/10.1007/978-981-15-8791-7_16

1 Introduction

The World Health Organization has defined diabetes mellitus (DM) as a chronic disease caused by inherited and/or acquired deficiency in the production of insulin by the pancreas, or by the ineffectiveness of the insulin produced [1]. It is a metabolic disorder characterized by hyperglycaemia, eventually leading to the damage of multiple body systems. Diabetes is categorized into two types, i.e., insulin dependent (IDDM) caused by impaired production of insulin and non-insulin dependent (NIDDM) commonly associated with the inability of the cells to respond to insulin. Another type of diabetes known as gestational diabetes is found to occur during pregnancy in women previously undiagnosed with the disease. DM is the major risk factor for morbidities like blindness, kidney failure, heart attacks and limb amputation. There has been a record increase in diabetic patients in India, from 31.7 million in 2000 to 72.96 million in 2019 (https://www.livemint.com/science/health/government-survey-found-11-8-prevalence-of-diabetes-in-india-11570702665713.html.pdf). A consequent increase in hypertension leading to cardiovascular diseases (CVD) in 70% of diabetic patients has been recorded [2]. Various categories of antidiabetic medications with different modes of actions are commercially available, such as insulin analogues, sulphonylureas, biguanides, dipeptidyl peptidase-4 inhibitors, thiazolidiones, α-glucosidase inhibitors, etc. [3]. However, even with extensive research in DM, a single medicine to eradicate the disease from the roots remains elusive.

The use of traditional medicine in India can be traced back to 1000 BC. Ancient texts elaborated the importance of 'Ayurveda' as an integrated science focusing on maintaining a state of equilibrium between the mental, physical, spiritual and social aspects of human beings. This was achieved by following proper techniques, procedures, regimens, diet and medicines. In Ayurveda, the primary source of drugs is derived from plants; however, metal, mineral, as well as marine- and animal-derived materials are also used. The global use of complementary and alternative medicine (CAM) for the management of DM is about 72.8% and often used in addition to, rather than in place of, conventional medicine [4]. Herbal medicines are now gaining popularity worldwide due to the common perception that they are free from side effects due to their natural origin. However, in reality, herbal remedies are complex potent mixtures of organic chemicals that are known to react with conventional drugs, sometimes possessing life-threatening consequences [5]. These herb–drug interactions during co-administration often affect pharmacokinetic and/or pharmacodynamic profile of the hypoglycaemic agents by either potentiating or antagonizing their pharmacological effects [6, 7]. The associated risks are of concern particularly due to the long-term nature of therapy for diabetic patients. Most herbal remedies when compared to conventional medicines possess several uncharacterized bioactive entities, each of which may or may not provide therapeutic activity. An additional complexity is the changing constitution of natural products subjected to part of the plant used, seasonality and growth conditions. This chapter aims to provide an overview of the studies investigating interactions between antidiabetic herbs and conventional medicine, identifying of both negative and positive aspects of these interactions, with a focus on mechanisms, clinical evidence, importance and management strategies.

2 Search Criteria

The chapter includes articles until 2020 reviewing indigenous and traditional herbal systems such as Ayurveda, Chinese, African, Unani and Greeko-Arab herbal system. Studies were included from Scopus, Pubmed, Science Direct, SciElo, Research gate, Google Scholar, by using the following keywords: ethnomedicinal herbs for diabetes, traditional medicinal systems for diabetes, herbal drugs for diabetes, phytochemicals as hypoglycaemic agents. An attempt was made to review important literature from the ancient time to modern era. The current trend for the purchase and use of herbal medicines was added by searching online retail stores like Amazon and Indiamart.

3 Common Antidiabetic Drugs

Various pharmaceutical agents are presently being used for the treatment of diabetes through different modes of action such as stimulation of insulin secretion by long acting (6–24 h) sulfonylureas and short acting (>6 h) meglitinides, enhancing the action of insulin agonists like peroxisome proliferator-activated receptor (PPAR) by thiazolidindiones, reduction of hepatic glucose output and enhancement of the peripheral uptake of glucose by biguanides, and delaying complex carbohydrate digestion by α-glucosidase inhibitors. Additionally, combination therapies using the above-mentioned classes of diabetic agents may be also used to minimize side effects and broaden therapeutic targets [8, 9].

4 Antidiabetic Plants in Traditional Medicines

The NAPRALERT database has listed over 1300 species including higher plants, algae and fungi representing 750 genera in 190 families [10]. Over half of these are used ethnopharmacologically in traditional medicine particularly against Type II DM as a normative basis for the maintenance of good health [11]. The increasing dependence on such medicinal plants in industrialized societies has led to the extraction and identification of potentially bioactive molecules such as complex carbohydrates, alkaloids, glycopeptides, terpenoids, peptides, amines, steroids, flavonoids, lipids, coumarins, sulphur compounds and inorganic ions [11]. These compounds may act by directly effecting insulin secretion, activation of glycogenesis and hepatic glycolysis, pancreatic beta cell potassium channel blocker activity, cAMP activation, adrenomimeticism modulation of glucose absorption from the intestine [12, 13].

5 Mechanisms of Herb–Drug Interaction

Treatment of patients with Type II DM is often initiated with oral antidiabetic drug monotherapy but as the diseases progresses, combination therapy with different oral antidiabetics may be required. Moreover, due to different comorbidities seen in DM, a multitargeted polypharmacotherapy including antihypertensive, lipid-lowering, and antiplatelet drugs is often considered especially for the elderly [13, 14]. Such patients are at increased risk of harmful drug–drug interactions and herb–drug interactions.

Clinical safety and efficacy of herb–drug interactions (HDI) are affected by additive/synergistic or antagonistic interactions among the herbal components and drug molecules. Interaction can occur when any two or more substances such as herbs, drugs, nutrients or foods are consumed at the same time or within a short interval, leading to physiological effects which differ from a simple summation of those produced by either substance when administered alone. Such interactions are known to alter both pharmacokinetic and pharmacodynamic properties of the drug causing an increase or decrease in the severity or effectiveness of the final outcome (Fig. 16.1).

5.1 Pharmacokinetic Herb–Drug Interactions

Pharmacokinetic (PK) interactions arise from altered absorption [15], interference in the patterns of distribution [16] and competition for metabolic and excretory pathways [17, 18]. However, the major underlying mechanisms of pharmacokinetic

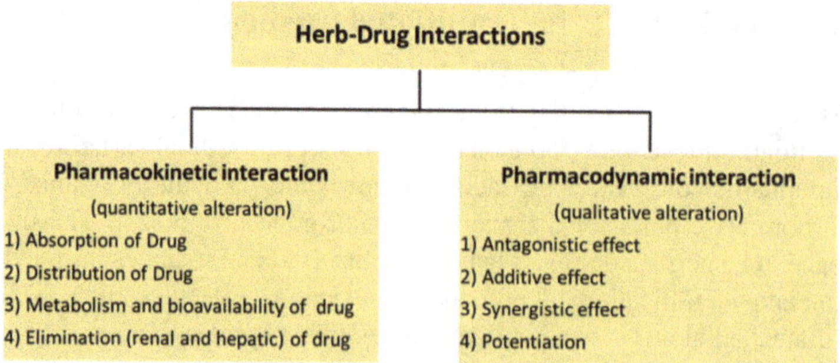

Fig. 16.1 Mechanisms of action of herb–drug interactions

drug–herb interaction are the induction and inhibition of intestinal and hepatic metabolic enzyme of cytochrome P450 (CYP450) isoenzyme family as well as drug transporters and efflux proteins [19]. The CYP is a family of monooxygenase enzymes catalysing phase I metabolic processes including oxidation, oxidative deamination, hydroxylation and S-and O-demethylation [20]. As some herbs and drugs may be substrates of the same CYP isoenzymes, they may inhibit or induce their activity when ingested concomitantly [21], for example, pioglitazone, repaglinide and rosiglitazone for CYP2C8; glibenclamide, glimepiride, glipizide, nateglinide and rosiglitazone for CYP2C9; proguanil for CYP2C19; and pioglitazone and repaginate for CYP3A4 [22]. The herb, St John's wort inhibits CYP2C, CYP3A and several ABC efflux transporters [23, 24]. *Ginkgo biloba* inhibits CYP3A4, CYP2C9 and CYP2C19 [25]. The major reason for drug–drug, food–drug, and herb–drug interactions is overlapping substrate specificity in the biotransformation pathways [26].

Drug absorption may also be impaired in the presence of fibres, gums and mucilage as seen in the case of flaxseed, psyllium, aloe gel and rhubarb. Protein-bound drugs may be displaced by components found in herbal preparations leading to increased activity of the drug [21]. Many herbs are known to interfere with the renal clearance by causing changes in diuretic activity, inhibition of tubular secretion, tubular reabsorption or glomerular filtration [17].

5.2 *Pharmacodynamic Herb–Drug Interactions*

Pharmacodynamic HDIs modify the drug/herb actions in a qualitative manner by the ability of different chemical groups to interact with receptor sites and alter the physiological environment [27]. Such interactions result in antagonistic, additive, synergistic or potentiation effects. For example, crude herbal medicines possess high antioxidant potential which could assist in reducing oxidative stress, a key pathogenic factor of diabetes [28]. There is a tendency to study and document negative or harmful interactions due to safety considerations, however, the additive/synergistic effects induced by HDIs resulting in an enhancement of desired pharmacological must also be considered upon co-administration [29]. For example, agrimony exhibits blood glucose increasing effect in the presence of antidiabetic drugs [30]. Antagonistic HDIs may result in decreased efficacy and therapeutic failure [29]. For example, ephedra or caffeine-containing herbs combined with other bioactives in herbal weight loss products, produce additive cardiovascular effects that may antagonize the effects of antihypertensive medications [29, 31].

6 Antidiabetic–Herb Interactions

Various types of HDIs are possible. However, the most common of these are the ones which result in a rise or fall in blood glucose levels, thereby disturbing the control of diabetes. Herb–drug interactions related to antidiabetic drugs are summarized in (Table 16.1), while a detailed discussion of Indian herbs used in the control of DM is described below.

6.1 *Karela (*Momordica charantia*)*

Karela or bitter melon is a tropical and subtropical vegetable included in the Asian diet and also traditionally used in the treatment of diabetes, abdominal pain, jaundice, cough, respiratory diseases, skin diseases, wounds, ulcer, gout and rheumatism [53]. It possesses a large number of chemical constituents such as sterols, glucoside mixtures, polypeptide P, momordin, pyrimidine nucleoside, vicine and charantin polypeptides [54]. These compounds increase the blood glucose-lowering effects of antidiabetics [29].

Evidence of Interaction Indication of the additive effect of karela and antidiabetics was first documented by Aslam and Stockley [55] in a diabetic person whose diabetes was poorly controlled by diet regulation and chlorpropamide. However, introduction of karela curry along with conventional oral antidiabetics had better blood glucose-lowering effect. With respect to glucose tolerance, potentiation was noted in Type II diabetics consuming karela with antidiabetic medication such as tolbutamide, chlorpropamide, glibenclamide, glymidine [56] and metformin [57]. Significant improvement in glucose tolerance in Type II diabetics was seen on inclusion of karela in the diet even without taking antidiabetics [58]. Non-diabetic children given karela tea reported hypoglycaemic coma and seizures [59].

Mechanism of Action Karela primarily acts by exhibiting sulfonylurea like activity [53], improving glucose tolerance and decreasing insulin resistance [56]. Other suggestive modes of interaction include increasing tissue glucose uptake and by increasing insulin sensitivity [60]. Inhibition of CYP2C9 with IC50 values between 63.4 and 425.9 μg/mL by *M. charantia* was reported [43]. Changes in phase I and phase II drug-metabolizing enzyme activities in streptozotocin (STZ)-induced diabetic animals are associated with enhanced expression of CYP2C9 and glutathione *S*-transferase (GST) isozymes [61]. The former is an enzyme for antidiabetic substrates glibenclamide, glimepiride, glipizide, nateglinide and rosiglitazone.

Management Even though the glucose-lowering activity of karela is well established, its efficacy is minimal. Patients often use karela in curries, commercially available aqueous extracts or in tablet or powder form [29].

Table 16.1 Herb–drug interactions related to antidiabetic drugs

Interacting herb	Co-administered antidiabetic drug	Observation	References
Hypericum perforatum (St. Johns wort)	Repaglinide	No effect on reduction of blood glucose and elevation of insulin levels by repaglinide	[32]
	Metformin	The herb causes decreased renal clearance of metformin while also decreasing the area under glucose concentration–time curve. Enhanced insulin secretion leads to improved glucose tolerance independent of insulin sensitivity in males	[33]
Lycium	Glibenclamide, glimepiride, glipizide, nateglinide, rosiglitazone	Slightly increased efficacy via additive effects with antidiabetics in general	[29, 34, 35]
Allium sativum (garlic)	Metformin	Increases the C_{max} and AUC (area under the curve) of metformin, potentiates hypoglycaemic effect, attenuates tubular toxicity induced by gentamycin	[36–38]
Zingiber officinale (ginger)	Glibenclamide	Reduction in blood glucose level greater than glibenclamide alone. Herb with sub-optimal dose of the drug produces effects as a full therapeutic dose of glibenclamide	[39]
	Metformin	Reduction in hyperglycaemia and improved renal dysfunction at reduced drug dose. Combination amends gentamicin nephrotoxicity	[40]
Gymnema sylvestre (Gymnema)	Metformin	Decreases the bioavailability of metformin, induces repair and regeneration of pancreatic beta cells, increases glucose uptake in the liver, antagonistic increase in serum glucose when compared to metformin alone	[41, 42]
Cassia fistula and *Cassia occidentalis*	Glibenclamide, glimepiride, glipizide, nateglinide, rosiglitazone	Additive effect by inhibiting enzyme activities of CYP2C9, 1A2, 2D6 and 3A4); inhibit glucose absorption from the small intestine	[43]
Radix astragali (Astragalus)	Pioglitazone	No effect on PK of pioglitazone	[44]
Scutellaria baicalensis (Chinese skullcap)	Metformin	Significant increase in pancreatic levels and reduction of hepatic levels of triglycerides and cholesterol	[45]
Aloe vera	Pioglitazone, repaglinide	Additive effects by inhibiting CYP3A4 and CYP2D6 along with insulin-sensitizing effects	[46, 47]

(continued)

Table 16.1 (continued)

Interacting herb	Co-administered antidiabetic drug	Observation	References
	Glibenclamide	Causes insulin release from pancreatic β cells by inhibiting ATP sensitive K^+ channels and depolarization of cell membrane	[48]
Opuntia aciculate (prickly pear cactus)	Glipizide, metformin	Additive hypoglycaemic effect in type II DM	[49]
Pongamia pinnata bark	Glibenclamide	Synergistic action on hypoglycaemic activity and maintenance of insulin secretory response to the drug	[50]
Flax seed	Metformin	Increase in liver glucose-6-phosphate dehydrogenase enzyme activity and normal glycogenesis, formation of new islets	[51]
Elderberry	Glibenclamide	Enhancement of insulin secretion, increased 2-deoxy-glucose transport, glucose oxidation and glycogenesis	[29]
Medicago sativa (alfalfa)		Stimulation of insulin secretion from the BRIN-BD11	[30]
Sesame oil	Gliberclamide	Additive/synergistic effect causing reduction in glycosylated Hb and blood glucose level. Improvement in enzymatic and non-enzymatic anti-oxidant levels	[52]

6.2 *Cinnamon (*Cinnamomum zeylanicum*)*

Cinnamon bark possesses insulin mimetic activity, inhibits pancreatic α-amylase and α-glucosidase [62], reduces insulin resistance [63], enhances glucose uptake [64], delays gastric emptying [65], stimulates glycogenesis [66] and inhibits gluconeogenesis [67].

Evidence of Interaction Studies on the concurrent use of cinnamon and metformin in diabetic rats have shown increased serum level of HDL-cholesterol significantly, more than metformin alone [68]. While cinnamon and metformin alone reduced activities of ALT, AST and LDH, greatest activities were observed with diabetic and metformin-diabetic rats which received high dose of cinnamon. In addition, researchers also reported induction of DNA damage in hepatocytes when co-administration of the herb with metformin was given, indicating an increase in oxidative stress together with poor glycaemic control [68]. Geil et al. [69] cautioned adjustment of dose of sulfonylureas in patients taking cinnamon concurrently.

Mechanism of Action Studies on cinnamaldehyde extracted from cinnamon bark have shown significant reduction of plasma glucose and HbA1c levels in streptozotocin induced diabetic rats [70], possibly by improving insulin sensitivity or by reducing the absorption of carbohydrates in the small intestine [64]. In addition, the hypoglycaemic activity may also be attributed to insulin receptor signalling, activation and regulation of enzymes involved in carbohydrate metabolism, stimulation of cellular glucose uptake, increased glucose transporter-4 receptor synthesis and delay gastric emptying [62]. The proanthocyanidin and cinnamtannin B1 stimulate phosphorylation of the insulin receptor β-subunit on adipocytes. Vascular complications arising due to oxidative stress may be overcome by polyphenols in cinnamon including epicatechin, camphene, eugenol, gamma-terpinene, phenol, salicylic acid and tannins [71].

Management The daily dose recommended is 1–4 g of cinnamon powder or 0.05–0.2 g of cinnamon oil. These doses are not recommended during pregnancy as it may induce uterine contractions and premature labour [72].

6.3 *Fenugreek (*Trigonella foenum-graecum*)*

Commonly used as a spice in South Asia, fenugreek is said to possess hypoglycaemic and hypocholesterolemic properties. The seeds, husk and cotyledons are rich in saponins like diosgenin; alkaloids and trigonelline inhibit intestinal glucose uptake [73].

Evidence of Interaction When taken along with metformin, fenugreek produces a huge decrease in plasma glucose level in patients of Type II diabetes [74]. Blended treatment with glibenclamide demonstrated reduction in lipid peroxidation and a greater antioxidant activity compared to the individual components [74]. It was noted that glipizide monotherapy was more efficacious in controlling FBG and HbA1c levels than fenugreek monotherapy or in combination with fenugreek. However, fenugreek monotherapy was found to be more effective in regulation of dyslipidaemia than in combination with glipizide [75].

Mechanism of Action The seed extracts are known to exhibit antidiabetic potential by decreasing the time of gastric emptying, glucose absorption rate, reducing glucose uptake, lowering blood glucose, regenerating and protecting β cells of pancreas [11], elevating serum insulin, stimulating glycogen synthetase activity [76], reducing proinflammatory cytokines and pancreatic enzymes, promoting restoration of depleted glycogen, correcting serum lipid profiles [76], promoting insulin sensitivity and recovering the level of HbA1c by utilization of glucose in peripheral tissues [11]. The increase in glucose-stimulated insulin release from islet cells is attributed to 4-hydroxyleucine, a novel amino acid from fenugreek seed while 4-hydroxyisoleucine possesses insulinotropic effects and enhances insulin sensitivity in peripheral tissues [11]. Fenugreek seed extracts are also known to phosphorylate proteins involved in signal transduction such as insulin receptor, IRS-1 and p85 subunit of PI3-K [77]. Fenugreek seed oil produces immunomodulatory- and insulin-stimulating [78].

Management Experiments have shown that the degummed fraction of fenugreek seeds has significantly lower hypoglycaemic effects when compared to raw seeds, seed powder, cooked seeds and gum isolate of the seeds. Hence, for a full protective range of activities, consumption of whole seeds is recommended. The fibre fraction alone provides a delayed response indicating that it is not the only component responsible for hypoglycaemic effects [79]. There may be other uncategorized responses as well.

6.4 *Ginseng (***Panax ginseng** *and* **Panax quinquefolium***)*

The antidiabetic effect of ginseng is largely due to ginsenosides such as Rb1, Rb2, Rc, Re, Rg3, Rh2, and compound K [80] which affect insulin dependent and insulin independent pathways. Among these, compound K, a final metabolite of protopanaxadiol ginsenosides produces insulin secreting action similar to metformin, eliciting additive effects up on combined treatment. Panax notoginseng saponins (PNS) produce anti-hyperglycaemic and anti-obesity effects by improving sensitivity to insulin.

Evidence of Interaction Supplementation of diet with Korean Ginseng (KG) along with normal therapy in Type II diabetics showed good glycaemic control, improved plasma glucose levels and enhanced plasma insulin regulation [81]. Extrapolation of these studies led to the understanding that for best glycaemic control in diabetic individuals, ginseng must be taken 40 min before a meal. However, for non-diabetic subjects, unintended hypoglycaemia can be avoided by taking the herb along with meals [82]. For American Ginseng (AG), reduction in postprandial glycaemia was reported irrespective of dose and time of administration [83], significantly increasing glycogen and HDL and lowering plasma cholesterol and LDL. In non-diabetics, AG taken at least 40 min before meals owed enhanced antioxidative and antihyperglycaemic effects [84]. Administration of the ginsenoside, malonyl ginsenosides (MGR) to high fat diet/streptozotocin diabetic rats significantly reduced fasting blood glucose levels, improved glucose tolerance and enhanced insulin sensitivity, without affecting body weight [85]. Co-administration of the bulky tri-glycosylated ginsenosides Rc with valsartan reduced the latter's distribution to the liver [86]. The lack of HDI between Rc and valsartan [87] is useful in understanding treatment regimens of patients taking antihypertensive, antidiabetics, anti-tumour drugs such as valsartan and repaglinide, substrates of OATPs transporters.

Mechanism of Action Ginseng stimulates insulin production, prevents β cell loss, stimulates insulin release [88], improves insulin sensitivity [89], decreases insulin resistance, increases glucose uptake [89], suppresses hepatic glucose production

[80], improves regulation of plasma glucose and plasma insulin and also improves expression of peroxisome proliferator-activated receptors (PPARγ) [90]. Induction of CYP3A activity and CYP3A4 by ginseng has been reported [91].

Ginseng polysaccharides may affect fasting blood glucose levels by stimulation of insulin secretion, promotion of glycogenesis [89], reduction in serum MDA levels, increase in serum insulin and superoxide dismutase activity. PPD-type ginsenoside like Rb1, Rb2, Rc, Rd, Rg3, compound K and Rh2 was found to inhibit OATP1B1 and OATP1B3, while PPT-type ginsenosides such as Re, Rg1, Rh1 and PPT did not inhibit OATP1B1 transport activity. OATPs assist in intestinal absorption and hepatic uptake of various therapeutic reagents thereby playing an important role in drug pharmacokinetics and biliary excretion [92].

Management Korean and American ginseng and also other species of ginseng possess blood glucose-lowering activity. However, this effect is not significant in patients with well controlled diabetes [29].

6.5 Andrographis paniculata *(Green Chireta)*

The antidiabetic agent in andrographis is the diterpene lactone compound, andrographolide [93]. The exact mechanism of action of andrographolide is still unknown however it has been found to reduce preprandial and postprandial blood glucose levels, increase expression of GLUT-4 protein in muscle tissues [93, 94] and lower HOMA-IR index [95]. Studies have also revealed that the pure biological active compound is not always better than their extract in reducing blood glucose levels as other component in the extract may play an important role as well [93]. Nonetheless, andrographis exhibits additive pharmacological effects when used in combination with antidiabetic medications.

Evidence of Interaction Studies on the efficacy of andrographolide in the presence of metformin in high fructose-fat feed diabetic rats highlighted the relationship between GLUT4 protein expression and the transportation of glucose into muscle tissue [94]. A significant decrease in the activity of metformin was reported. Ethanolic extract of andrographis used in combination with curcuminoid fraction of *Curcuma xanthorrhiza* indicates better hypoglycaemic activity than when administered singly or with ethyl acetate soluble fraction of propolis [96]. Hence the former is the chosen therapy using only herbal medicine. A combination of high levels of two active constituents such as andrographolide from *Andrograpis paniculata* and asiaticoside-enriched extract of *Centella asiatica* decreased blood glucose levels by 68.13% when administered in a ratio of 70:30. In addition, significant decrease in the levels of cholesterol, LDL, triglycerides was also noted [96].

Mechanism of Action Andrographolide enhanced uptake of glucose in muscle of STZ-diabetic rats in a concentration-dependent manner, an outcome which could be attributed to an increase in the expression of GLUT4 mRNA and protein levels [97]. Ethanolic extract of the herb was found to decrease TBARS, while significantly increasing SOD and CAT activities. Potent inhibition of CYP3A4, CYP2C9, CYP2C19 and CYP2D6 was also reported [98]. Glibenclamide, glimepiride, glipizide, nateglinide and rosiglitazone are substrates of CYP2C9, and pioglitazone and repaglinide are substrates of CYP3A4.

Management Studies on the exact mechanism of action of andrographis are inconclusive. Hence applying the evidence in a clinical setup is not advisable. Patients using combination therapy are advised about the potential additive effects.

7 Conclusion

The use of herbal therapies is increasing around the world as they are perceived to be free from side effects, economical and are ethnomedicinally significant. The aim of this chapter is not to suppress the use of traditional herbal remedies but to inculcate rational, safe, and effective usage. Often, herbs are prescribed by physicians to fill the lacunae where conventional medicines seem unsuitable. It becomes imperative to maintain a database for natural products, collecting and preserving clinical evidence and documenting their PK and PD properties. A thorough understanding of the potential herb–drug interactions leading to ideal treatment strategies ensures rational use of herbal medicines.

Patients with diabetes often exhibit various comorbidities that necessitate the use of multiple medications or polypharmacotherapy. These are often lifelong increasing the chance of drug–drug, drug–herb and herb–herb interactions. The results presented in this chapter clearly highlight favourable and unfavourable responses to herbs when given in combination with conventional antidiabetic drugs. A marked alteration in their pharmacokinetic and/or pharmacodynamic properties is often encountered. The interactions are highly complex due to variable targets, multicomponent nature of herbal medicine, batch variation dependent on climatic fluctuations, processing criteria, etc. Combination therapy may result in additive or antagonistic effects. Hypoglycaemia may be encountered if the HDI results in enhancement of glucose lowering property. Diabetic patients have to closely monitor their blood glucose levels in such cases. Even though interactions with non-diabetic medicines are not reviewed in this chapter, caution must be exercised especially in persons suffering from chronic illnesses and the elderly.

Despite a few adverse HDI, combination therapy for antidiabetic treatment has several positive clinical ramifications, enabling reduction in the dose of the prescribed drug, and consequently lowering chances of side effects. As the drug properties vary for drugs even from the same class, drug interaction cannot be generalized and needs to be experimentally tested for each drug to establish clinical correlation. Such data is critical for developing future clinical guidelines to ensure better health care consequences.

References

1. Alberti K, Zimmet PZ, Consultation WHO (1998) Definition, diagnosis and classification of diabetes mellitus and its complications part 1: diagnosis and classification of diabetes mellitus – provisional report of a WHO consultation. Diabet Med 15(7):539–553. https://doi.org/10.1002/(sici)1096-9136(199807)15:7<539::Aid-dia668>3.0.Co;2-s
2. Dhaliwall C, Erinmacpherson R (2015) Effectiveness of telephone-delivered interventions for increasing physical activity levels in persons with type 2 diabetes or hypertension: a systematic review. J Crit Rev 2(4):6–11
3. Li W, Yuan G, Pan Y, Wang C, Chen H (2017) Network pharmacology studies on the bioactive compounds and action mechanisms of natural products for the treatment of diabetes mellitus: a review. Front Pharmacol 8:15–24. https://doi.org/10.3389/fphar.2017.00074
4. Manya K, Champion B, Dunning T (2012) The use of complementary and alternative medicine among people living with diabetes in Sydney. BMC Complement Altern Med 12. https://doi.org/10.1186/1472-6882-12-2
5. Bent S, Goldberg H, Padula A, Avins AL (2005) Spontaneous bleeding associated with Ginkgo biloba – a case report and systematic review of the literature. J Gen Intern Med 20(7):657–661. https://doi.org/10.1007/s11606-005-0114-4
6. Hafner-Blumenstiel V (2011) Herbal drug-drug interaction and adverse drug reactions (Phytopharmaka als Ursache von unerwunschten Arzneimittelwirkungen.). Therapeutische Umschau Revue Therapeutique 68(1):54–57. https://doi.org/10.1024/0040-5930/a000120
7. Tarirai C, Viljoen AM, Hamman JH (2010) Herb-drug pharmacokinetic interactions reviewed. Expert Opin Drug Metab Toxicol 6(12):1515–1538. https://doi.org/10.1517/17425255.2010.529129
8. Del Prato S, Bianchi C, Marchetti P (2007) Beta-cell function and anti-diabetic pharmacotherapy. Diabetes Metab Res Rev 23(7):518–527. https://doi.org/10.1002/dmrr.770
9. Jovanovic L, Hassman DR, Gooch B, Jain R, Greco S, Khutoryansky N, Hale PM, Hale PM, Repaglinide Pioglitazone Study G (2004) Treatment of type 2 diabetes with a combination regimen of repaglinide plus pioglitazone. Diabetes Res Clin Pract 63(2):127–134. https://doi.org/10.1016/j.diabres.2003.09.004
10. Marles RJ, Farnsworth NR (1995) Antidiabetic plants and their active constituents. Phytomedicine 2(2):137–189. https://doi.org/10.1016/s0944-7113(11)80059-0
11. Chang CLT, Lin Y, Bartolome AP, Chen Y-C, Chiu S-C, Yang W-C (2013) Herbal therapies for type 2 diabetes mellitus: chemistry, biology, and potential application of selected plants and compounds. Evid Based Complement Alternat Med. https://doi.org/10.1155/2013/378657
12. Patel DK, Prasad SK, Kumar R, Hemalatha S (2012) An overview on antidiabetic medicinal plants having insulin mimetic property. Asian Pac J Trop Biomed 2(4):320–330. https://doi.org/10.1016/s2221-1691(12)60032-x
13. Gaede P, Lund-Andersen H, Parving H-H, Pedersen O (2008) Effect of a multifactorial intervention on mortality in type 2 diabetes. N Engl J Med 358(6):580–591. https://doi.org/10.1056/NEJMoa0706245
14. Caughey GE, Roughead EE, Vitry AI, McDermott RA, Shakib S, Gilbert AL (2010) Comorbidity in the elderly with diabetes: identification of areas of potential treatment conflicts. Diabetes Res Clin Pract 87(3):385–393. https://doi.org/10.1016/j.diabres.2009.10.019
15. Mochiki E, Yanai M, Ohno T, Kuwano H (2010) The effect of traditional Japanese medicine (Kampo) on gastrointestinal function. Surg Today 40(12):1105–1111. https://doi.org/10.1007/s00595-010-4388-8
16. Rodriguez-Landa JF, Contreras CA (2003) A review of clinical and experimental observations about antidepressant actions and side effects produced by *Hypericum perforatum* extracts. Phytomedicine 10(8):688–699. https://doi.org/10.1078/0944-7113-00340
17. Bagnis CI, Deray G, Baumelou A, Le Quintrec M, Vanherweghem JL (2004) Herbs and the kidney. Am J Kidney Dis 44(1):1–11. https://doi.org/10.1053/j.ajkd.2004.02.009

18. Yang X-X, Hu Z-P, Duan W, Zhu Y-Z, Zhou S-F (2006) Drug-herb interactions: eliminating toxicity with hard drug design. Curr Pharm Des 12(35):4649–4664. https://doi.org/10.2174/138161206779010440
19. Nowack R (2008) Review article: cytochrome P450 enzyme, and transport protein mediated herb-drug interactions in renal transplant patients: grapefruit juice, St John's Wort - and beyond. Nephrology 13(4):337–347. https://doi.org/10.1111/j.1440-1797.2008.00940.x
20. Karyekar CS, Eddington ND, Dowling TC (2002) Effect of St. John's Wort extract on intestinal expression of cytochrome P4501A2: studies in LS180 cells. J Postgrad Med 48(2):97–100
21. Wang J-F, Chou K-C (2010) Molecular modeling of cytochrome P450 and drug metabolism. Curr Drug Metab 11(4):342–346. https://doi.org/10.2174/138920010791514180
22. Gupta RC, Chang D, Nammi S, Bensoussan A, Bilinski K, Roufogalis BD (2017) Interactions between antidiabetic drugs and herbs: an overview of mechanisms of action and clinical implications. Diabetol Metab Syndr 9. https://doi.org/10.1186/s13098-017-0254-9
23. Whitten DL, Myers SP, Hawrelak JA, Wohlmuth H (2006) The effect of St John's wort extracts on CYP3A: a systematic review of prospective clinical trials. Br J Clin Pharmacol 62 (5):512–526. https://doi.org/10.1111/j.1365-2125.2006.02755.x
24. Zhou S-F, Lai X (2008) An update on clinical drug interactions with the herbal antidepressant St. John's wort. Curr Drug Metab 9(5):394–409. https://doi.org/10.2174/138920008784746391
25. Fan L, Mao XQ, Tao GY, Wang G, Jiang F, Chen Y, Li Q, Zhang W, Lei HP, Hu DL, Huang YF, Wang D, Zhou HH (2009) Effect of *Schisandra chinensis* extract and *Ginkgo biloba* extract on the pharmacokinetics of talinolol in healthy volunteers. Xenobiotica 39(3):249–254. https://doi.org/10.1080/00498250802687657
26. Colalto C (2010) Herbal interactions on absorption of drugs: mechanisms of action and clinical risk assessment. Pharmacol Res 62(3):207–227. https://doi.org/10.1016/j.phrs.2010.04.001
27. Izzo AA, Ernst E (2009) Interactions between herbal medicines and prescribed drugs an updated systematic review. Drugs 69(13):1777–1798. https://doi.org/10.2165/11317010-000000000-00000
28. Nasri H, Shirzad H, Baradaran A, Rafieian-kopaei M (2015) Antioxidant plants and diabetes mellitus. J Res Med Sci 20(5):491–502
29. Williamson EM, Driver S, Baxter K, Lee CR (2013) Stockley's herbal medicines interactions. Pharmaceutical Press, London
30. Gray AM, Flatt PR (1998) Actions of the traditional anti-diabetic plant, agrimony eupatoria (agrimony): effects on hyperglycaemia, cellular glucose metabolism and insulin secretion. Br J Nutr 80(1):109–114. https://doi.org/10.1017/s0007114598001834
31. Scott GN, Elmer GW (2002) Update on natural product-drug interactions. Am J Health Syst Pharm 59(4):339–347. https://doi.org/10.1093/ajhp/59.4.339
32. Fan L, Zhou G, Guo D, Liu Y-L, Chen W-Q, Liu Z-Q, Tan Z-R, Sheng D, Zhou H-H, Zhang W (2011) The Pregnane X receptor agonist St John's Wort has no effects on the pharmacokinetics and pharmacodynamics of Repaglinide. Clin Pharmacokinet 50(9):605–611. https://doi.org/10.2165/11587310-000000000-00000
33. Stage TB, Pedersen RS, Damkier P, Christensen MMH, Feddersen S, Larsen JT, Hojlund K, Brosen K (2015) Intake of St John's wort improves the glucose tolerance in healthy subjects who ingest metformin compared with metformin alone. Br J Clin Pharmacol 79(2):298–306. https://doi.org/10.1111/bcp.12510
34. Lam AY, Elmer GW, Mohutsky MA (2001) Possible interaction between warfarin and Lycium barbarum L. Ann Pharmacother 35(10):1199–1201
35. Zhao H, Alexeev A, Chang E, Greenburg G, Bojanowski K (2005) *Lycium barbarum* glycoconjugates: effect on human skin and cultured dermal fibroblasts. Phytomedicine 12 (1-2):131–137. https://doi.org/10.1016/j.phymed.2003.08.002
36. Chourey S, Narsinghani T, Soni L (2011) Effect of *Allium sativum* on the pharmacokinetic of metformin in rat plasma: a herb–drug interaction study. Der Pharm Chem 3:287–291

37. Rafieian-Kopaei M, Baradaran A, Merrikhi A, Nematbakhsh M, Madihi Y, Nasri H (2013) Efficacy of co-administration of garlic extract and metformin for prevention of gentamicin-renal toxicity in Wistar rats: a biochemical Study. Int J Prev Med 4(3):258–264
38. Tripathi P, Gupta P, Lal V (2013) Effect of co-administration of *Allium sativum* extract and metformin on blood glucose of Streptozotocin induced diabetic rats. J Intercult Ethnopharmacol 2:81–84
39. al Omari I, Afifi F, Salhab A (2012) Therapeutic effect and possible herb drug interactions of ginger (*Zingiberofficinale roscoe*, Zingiberaceae) crude extract with glibenclamide and insulin. Pharmacognosy Comm 2:12–20
40. Arshad M, Ashoka S, Samuel SM, Shabaraya AR (2013) Evaluation of the anti-diabetic activity of glibenclamide in combination with ginger in Streptozotocin induced diabetes. Int J Pharm Chem Sci 2(3):1339–1344
41. Shravan KD, Ramakrishna R, Kannappan (2013) Effect of Gymnema sylvestre on the pharmacokinetics and pharmacodynamics of metformin in diabetic rats. An Int J 1(7):657–664
42. Srujan K, Ramakrishna R, Shravan KD (2014) Effect of *Gymnema sylvestre* extract on the pharmacokinetics and pharmacodynamics of oral hypoglycemic drug – metformin in hyperglycemic rats. J Pharm Res 8(4):614–620
43. Appiah-Opong R, Commandeur JNM, Axson C, Vermeulen NPE (2008) Interactions between cytochromes P450, glutathione S-transferases and Ghanaian medicinal plants. Food Chem Toxicol 46(12):3598–3603. https://doi.org/10.1016/j.fct.2008.09.002
44. Shi Z, Gao J, Yuan Y, Zhu S, Yao M (2014) Effect of raw radix Rehmanniae on the pharmacokinetics of pioglitazone in rats. Pak J Pharm Sci 27(3):537–539
45. Waisundara VY, Hsu A, Huang D, Tan BK-H (2008) *Scutellaria baicalensis* enhances the antidiabetic activity of metformin in streptozotocin-induced diabetic Wistar rats. Am J Chin Med 36 (3):517–540. https://doi.org/10.1142/s0192415x08005953
46. Rehman SU, Choi MS, Choe K, Yoo HH (2015) Interactions between herbs and antidiabetics: an overview of the mechanisms, evidence, importance, and management. Arch Pharm Res 38 (7):1281–1298. https://doi.org/10.1007/s12272-014-0517-z
47. Yongchaiyudha S, Rungpitarangsi V, Bunyapraphatsara N, Chokechaijaroenporn O (1996) Antidiabetic activity of *Aloe vera* L. juice. I. Clinical trial in new cases of diabetes mellitus. Phytomedicine 3(3):241–243. https://doi.org/10.1016/s0944-7113(96)80060-2
48. Bunyapraphatsara N, Yongchaiyudha S, Rungpitarangsi V, Chokechaijaroenporn O (1996) Antidiabetic activity of *Aloe vera* L. juice II. Clinical trial in diabetes mellitus patients in combination with glibenclamide. Phytomedicine 3(3):245–248. https://doi.org/10.1016/s0944-7113(96)80061-4
49. Sobieraj DM, Freyer CW (2010) Probable Hypoglycemic adverse drug reaction associated with prickly pear cactus, Glipizide, and metformin in a patient with type 2 diabetes mellitus. Ann Pharmacother 44(7-8):1334–1337. https://doi.org/10.1345/aph.1P148
50. Heroor S, Beknal A, Mahurkar N (2015) Effect of Pongamia bark extract on antihyperglycemic effect of glibenclamide for possible herb-drug interaction. Ind J Pharm Edu Res 49(1):76–81. https://doi.org/10.5530/ijper.49.1.11
51. Dusane MB, Joshi BN (2013) Beneficial effect of flax seeds in streptozotocin (STZ) induced diabetic mice: isolation of active fraction having islet regenerative and glucosidase inhibitory properties. Can J Physiol Pharmacol 91(5):325–331. https://doi.org/10.1139/cjpp-2011-0428
52. Sankar D, Ali A, Sambandam G, Rao R (2011) Sesame oil exhibits synergistic effect with antidiabetic medication in patients with type 2 diabetes mellitus. Clin Nutr 30(3):351–358. https://doi.org/10.1016/j.clnu.2010.11.005
53. Grover JK, Yadav SP (2004) Pharmacological actions and potential uses of *Momordica charantia*: a review. J Ethnopharmacol 93(1):123–132. https://doi.org/10.1016/j.jep.2004.03.035
54. Basch E, Gabardi S, Ulbricht C (2003) Bitter melon (*Momordica charantia*): a review of efficacy and safety. Am J Health Syst Pharm 60(4):356–359. https://doi.org/10.1093/ajhp/60.4.356

55. Aslam M, Stockley IH (1979) Interaction between curry ingredient (karela) and drug (chlorpropamide). Lancet (London, England) 1(8116):607–607. https://doi.org/10.1016/s0140-6736(79)91028-6
56. Leatherdale BA, Panesar RK, Singh G, Atkins TW, Bailey CJ, Bignell AH (1981) Improvement in glucose tolerance due to *Momordica charantia* (karela). Br Med J (Clin Res Ed) 282(6279):1823–1824. https://doi.org/10.1136/bmj.282.6279.1823
57. Tongia A, Tongia SK, Dave M (2004) Phytochemical determination and extraction of *Momordica charantia* fruit and its hypoglycemic potentiation of oral hypoglycemic drugs in diabetes mellitus (NIDDM). Indian J Physiol Pharmacol 48(2):241–244
58. Srivastava Y, Venkatakrishna-Bhatt H, Verma Y, Venkaiah K, Raval BH (1993) Antidiabetic and adaptogenic properties of *Momordica charantia* extract: an experimental and clinical evaluation. Phytother Res 7(4):285–289. https://doi.org/10.1002/ptr.2650070405
59. Hulin A, Wavelet M, Desbordes JM (1988) Intoxication aigue¨ par Momordica charantia (Sorrossi) (a` propos de deux cas). Me´decine d'Afrique Noire 35:671–674
60. Sridhar MG, Vinayagamoorthi R, Suyambunathan VA, Bobby Z, Selvaraj N (2008) Bitter gourd (*Momordica charantia*) improves insulin sensitivity by increasing skeletal muscle insulin-stimulated IRS-1 tyrosine phosphorylation in high-fat-fed rats. Br J Nutr 99(4):806–812. https://doi.org/10.1017/s000711450783176x
61. Raza H, Ahmed I, John A, Sharma AK (2000) Modulation of xenobiotic metabolism and oxidative stress in chronic streptozotocin-induced diabetic rats fed with *Momordica charantia* fruit extract. J Biochem Mol Toxicol 14(3):131–139. https://doi.org/10.1002/(sici)1099-0461(2000)14:3<131::Aid-jbt2>3.0.Co;2-q
62. Adisakwattana S, Lerdsuwankij O, Poputtachai U, Minipun A, Suparpprom C (2011) Inhibitory activity of cinnamon bark species and their combination effect with Acarbose against intestinal alpha-glucosidase and pancreatic alpha-amylase. Plant Foods Hum Nutr 66(2):143–148. https://doi.org/10.1007/s11130-011-0226-4
63. Qin B, Nagasaki M, Ren M, Bajotto G, Oshida Y, Sato Y (2004) Cinnamon extract prevents the insulin resistance induced by a high-fructose diet. Horm Metab Res 36(2):119–125. https://doi.org/10.1055/s-2004-814223
64. Kim W, Khil LY, Clark R, Bok SH, Kim EE, Lee S, Jun HS, Yoon JW (2006) Naphthalenemethyl ester derivative of dihydroxyhydrocinnamic acid, a component of cinnamon, increases glucose disposal by enhancing translocation of glucose transporter 4. Diabetologia 49(10):2437–2448. https://doi.org/10.1007/s00125-006-0373-6
65. Hlebowicz J, Darwiche G, Bjorgell O, Almer L-O (2007) Effect of cinnamon on postprandial blood glucose, gastric emptying, and satiety in healthy subjects. Am J Clin Nutr 85(6):1552–1556
66. Khan A, Safdar M, Khan MMA, Khattak KN, Anderson RA (2003) Cinnamon improves glucose and lipids of people with type 2 diabetes. Diabetes Care 26(12):3215–3218. https://doi.org/10.2337/diacare.26.12.3215
67. Anand P, Murali KY, Tandon V, Murthy PS, Chandra R (2010) Insulinotropic effect of cinnamaldehyde on transcriptional regulation of pyruvate kinase, phosphoenolpyruvate carboxykinase, and GLUT4 translocation in experimental diabetic rats. Chem Biol Interact 186(1):72–81. https://doi.org/10.1016/j.cbi.2010.03.044
68. Ashoor LA, Qusti SY (2010) Potential interactions between cinnamon and metformin treatment in diabetic rats. Biosci Biotechnol Res Asia 7:607–616
69. Geil P, Shane-McWhorter L (2008) Dietary supplements in the management of diabetes: potential risks and benefits. J Am Diet Assoc 108(4):S59–S65. https://doi.org/10.1016/j.jada.2008.01.020
70. Babu PS, Prabuseenivasan S, Ignacimuthu S (2007) Cinnamaldehyde – a potential antidiabetic agent. Phytomedicine 14(1):15–22. https://doi.org/10.1016/j.phymed.2006.11.005
71. Dugoua J-J, Seely D, Perri D, Cooley K, Forelli T, Mills E, Koren G (2007) From type 2 diabetes to antioxidant activity: a systematic review of the safety and efficacy of common

and cassia cinnamon bark. Can J Physiol Pharmacol 85(9):837–847. https://doi.org/10.1139/y07-080
72. Ernst PE (2002) Herbal medicinal products during pregnancy: are they safe? Bjog-an Int J Obstet Gynaecol 109(3):227–235
73. Tharaheswari M, Kumar J, Reddy J, Sridharan S, Sudha Rani S (2014) Fenugreek seed extract stabilize plasma lipid levels in type 2 diabetes by modulating PPARs and GLUT4 in insulin target tissues. Am J Phytomed Clin Ther 2
74. Neha S, Anand K, Sunanda P (2019) Administration of fenugreek seed extract produces better effects in glibenclamide-induced inhibition in hepatic lipid peroxidation: an in vitro Study. Chin J Integr Med 25(4):278–284. https://doi.org/10.1007/s11655-015-1793-z
75. Singh A, Rai J, Mahajan D (2016, 2016) Comparative evaluation of glipizide and fenugreek (*Trigonella foenum-graecum*) seeds as monotherapy and combination therapy on glycaemic control and lipid profile in patients with type 2 diabetes mellitus. Int J Basic Clin Pharmacol:942–950. https://doi.org/10.18203/2319-2003.ijbcp20161549
76. Bera TK, Ali KM, Jana K, Ghosh A, Ghosh D (2013) Protective effect of aqueous extract of seed of *Psoralea corylifolia* (Somraji) and seed of *Trigonella foenum-graecum* L. (Methi) in streptozotocin-induced diabetic rat: a comparative evaluation. Pharm Res 5(4):277–285. https://doi.org/10.4103/0974-8490.118840
77. Vijayakumar MV, Singh S, Chhipa RR, Bhat MK (2005) The hypoglycaemic activity of fenugreek seed extract is mediated through the stimulation of an insulin signalling pathway. Br J Pharmacol 146(1):41–48. https://doi.org/10.1038/sj.bjp.0706312
78. Khan V, Najmi AK, Akhtar M, Aqil M, Mujeeb M, Pillai KK (2012) A pharmacological appraisal of medicinal plants with antidiabetic potential. J Pharm Bioallied Sci 4(1):27–42. https://doi.org/10.4103/0975-7406.92727
79. Neelakantan N, Narayanan M, de Souza RJ, van Dam RM (2014) Effect of fenugreek (*Trigonella foenum-graecum* L.) intake on glycemia: a meta-analysis of clinical trials. Nutr J 13. https://doi.org/10.1186/1475-2891-13-7
80. Kim JH, Yi Y-S, Kim M-Y, Cho JY (2017) Role of ginsenosides, the main active components of *Panax ginseng*, in inflammatory responses and diseases. J Ginseng Res 41(4):435–443. https://doi.org/10.1016/j.jgr.2016.08.004
81. Vuksan V, Sung M-K, Sievenpiper JL, Stavro PM, Jenkins AL, Di Buono M, Lee K-S, Leiter LA, Nam KY, Arnason JT, Choi M, Naeem A (2008) Korean red ginseng (*Panax ginseng*) improves glucose and insulin regulation in well-controlled, type 2 diabetes: results of a randomized, double-blind, placebo-controlled study of efficacy and safety. Nutr Metab Cardiovasc Dis 18(1):46–56. https://doi.org/10.1016/j.numecd.2006.04.003
82. Vuksan V, Sievenpiper JL, Koo VYY, Francis T, Beljan-Zdravkovic U, Xu Z, Vidgen E (2000) American ginseng (*Panax quinquefolius* L) reduces postprandial glycemia in nondiabetic subjects and subjects with type 2 diabetes mellitus. Arch Intern Med 160(7):1009–1013. https://doi.org/10.1001/archinte.160.7.1009
83. Vuksan V, Stavro MP, Sievenpiper JL, Beljan-Zdravkovic U, Leiter LA, Josse RG, Xu Z (2000) Similar postprandial glycemic reductions with escalation of dose and administration time of American ginseng in type 2 diabetes. Diabetes Care 23(9):1221–1226. https://doi.org/10.2337/diacare.23.9.1221
84. Vuksan V, Stavro MP, Sievenpiper JL, Koo VYY, Wong E, Beljan-Zdravkovic U, Francis T, Jenkins AL, Leiter LA, Josse RG, Xu Z (2000) American ginseng improves glycemia in individuals with normal glucose tolerance: effect of dose and time escalation. J Am Coll Nutr 19(6):738–744. https://doi.org/10.1080/07315724.2000.10718073
85. Liu Z, Li W, Li X, Zhang M, Chen L, Y-n Z, G-z S, C-c R (2013) Antidiabetic effects of malonyl ginsenosides from *Panax ginseng* on type 2 diabetic rats induced by high-fat diet and streptozotocin. J Ethnopharmacol 145(1):233–240. https://doi.org/10.1016/j.jep.2012.10.058
86. Jeon J-H, Lee S, Lee W, Jin S, Kwon M, Shin CH, Choi M-K, Song I-S (2020) Herb-drug interaction of red ginseng extract and Ginsenoside Rc with valsartan in rats. Molecules 25(3). https://doi.org/10.3390/molecules25030622

87. Jin S, Jeon J-H, Lee S, Kang WY, Seong SJ, Yoon Y-R, Choi M-K, Song I-S (2019) Detection of 13 Ginsenosides (Rb1, Rb2, Rc, Rd, re, Rf, Rg1, Rg3, Rh2, F1, compound K, 20(S)-Protopanaxadiol, and 20(S)-Protopanaxatriol) in human plasma and application of the analytical method to human pharmacokinetic studies following two week-repeated administration of red ginseng extract. Molecules 24(14). https://doi.org/10.3390/molecules24142618
88. Kim K, Kim HY (2008) Korean red ginseng stimulates insulin release from isolated rat pancreatic islets. J Ethnopharmacol 120(2):190–195. https://doi.org/10.1016/j.jep.2008.08.006
89. Lee H-M, Lee O-H, Kim K-J, Lee B-Y (2012) Ginsenoside Rg1 promotes glucose uptake through activated AMPK pathway in insulin-resistant muscle cells. Phytother Res 26 (7):1017–1022. https://doi.org/10.1002/ptr.3686
90. Shang W, Yang Y, Jiang B, Jin H, Zhou L, Liu S, Chen M (2007) Ginsenoside Rb-1 promotes adipogenesis in 3T3-L1 cells by enhancing PPAR gamma(2) and C/EBP alpha gene expression. Life Sci 80(7):618–625. https://doi.org/10.1016/j.lfs.2006.10.021
91. Jo JJ, Lee S (2017) Investigation of herb-drug interactions between Korean red ginseng extract and five CYP substrates by LC-MS/MS. Mass Spectrometry Lett 8:4. https://doi.org/10.5478/msl.2017.8.4.98
92. Yamashiro W, Maeda K, Hirouchi M, Adachi Y, Hu Z, Sugiyama Y (2006) Involvement of transporters in the hepatic uptake and biliary excretion of valsartan, a selective antagonist of the angiotensin II AT1-receptor, in humans. Drug Metab Dispos 34(7):1247–1254. https://doi.org/10.1124/dmd.105.008938
93. Nugroho AE, Andrie M, Warditiani NK, Siswanto E, Pramono S, Lukitaningsih E (2012) Antidiabetic and antihiperlipidemic effect of *Andrographis paniculata* (Burm. f.) Nees and andrographolide in high-fructose-fat-fed rats. Ind J Pharmacol 44(3):377–381. https://doi.org/10.4103/0253-7613.96343
94. Eka SS, Agung EN, Suwijiyo P (2011) The antidiabetics of combination metformin and purified extract of *Andrographis paniculata* (Burn). F. Ness in high fructose-fat fed rats. Majalah Obat Tradisional 16(3):124–132
95. Nugroho AE, Adriawan IR, Andrie M, Susilowati R, Pramono S (2014) Evaluasi efek anti-diabetes melitus ekstrak terpurifikasi *Andrographis paniculata* (Burm. f.) Nees dan andrografolid dengan parameter indeks HOMA-IR. Majalah Obat Trad 19:19–23
96. Nugroho AE, Kusumaramdani G, Widyaninggar A, Anggoro DP, Pramono S (2014) Antidiabetic effect of combinations of n-hexane insoluble fraction of ethanolic extract of *Andrographis paniculata* with other traditional medicines. Int Food Res J 21:785–789
97. Yu BC, Hung CR, Chen WC, Cheng JT (2003) Antihyperglycemic effect of andrographolide in streptozotocin-induced diabetic rats. Planta Med 69(12):1075–1079
98. Pan Y, Abd-Rashid BA, Ismail Z, Ismail R, Mak JW, Pook PCK, Er HM, Ong CE (2011) In vitro determination of the effect of *Andrographis paniculata* extracts and andrographolide on human hepatic cytochrome P450 activities. J Nat Med 65(3-4):440–447. https://doi.org/10.1007/s11418-011-0516-z

Chapter 17
Role of Micronutrients and Trace Elements in Diabetes Mellitus: A Review

Amar Godavari and Nagerathinam Manickamoorthi

Abstract The present review tries to understand the relationship between minerals, trace elements, and blood glucose levels in diabetic patients in chronic hyperglycemic conditions, a state where there is impaired metabolism of carbohydrates, proteins, and fats, which is caused by abnormal insulin secretion or insulin action or both. Diabetes mellitus is a metabolic syndrome characterized by increased blood sugar levels and glucose intolerance due to insulin resistance and deficiency. Many research trials have shed light on the relationship between trace elements and diabetes mellitus. Certain trace elements like copper, chromium, magnesium, vanadium, zinc, iron, and selenium are known to potentiate the insulin action. The mechanism behind this includes the activation of receptor sites of insulin, by involving in glucose metabolism enzyme systems serving as components or cofactors in it. Many studies have convincingly shown a link between micronutrients and glucose homeostasis. There is also evidence that the deficiency and efficiency of these micronutrients might have precise role in the pathogenesis and development of this disease.

Keywords Diabetes mellitus · Diabetes review · Micronutrients · Vitamins · Trace elements

1 Introduction

Diabetes is a serious, long-term condition that occurs when the body cannot produce any or enough insulin or cannot effectively use the insulin it produces. Diabetes mellitus (DM) is a chromic metabolic disorder and the takes fourth place being leading cause of death. It is estimated that 463.0 million adults aged between 20 and 79 years worldwide have diabetes. According to the current trend in the prevalence

A. Godavari (✉) · N. Manickamoorthi
Vels Institute of Science, Technology and Advanced Studies (Vels University), Chennai, Tamil Nadu, India

H. Chen, M. Zhang (eds.), *Structure and Health Effects of Natural Products on Diabetes Mellitus*, https://doi.org/10.1007/978-981-15-8791-7_17

of diabetes by 2030, a projected 578.4 million, and by 2045, 700.2 million adults aged between 20 and 79 years will be affected by diabetes [1]. DM is characterized by high blood glucose levels, hyperinsulinemia, insulin resistance, β-cell dysfunction, and failure [2]. It is further complicated by the oxidative stress forming imbalance between antioxidants and free radicals. Antioxidant properties of certain micronutrients may help in diabetic conditions [2, 3]. Diabetic patients are susceptible to multiple micronutrient deficiencies [4].

Nutrients are required by the body to carry out day to day functions normally. Broadly, nutrients are classified as macronutrients and micronutrients. Micronutrients include four major classes: macrominerals, trace minerals, vitamins, and organic acids. Macronutrients are composed of chloride, calcium, phosphorus, magnesium, sodium, and potassium. The trace minerals include chromium, iron, manganese, cobalt, copper, sulfur, boron, zinc, iodine, selenium, fluoride, and molybdenum. Vitamin types include A, D, C, E, K, and B complex [5].

Macronutrients react with vitamins and produce hormones whereby metabolic processes progress at faster rates. Micronutrients are important for immune regulations, tissue and cellular functions, nerve transmission, membrane regulations, muscle contractions, and enzyme actions [6].

Many researches have witnessed the effect of micronutrients with diabetes. Action of micronutrients on lowering the blood glucose levels has been observed. Their role of mechanism may be like acting as cofactors for enzyme systems in carbohydrate metabolism [7] and by increasing insulin sensitivity [8].

Certain micronutrients are also responsible for onset of this disease and their progress. Recommendations of vitamins and minerals in pharmacological doses like medications for therapeutic purpose should be clinically tested for their safety and efficacy for human health [4]. In this review, we will throw light on the positive and negative correlation of these micronutrients with diabetes along with the effect of deficiency and toxicity of excess consumption of these micronutrients.

2 Zinc

Insulin resistance determined by homeostasis model assessment (HOMA) marker in a group of 365 Spanish school children showed that children with zinc deficiency had higher HOMA values than those with normal zinc values. The greater insulin resistance value decreased as zinc serum levels increased. This study predicted that zinc can promote the health by lowering the insulin resistance [9].

3 Magnesium

Lower levels of magnesium in body have effect on DM [10]. Many studies showed inverse relationship among intake of magnesium and its serum levels, leading to insulin resistance and glucose metabolism [11, 12] Magnesium supplementation has shown to induce insulin secretion and its action [13]. Oral magnesium supplementation may improve insulin sensitivity and blood glucose control in patients with magnesium deficiency [11, 14].

4 Iron

Iron acts as cofactor for several enzymes and transports oxygen in the body. Higher intake of heme iron and iron stored in body was susceptible for greater risk for type 2 diabetes mellitus (T2DM). Dietary intake of non-heme iron and supplement of iron were not in danger of getting T2DM [15].

5 Copper

Imbalance in copper ions showed an increase in complications of diabetic patients. The antioxidant property of this important trace element deals with the increased free radical production thereby reducing the oxidative damage [16].

6 Chromium

Many trials have been carried out which enlighten the fact that supplementation of chromium, an essential trace element has hypoglycemic activity to great extent. The proposed mechanism is the activation of insulin receptors. It also has positive effects on body weight of both diabetic and healthy volunteers [17–19]. Safety of chromium was also evaluated. Controlled clinical trials have shown that treatment with chromium at doses up to 1000 g/day and for periods for up to 64 months had no lethal effects [20].

7 Other Trace Elements

Studies have shown that supplementation of vanadyl sulfate has shown reduction in blood glucose levels and glycosylated hemoglobin levels [21, 22]. In a study of randomized double-blind placebo-controlled trial, selenium supplementation showed an increased risk for DM [23].

8 Combined Trace Elements

Association between dietary intakes of zinc, copper, and iron among Japanese population showed that subjects with intakes of iron and copper were at higher risk for Type 2 DM (T2DM), while subjects with zinc had reduced risk for T2DM [24]. Combined supplementation of vitamin E and C along with zinc and magnesium reduced urinary albumin output in T2DM patients [25]. Combination of zinc sulfate with multivitamin/mineral (vitamin A, vitamin D3, vitamin E, magnesium, manganese, copper, and selenium) given as a supplementation exhibited beneficial effects in the glucose metabolism of T2DM patients [26].

9 Vitamin K

Vitamin K complex constitutes phylloquinone and menaquinone [27]. Study was done which correlated between dietary intake of vitamin K and diabetic markers, which showed no changes in baseline values. But after 1 year follow up study, the subjects had low glucagon-like peptide, IL-6, and glucose-dependent insulinotropic peptides in plasma concentrations. Lower risk for DM was with high intake of vitamin K [28, 29].

10 Vitamin E

Vitamin E is present in the form of α-tocopherol in human plasma [27]. A decrease in plasma concentration of tocopherol was seen in diabetic subjects. Vitamin E has antioxidant activity which may be the underlying mechanism for its effect on DM [30, 31].

11 Vitamin D

Vitamin D or calciferol plays an important role in glucose metabolism. Vitamin D receptors present in pancreatic β-cell activate the calcium channels, thereby producing pro-insulin and its conversion [32, 33]. It also improves the insulin sensitivity by stimulation of insulin receptor expression and PPAR-γ activation [34–36]. Many studies have shown associations between low levels of vitamin D and DM [37–42]. Association between vitamin D levels, insulin resistance, and beta-cell dysfunction are also experimented [43–47].

12 Vitamin C

Vitamin C or ascorbic acid is a powerful antioxidant. In DM, hyperglycemic conditions cause oxidative stress by producing high levels of free radicals. This vitamin brings down the levels of blood glucose by decreasing the free radicals [48]. In diabetic patients the levels of vitamin C and vitamin E levels were found to be low compared to healthy subjects [49, 50]. Inverse correlation was observed between plasma concentration of vitamin C and glycosylated hemoglobin, fasting and post-prandial blood glucose levels [51, 52].

13 Vitamin B Complex

Biotin also called as vitamin H or B7, supplemented as an adjuvant along with insulin therapy has shown to improve glycemic control and also serum lipids concentrations in type 1 diabetes mellitus (T1DM) patients with no side effects [53]. Antidiabetic activity of biotin and chromium picolinate combined supplementation was shown in rat model. There was increase in the serum insulin level at the end of the study. At the cellular level, there is an increase in PPAR-γ in adipose tissue and phosphorylated insulin receptor substrate expression of liver, kidney, and muscle tissues when compared to diabetic rats [54].

Thiamine or vitamin B1 supplementation for 6 weeks to diabetic individuals in randomized, double-blind pilot study prevented rise in fasting blood glucose levels and insulin levels when compared to placebo study, but no reduction was found in glucose levels [55]. In a study thiamine supplementation for 1 month showed lowered levels of glucose and leptin in DM patients [56].

Supplementation of folate or vitamin B9 has shown to control blood glucose levels, decrease in fasting blood glucose levels, drop in glycosylated hemoglobin, insulin resistance, and serum insulin levels in patients with T2DM [57]. Folate, vitamin B12, and pyridoxine supplementation has positive effects on diabetic retinopathy symptoms [58]. Regular usage of metformin may cause vitamin B12 deficiency; supplementation with folate in diabetic patients on metformin had improvements in total antioxidant capacity, homocysteine levels, and malondialdehyde [59]. Vitamin B12 deficiency is considered as a risk factor for DM and its complications mainly related to peripheral neuropathy associated with hyperhomocisteinemia in diabetic patients [60, 61]. Based on many reviews, it can be concluded that diabetic patients are at a higher risk of developing vitamin B12 deficiency when using metformin and its resultant hyperhomocysteinemia leads to neuropathy and other complications. Thus it is being recommended to supplement cobalamine to those diabetic patients treated using metformin [62–67].

T1DM was caused by autoimmune destruction of β-cells in pancreas thereby affecting the insulin secretion in children [68]. Negative effect of nicotinamide or vitamin B3 was found which are associated with oxidative stress and insulin resistance on the development of T2DM complications [69].

The active form of this vitamin B6 is pyridoxal-5′-phospate (PLP). Recently diagnosed diabetic patients contain low PLP levels than healthy subjects [70]. An investigational model of supplementation of pyridoxamine indicated reduced insulin concentration and insulin sensitivity with no effect on blood glucose levels [71].

14 Conclusion

For management of diabetes mellitus, we should understand the role of nutrients in glucose homeostasis, the impact of deficiency of micronutrients and the effect of their supplementation for prevention of diabetes. These micronutrients supplementation may reduce the peripheral consequences of DM which are treated with the present drugs for DM. The importance of procuring necessary quantities of micronutrients, trace elements, and vitamins from food sources should be explained to diabetic patients and pre-diabetic people. Multivitamin and multimineral supplementation is highly recommended for certain group such as aged, pregnant or lactating women, vegeterians and diet restricted people. Macronutrients are required in larger amounts and micronutrients in little amount for maintaining the normal functioning of the body. Nutrition therapy tries to maintain normal blood glucose levels and prevent the complications related to it.

References

1. Huang Y (2017) IDF diabetes atlas, 8th edn. International Diabetes Federation, Brussels Belgium
2. Rajalakshmy P (2019) Role of micronutrients on type II diabetes mellitus. Acta Sci Nutr Health 3:44–47. https://doi.org/10.31080/ASNH.2019.03.0444
3. Khalid S, Nahla B, Salini SJJS (2014) Variation in macro and trace elements in progression of type 2 diabetes. Sci World J 2014:1–9

4. Chehade JM, Sheikh-Ali M, Mooradian ADJDS (2009) The role of micronutrients in managing diabetes. Diabetes Spectr 22(4):214–218
5. Siddiqui K, Bawazeer N, Joy SS (2014) Variation in macro and trace elements in progression of type 2 diabetes. Sci World J:9. https://doi.org/10.1155/2014/461591
6. Matsumura M, Nakashima A, Tofuku Y (2000) Electrolyte disorders following massive insulin overdose in a patient with type 2 diabetes. Intern Med 39(1):55–57. https://doi.org/10.2169/internalmedicine.39.55
7. Vincent JB (2000) Quest for the molecular mechanism of chromium action and its relationship to diabetes. Nutr Rev 58(3):67–72
8. Kruse-Jarres JD, Rukgauer M (2000) Trace elements in diabetes mellitus. Peculiarities and clinical validity of determinations in blood cells. J Trace Elem Med Biol 14(1):21–27. https://doi.org/10.1016/s0946-672x(00)80019-x
9. Ortega RM, Rodriguez-Rodriguez E, Aparicio A, Jimenez AI, Lopez-Sobaler AM, Gonzalez-Rodriguez LG, Andres P (2012) Poor zinc status is associated with increased risk of insulin resistance in Spanish children. Br J Nutr 107(3):398–404. https://doi.org/10.1017/s0007114511003114
10. Kaur B, Henry J (2014) Micronutrient status in type 2 diabetes: a review. Adv Food Nutr Res 71:55–100. https://doi.org/10.1016/b978-0-12-800270-4.00002-x
11. Mooren FC, Kruger K, Volker K, Golf SW, Wadepuhl M, Kraus A (2011) Oral magnesium supplementation reduces insulin resistance in non-diabetic subjects—a double-blind, placebo-controlled, randomized trial. Diabetes Obes Metab 13(3):281–284. https://doi.org/10.1111/j.1463-1326.2010.01332.x
12. Guerrero-Romero F, Tamez-Perez HE, Gonzalez-Gonzalez G, Salinas-Martinez AM, Montes-Villarreal J, Trevino-Ortiz JH, Rodriguez-Moran M (2004) Oral magnesium supplementation improves insulin sensitivity in non-diabetic subjects with insulin resistance. A double-blind placebo-controlled randomized trial. Diabetes Metab 30(3):253–258. https://doi.org/10.1016/s1262-3636(07)70116-7
13. Barbagallo M, Dominguez LJ, Galioto A, Ferlisi A, Cani C, Malfa L, Pineo A, Busardo A, Paolisso G (2003) Role of magnesium in insulin action, diabetes and cardio-metabolic syndrome X. Mol Asp Med 24(1–3):39–52. https://doi.org/10.1016/s0098-2997(02)00090-0
14. Yokota K, Kato M, Lister F, Ii H, Tajima N (2004) Clinical efficacy of magnesium supplementation in patients with type 2 diabetes. J Am Coll Nutr 23(5):506S–509S
15. Bao W, Rong Y, Rong S, Liu LG (2012) Dietary iron intake, body iron stores, and the risk of type 2 diabetes: a systematic review and meta-analysis. BMC Med 10:13. https://doi.org/10.1186/1741-7015-10-119
16. Lefevre M, Keen CL, Lönnerdal B, Hurley LS, Schneeman BO (1986) Copper deficiency-induced hypercholesterolemia: effects on HDL subfractions and hepatic lipoprotein receptor activity in the rat. J Nutr 116(9):1735–1746. https://doi.org/10.1093/jn/116.9.1735
17. Amato P, Morales AJ, Yen SSC (2000) Effects of chromium picolinate supplementation on insulin sensitivity, serum lipids, and body composition in healthy, nonobese, older men and women. J Gerontol Ser A Biol Sci Med Sci 55(5):M260–M263. https://doi.org/10.1093/gerona/55.5.M260
18. Volpe SL, Huang HW, Larpadisorn K, Lesser II (2001) Effect of chromium supplementation and exercise on body composition, resting metabolic rate and selected biochemical parameters in moderately obese women following an exercise program. J Am Coll Nutr 20(4):293–306. https://doi.org/10.1080/07315724.2001.10719050
19. Campbell WW, Joseph LJO, Anderson RA, Davey SL, Hinton J, Evans WJ (2002) Effects of resistive training and chromium picolinate on body composition and skeletal muscle size in older women. Int J Sport Nutr Exerc Metab 12(2):125–135. https://doi.org/10.1123/ijsnem.12.2.125
20. Jeejeebhoy KN (1999) The role of chromium in nutrition and therapeutics and as a potential toxin. Nutr Rev 57(11):329–335

21. Goldfine AB, Patti ME, Zuberi L, Goldstein BJ, LeBlanc R, Landaker EJ, Jiang ZY, Willsky GR, Kahn CR (2000) Metabolic effects of vanadyl sulfate in humans with non-insulin-dependent diabetes mellitus: in vivo and in vitro studies. Metab Clin Exp 49(3):400–410. https://doi.org/10.1016/s0026-0495(00)90418-9
22. Cusi K, Cukier S, DeFronzo RA, Torres M, Puchulu FM, Redondo JCP (2001) Vanadyl sulfate improves hepatic and muscle insulin sensitivity in type 2 diabetes. J Clin Endocrinol Metab 86 (3):1410–1417. https://doi.org/10.1210/jc.86.3.1410
23. Stranges S, Marshall JR, Natarajan R, Donahue RP, Trevisan M, Combs GF, Cappuccio FP, Ceriello A, Reid ME (2007) Effects of long-term selenium supplementation on the incidence of type 2 diabetes - a randomized trial. Ann Intern Med 147(4):217–223. https://doi.org/10.7326/0003-4819-147-4-200708210-00175
24. Eshak ES, Iso H, Maruyama K, Muraki I, Tamakoshi A (2018) Associations between dietary intakes of iron, copper and zinc with risk of type 2 diabetes mellitus: a large population-based prospective cohort study. Clin Nutr 37(2):667–674. https://doi.org/10.1016/j.clnu.2017.02.010
25. Farvid MS, Jalali M, Siassi F, Hosseini M (2005) Comparison of the effects of vitamins and/or mineral supplementation on glomerular and tubular dysfunction in type 2 diabetes. Diabetes Care 28(10):2458–2464. https://doi.org/10.2337/diacare.28.10.2458
26. Gunasekara P, Hettiarachchi M, Liyanage C, Lekamwasam S (2011) Effects of zinc and multimineral vitamin supplementation on glycemic and lipid control in adult diabetes. Diabetes Metab Syndr Obes Targets Therapy 4:53–60. https://doi.org/10.2147/dmso.S16691
27. Valdes-Ramos R, Laura GLA, Elina MCB, Donaji BAA (2015) Vitamins and type 2 diabetes mellitus. Endocr Metab Immune Disord Drug Targets 15(1):54–63. https://doi.org/10.2174/1871530314666141111103217
28. Beulens JWJ, van der A DL, Grobbee DE, Sluijs I, Spijkerman AMW, van der Schouw YT (2010) Dietary Phylloquinone and Menaquinones intakes and risk of type 2 diabetes. Diabetes Care 33(8):1699–1705. https://doi.org/10.2337/dc09-2302
29. Ibarrola-Jurado N, Salas-Salvado J, Martinez-Gonzalez MA, Bullo M (2012) Dietary phylloquinone intake and risk of type 2 diabetes in elderly subjects at high risk of cardiovascular disease. Am J Clin Nutr 96(5):1113–1118. https://doi.org/10.3945/ajcn.111.033498
30. Polidori MC, Mecocci P, Stahl W, Parente B, Cecchetti R, Cherubini A, Cao P, Sies H, Senin U (2000) Plasma levels of lipophilic antioxidants in very old patients with type 2 diabetes. Diabetes Metab Res Rev 16(1):15–19. https://doi.org/10.1002/(sici)1520-7560(200001/02)16:1<15::Aid-dmrr71>3.3.Co;2-2
31. Ebesunun MO, Obajobi EO (2012) Elevated plasma homocysteine in type 2 diabetes mellitus: a risk factor for cardiovascular diseases. Pan Afr Med J 12:48
32. Mitri J, Pittas AG (2014) Vitamin D and Diabetes. Endocrinol Metabol Clin North Am 43 (1):205–20+. https://doi.org/10.1016/j.ecl.2013.09.010
33. Mathieu C, Gysemans C, Giulietti A, Bouillon R (2005) Vitamin D and diabetes. Diabetologia 48(7):1247–1257. https://doi.org/10.1007/s00125-005-1802-7
34. Mitri J, Muraru MD, Pittas AG (2011) Vitamin D and type 2 diabetes: a systematic review. Eur J Clin Nutr 65(9):1005–1015. https://doi.org/10.1038/ejcn.2011.118
35. van Etten E, Mathieu C (2005) Immunoregulation by 1,25-dihydroxyvitamin D-3: basic concepts. J Steroid Biochem Mol Biol 97(1–2):93–101. https://doi.org/10.1016/j.jsbmb.2005.06.002
36. Guo JJ, Xiao ZD, Xue X, Liu X, Lu Y, Yin X, Ma K (2013) 25-Hydroxyvitamin D is closely related with the function of the pancreatic islet beta cells. Pak J Med Sci 29(3):809–813. https://doi.org/10.12669/pjms.293.2982
37. Knekt P, Laaksonen M, Mattila C, Harkanen T, Marniemi J, Heliovaara M, Rissanen H, Montonen J, Reunanen A (2008) Serum vitamin D and subsequent occurrence of type 2 diabetes. Epidemiology 19(5):666–671. https://doi.org/10.1097/EDE.0b013e318176b8ad
38. Pittas AG, Dawson-Hughes B, Li T, Van Dam RM, Willett WC, Manson JE, Hu FB (2006) Vitamin D and calcium intake in relation to type 2 diabetes in women. Diabetes Care 29 (3):650–656. https://doi.org/10.2337/diacare.29.03.06.dc05-1961

39. Liu S, Song Y, Ford ES, Manson JE, Buring JE, Ridker PM (2005) Dietary calcium, vitamin D, and the prevalence of metabolic syndrome in middle-aged and older US women. Diabetes Care 28(12):2926–2932. https://doi.org/10.2337/diacare.28.12.2926
40. Afzal S, Brondum-Jacobsen P, Bojesen SE, Nordestgaard BG (2014) Vitamin D concentration, obesity, and risk of diabetes: a mendelian randomisation study. Lancet Diabetes Endocrinol 2 (4):298–306. https://doi.org/10.1016/s2213-8587(13)70200-6
41. Song YQ, Wang L, Pittas AG, Del Gobbo LC, Zhang CL, Manson JE, Hu FB (2013) Blood 25-Hydroxy vitamin D levels and incident type 2 diabetes. Diabetes Care 36(5):1422–1428. https://doi.org/10.2337/dc12-0962
42. Gagnon C, Lu ZX, Magliano DJ, Dunstan DW, Shaw JE, Zimmet PZ, Sikaris K, Grantham N, Ebeling PR, Daly RM (2011) Serum 25-Hydroxyvitamin D, calcium intake, and risk of type 2 diabetes after 5 years results from a national, population-based prospective study (the Australian diabetes, obesity and lifestyle study). Diabetes Care 34(5):1133–1138. https://doi.org/10.2337/dc10-2167
43. Kayaniyil S, Vieth R, Retnakaran R, Knight JA, Qi Y, Gerstein HC, Perkins BA, Harris SB, Zinman B, Hanley AJ (2010) Association of Vitamin D with insulin resistance and beta-cell dysfunction in subjects at risk for type 2 diabetes. Diabetes Care 33(6):1379–1381. https://doi.org/10.2337/dc09-2321
44. Broder AR, Tobin JN, Putterman C (2010) Disease-specific definitions of vitamin D deficiency need to be established in autoimmune and non-autoimmune chronic diseases: a retrospective comparison of three chronic diseases. Arthritis Res Ther 12(5):8. https://doi.org/10.1186/ar3161
45. Bellan M, Guzzaloni G, Rinaldi M, Merlotti E, Ferrari C, Tagliaferri A, Pirisi M, Aimaretti G, Scacchi M, Marzullo P (2014) Altered glucose metabolism rather than naive type 2 diabetes mellitus (T2DM) is related to vitamin D status in severe. Cardiovasc Diabetol 13:10. https://doi.org/10.1186/1475-2840-13-57
46. Kostoglou-Athanassiou I, Athanassiou P, Gkountouvas A, Kaldrymides P (2013) Vitamin D and glycemic control in diabetes mellitus type 2. Ther Adv Endocrinol Metab 4(4):122–128. https://doi.org/10.1177/2042018813501189
47. Stadlmayr A, Aigner E, Huber-Schonauer U, Niederseer D, Zwerina J, Husar-Memmer E, Hohla F, Schett G, Patsch W, Datz C (2015) Relations of vitamin D status, gender and type 2 diabetes in middle-aged Caucasians. Acta Diabetol 52(1):39–46. https://doi.org/10.1007/s00592-014-0596-9
48. Mandl J, Szarka A, Banhegyi G (2009) Vitamin C: update on physiology and pharmacology. Br J Pharmacol 157(7):1097–1110. https://doi.org/10.1111/j.1476-5381.2009.00282.x
49. Odum EP, Ejilemele AA, Wakwe VC (2012) Antioxidant status of type 2 diabetic patients in Port Harcourt, Nigeria. Niger J Clin Pract 15(1):55–58. https://doi.org/10.4103/1119-3077.94099
50. Sundaram RK, Bhaskar A, Vijayalingam S, Viswanathan M, Mohan R, Shanmugasundaram KR (1996) Antioxidant status and lipid peroxidation in type II diabetes mellitus with and without complications. Clin Sci 90(4):255–260. https://doi.org/10.1042/cs0900255
51. Carter P, Gray LJ, Talbot D, Morris DH, Khunti K, Davies MJ (2013) Fruit and vegetable intake and the association with glucose parameters: a cross-sectional analysis of the Let's prevent diabetes study. Eur J Clin Nutr 67(1):12–17. https://doi.org/10.1038/ejcn.2012.174
52. Mazloom Z, Hejazi N, Dabbaghmanesh M-H, Tabatabaei H-R, Ahmadi A, Ansar H (2011) Effect of vitamin C supplementation on postprandial oxidative stress and lipid profile in type 2 diabetic patients. Pak J Biol Sci 14(19):900–904
53. Hemmati M, Babaei H, Abdolsalehei M (2013) Survey of the effect of biotin on glycemic control and plasma lipid concentrations in type 1 diabetic patients in Kermanshah in Iran (2008–2009). Oman Med J 28(3):195–198. https://doi.org/10.5001/omj.2013.53
54. Sahin K, Tuzcu M, Orhan C, Sahin N, Kucuk O, Ozercan IH, Juturu V, Komorowski JR (2013) Anti-diabetic activity of chromium picolinate and biotin in rats with type 2 diabetes induced by

high-fat diet and streptozotocin. Br J Nutr 110(2):197–205. https://doi.org/10.1017/s0007114512004850
55. Shahmiri FA, Soares MJ, Zhao Y, Sherriff J (2013) High-dose thiamine supplementation improves glucose tolerance in hyperglycemic individuals: a randomized, double-blind cross-over trial. Eur J Nutr 52(7):1821–1824. https://doi.org/10.1007/s00394-013-0534-6
56. Gonzalez-Ortiz M, Martinez-Abundis E, Robles-Cervantes JA, Ramirez-Ramirez V, Ramos-Zavala MG (2011) Effect of thiamine administration on metabolic profile, cytokines and inflammatory markers in drug-naive patients with type 2 diabetes. Eur J Nutr 50(2):145–149. https://doi.org/10.1007/s00394-010-0123-x
57. Gargari BP, Aghamohammadi V, Aliasgharzadeh A (2011) Effect of folic acid supplementation on biochemical indices in overweight and obese men with type 2 diabetes. Diabetes Res Clin Pract 94(1):33–38. https://doi.org/10.1016/j.diabres.2011.07.003
58. Smolek MK, Notaroberto NF, Jaramillo AG, Pradillo LR (2013) Intervention with vitamins in patients with nonproliferative diabetic retinopathy: a pilot study. Clin Ophthalmol 7:1451–1458. https://doi.org/10.2147/opth.S46718
59. Aghamohammadi V, Gargari BP, Aliasgharzadeh A (2011) Effect of folic acid supplementation on Homocysteine, serum Total antioxidant capacity, and malondialdehyde in patients with type 2 diabetes mellitus. J Am Coll Nutr 30(3):210–215. https://doi.org/10.1080/07315724.2011.10719962
60. Ben Ahmed H, Bouzid K, Hassine M, Saadi O, Bahlous A, Abdelmoula J, Baccar H, Ben Mami Ben Miled F (2014) Prevalence of non-conventional cardiovascular risk factors in Tunisian diabetics (Prevalence des facteurs de risque cardiovasculaire non conventionnels chez les sujets diabetiques tunisiens.). Presse Med 43(1):e9–e16. https://doi.org/10.1016/j.lpm.2013.07.019
61. Molina M, Gonzalez R, Folgado J, Real JT, Martinez-Hervas S, Priego A, Lorente R, Chaves FJ, Ascaso JF (2013) Correlation between plasma concentrations of homocysteine and diabetic polyneuropathy evaluated with the Semmes-Weinstein monofilament test in patients with type 2 diabetes mellitus. Med Clin 141(9):382–386. https://doi.org/10.1016/j.medcli.2012.09.042
62. Calvo Romero JM, Ramiro Lozano JM (2012) Vitamin B(12) in type 2 diabetic patients treated with metformin. Endocrinol Nutr 59(8):487–490. https://doi.org/10.1016/j.endonu.2012.06.005
63. Kos E, Liszek MJ, Emanuele MA, Durazo-Arvizu R, Camacho P (2012) Effect of metformin therapy on vitamin D and vitamin B-12 levels in patients with type 2 diabetes mellitus. Endocr Pract 18(2):179–184. https://doi.org/10.4158/ep11009.Or
64. Moore EM, Mander AG, Ames D, Kotowicz MA, Carne RP, Brodaty H, Woodward M, Boundy K, Ellis KA, Bush AI, Faux NG, Martins R, Szoeke C, Rowe C, Watters DA, Investigators A (2013) Increased risk of cognitive impairment in patients with diabetes is associated with metformin. Diabetes Care 36(10):2981–2987. https://doi.org/10.2337/dc13-0229
65. Obeid R, Jung J, Falk J, Herrmann W, Geisel J, Friesenhahn-Ochs B, Lammert F, Fassbender K, Kostopoulos P (2013) Serum vitamin B12 not reflecting vitamin B12 status in patients with type 2 diabetes. Biochimie 95(5):1056–1061. https://doi.org/10.1016/j.biochi.2012.10.028
66. Pierce SA, Chung AH, Black KK (2012) Evaluation of vitamin B-12 monitoring in a veteran population on long-term, high-dose metformin therapy. Ann Pharmacother 46(11):1470–1476. https://doi.org/10.1345/aph.1R223
67. Reinstatler L, Qi YP, Williamson RS, Garn JV, Oakley GP (2012) Association of Biochemical B-12 deficiency with metformin therapy and vitamin B-12 supplements the National Health and nutrition examination survey, 1999–2006. Diabetes Care 35(2):327–333. https://doi.org/10.2337/dc11-1582
68. Orban T, Sosenko JM, Cuthbertson D, Krischer JP, Skyler JS, Jackson R, Yu LP, Palmer JP, Schatz D, Eisenbarth G, Diabet Prevention Trial Type S (2009) Pancreatic islet autoantibodies as predictors of type 1 diabetes in the diabetes prevention trial-type 1. Diabetes Care 32(12):2269–2274. https://doi.org/10.2337/dc09-0934

69. Zhou SS, Li D, Sun WP, Guo M, Lun YZ, Zhou YM, Xiao FC, Jing LX, Sun SX, Zhang LB, Luo N, Bian FN, Zou W, Dong LB, Zhao ZG, Li SF, Gong XJ, Yu ZG, Sun CB, Zheng CL, Jiang DJ, Li ZN (2009) Nicotinamide overload may play a role in the development of type 2 diabetes. World J Gastroenterol 15(45):5674–5684. https://doi.org/10.3748/wjg.15.5674
70. Ahn HJ, Min KW, Cho YO (2011) Assessment of vitamin B-6 status in Korean patients with newly diagnosed type 2 diabetes. Nutr Res Pract 5(1):34–39. https://doi.org/10.4162/nrp.2011.5.1.34
71. Unoki-Kubota H, Yamagishi S, Takeuchi M, Bujo H, Saito Y (2010) Pyridoxamine, an inhibitor of advanced Glycation end product (AGE) formation ameliorates insulin resistance in obese, type 2 diabetic mice. Protein Pept Lett 17(9):1177–1181. https://doi.org/10.2174/092986610791760423